BRAIN DISEASES AND METALLOPROTEINS

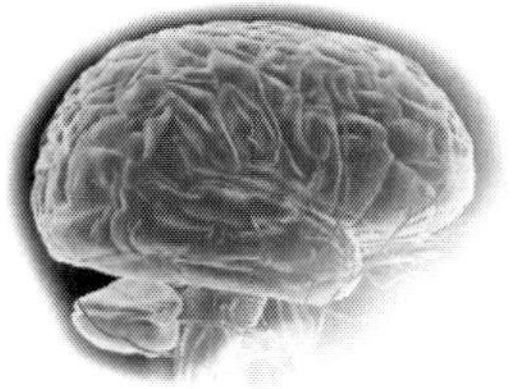

edited by David R. Brown

BRAIN DISEASES AND METALLOPROTEINS

Pan Stanford Publishing

Published by

Pan Stanford Publishing Pte. Ltd.
Penthouse Level, Suntec Tower 3
8 Temasek Boulevard
Singapore 038988

Email: editorial@panstanford.com
Web: www.panstanford.com

British Library Cataloguing-in-Publication Data
A catalogue record for this book is available from the British Library.

Brain Diseases and Metalloproteins

ISBN 978-981-4316-01-9 (Hardcover)
ISBN 978-981-4364-07-2 (eBook)

Printed in the USA

Contents

Chapter 1

Introduction

David R. Brown
Department of Biology and Biochemistry, University of Bath, Bath, BA2 7AY, United Kingdom

1.1 Brain Diseases

Brain diseases are one of the most frightening sets of conditions that exist. Mention of these diseases usually conjures up images of either retirement homes full of aging people lost to the world in the haze of dementia or debilitating conditions that strip away the very fabric of who we are. Dementias are truly dreadful conditions as not only are they mostly fatal but they are diseases of attrition and still, to this day, quite incurable. Brain diseases are usually neurodegenerative, in that they result in the loss of brains cells, especially neurons, the cells with which we think and hold the essence of our conscious selves. While there are brain diseases that are not neurodegenerative such as cancers like gliomas or tumors of children such as primary neuroectodermal tumors, this book will consider only neurodegenerative diseases.

Outside of the dreadful cost to individuals and their families, concern over neurodegenerative diseases is mounting because of the trend towards longer life. As the percentage of people over 60 increases across the world, the incidence of these diseases increases as well. This brings with it an increase social and economic burden that necessitates a greater effort to understand these diseases and

Brain Diseases and Metalloproteins
Edited by David R. Brown

ISBN 978-981-4316-01-9 (Hardcover), 978-981-4364-07-2 (eBook)
www.panstanford.com

find ways to manage, alleviate, or even prevent them. However, this is quite a daunting challenge. The most well-known and also the most common of these diseases, Alzheimer's disease was first described over 100 years ago, but it seems in many ways that we are still no closer to a cure for the disease, despite so much research and study. What could make a difference in this situation? The answer may lie with looking at these diseases in an altered way or going back to fundamental assumptions about how these diseases occur at the cellular level. While drug trials must continue, there is an increasing need to understand neurodegenerative disease at the basic level of cells and molecules. Basic research is therefore the key ingredient necessary to change assumptions and challenge our views of the cause of these diseases. Many of the contributors to this book are scientists addressing a fundamental question about proteins involved in these diseases.

Neurodegenerative diseases are a very large family. However, not all of them lead to dementia as dementia is the result of neuronal loss in areas of the brain associated with cognition or memory. The two most common neurodegenerative disorders are Alzheimer's disease and Parkinson's disease (Ross and Poirier, 2004). Other diseases include, motor neuron disease, amyotrophic lateral sclerosis (ALS) (Strong and Rosenfeld, 2003), Huntington's disease, dementia with lewy bodies, multiple sclerosis, spinocerebellar ataxia (Hontiand Vecsei, 2005), progressive supranuclear palsy (Lubarsky and Juncos, 2008), frontotemporal dementia (Kertesz, 2008) and a variety of less common diseases such as Menkes disease (Madsen and Gitlin, 2007). One of the most notorious family of neurodegenerative diseases is prion disease or transmissible spongiform encephalopathy (Prusiner, 1998). These diseases remain quite rare despite the scare of an epidemic of Variant Creutzfeldt–Jakob disease (vCJD). The total number of vCJD cases worldwide remains below 250. There are numerous forms of prion diseases in humans and also animals. Bovine spongiform encephalopathy (BSE) and scrapie have become well-known animal neurodegenerative diseases, while almost all other animal dementias are considered obscure (de Lehante, 1990).

Most neurodegenerative diseases of humans affect people in the later stages of life. Alzheimer's disease being the most common neurodegenerative disease in elderly people and is the cause of 70% of all cases of dementia. It has a frequency of 6% in those aged 70–74 and this value increases in frequency with age until people over 85 have a 42% chance of developing the disease (Alzheimer's Association, 2008). There are often early-onset forms where the patients are much younger than the expected age for the disease to occur. There are also inherited forms of Alzheimer's, Parkinson's and prion diseases. Some of the rarer diseases such as Menkes disease only exist as inherited forms. Currently, the only known neurodegenerative disease that can be transmitted between individuals is prion disease (vCJD, iatrogenic CJD) where a small number of cases have been transmitted through blood or other products such as growth factors or dura mata (Will, 2003). While many neurodegenerative diseases have "early onset" subsets where individuals are diagnosed with the disease many years earlier than the mode, in vCJD the age of onset is typically in the late 20s.

Outside of clinical aspects of neurodegenerative orders, molecular biology and biochemistry have provided insights into how such disease develops and the proteins and other molecules associated with disease onset and progression. Although there is no complete picture for any of the diseases, there are many common themes and significant bodies of information on changes associated with these. The most common theme linking the diseases is that there is often a protein or a fragment is involved that changes conformation or takes on a new role (Chiti and Dobson, 2006). Often these proteins are poorly understood in terms of their importance to a cell in health, but it is becoming an exaggeration to say that we do not know their function. These aggregating proteins are termed amyloidogenic and often change conformation to one rich in beta sheet structural elements and/or become resistant to proteases. These proteins can then form aggregates or fibrils that build up in deposits such as plaques which may have an intracellular or extracellular location depending on the disease.

Neurodegenerative disease in which a protein aggregates to form fibrils or other aggregates are termed amyloidoses. These diseases all result in deposition of abnormal protein isoforms within the brain. The abnormal isoforms of these proteins can form a variety of aggregate types. Chief among these are fibrils often referred to as amyloid because of their association with amyloid plaques. Fibrils represent a form of order polymerisation into long chains.

Fibrils are not the only form of aggregate found in disease. Indeed, despite extensive research on this form and the presence of fibrils considered evidence of disease presence, it is now largely thought that these aggregates are "tomb stones", products of the condition that play little actual role in disease progression. Most modern research now focuses on smaller aggregates termed oligomers which are likely to be more soluble and probably more toxic to neurons and other cells of the brain.

1.2 Metalloproteins

Proteins play many roles in the cell and are usually quite essential, carrying out vital functions such as enzymatic catalysis or forming parts of the cytoskeleton. The proteins we associated with neurodegenerative disease often have "normal" isoforms. This suggests that there are harmless forms of these proteins that usually play some other role in the cell. However, many of the proteins associated with neurodegeneration, apart from being amyloidogenic appear to be non-essential. Three of the best known amyloidogenic proteins as the prion protein, the amyloid precursor protein or alpha-synuclein and are associated with neurodegenerative diseases. Mice genetically altered to lack such proteins develop normally and show little abnormality. This may be because other proteins have similar functions and are still active in cells or because their functions are beneficial but not essential. Nevertheless, knowing more about what these disease associated proteins do in the healthy cell might give an insight into what converts them into their more deadly isoforms.

Proteins expressed in cells often require other factors to be active and carry out their functions. Examples of such factors are cofactors. Cofactors are usually metal ions that are bound or otherwise associated with the proteins. Proteins that bind metals are termed metalloproteins. One of the most surprising findings of recent years is that a considerable number of proteins associated with neurodegenerative diseases are metalloproteins. In particular many of these proteins bind copper. Some of these proteins are amyloidogenic proteins as well, but a number are not. Those proteins that interact with copper but are non-amyloidogenic are associated with inherited disease. The two leading examples of these are the

P type ATPases. Mutations in these proteins cause two diseases, Wilson's disease and Menkes disease. Both diseases are associated with changes in cellular metabolism that result in mishandling of copper. These diseases and the proteins involved are dealt with in Chapter 5. Another disease associated with mutations in a metal binding protein is familial amyotrophic lateral sclerosis or fALS. In this cases mutations in an antioxidant protein, superoxide dismutase (SOD-1) cause a gain of a toxic function that causes loss of neurons. Studies on this protein and its potential role in fALS are covered in Chapter 11.

More significant than diseases associated with mutations are sporadic neurodegenerative diseases that make up the majority of cases. Astonishing, three of these diseases are associated with proteins that have been shown to be metalloproteins. These proteins are the prion protein of prion diseases, both the amyloid precursor protein (APP) and amyloid-beta (Aβ) of Alzheimer's disease, and more recently alpha-synculein of Parkinson's disease. In all cases the normal isoform of the protein involved can bind copper but the functional consequences of this are still unclear. These three diseases have received considerable attention both as the focus of research and also public attention. In the case of Alzheimer's and Parkinson's disease this is clearly warranted because they are the two most common neurodegenerative diseases. Prion disease on the other hand is very rare but is important for two reasons. First the great public concern that is associated with a transmissible neurodegenerative disease and second because it is the only neurodegenerative disease with a reliable animal model. The majority of the chapters in this book deal with these diseases.

Metals and amyloidogenic proteins are a curious combination. As mentioned, metal binding to most proteins can be important to their activity. Copper binding to potentially amyloidogenic proteins could have a variety of consequences (Fig. 1.1). While the normal isoform of the protein might bind a metal specifically for a functional reason, amyloid proteins can bind metal non-specifically. In this regard, metal binding could have two potential effects. Thus binding could either inhibit aggregation or promote it. Binding of copper to the normal prion protein can prevent its conversion to the disease specific isoform (Bocharova *et al.*, 2005). However once the disease process of prion disease has initiated protein conversion to

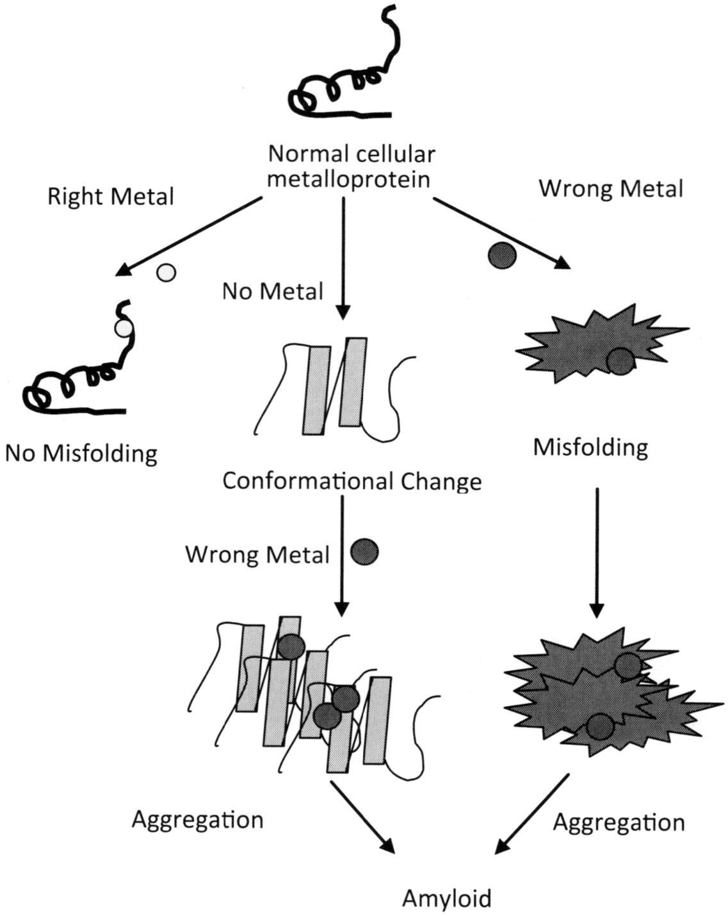

Figure 1.1 Some metalloproteins can become amyloid protein. This summary indicates how normal cellular metalloproteins can become amyloidogenic proteins. Under normal cellular conditions a metalloprotein binds its cofactor which protects it from misfolding or conformational changes. However, there are two potential alternative pathways if this does not occur. If the protein does not bind its normal cofactor it may undergo an inappropriate conformational change. Under those circumstances binding a metal could facilitate aggregation. Alternatively, binding an inappropriate metal during protein synthesis might also lead to misfolding (or conformational change). Both pathways would generate abnormal protein that could be amyloidogenic. See also Colour Insert.

the abnormal isoform, the presence of copper can stimulate this process. Copper can increase the infectivity of prions or increase the protease resistance of the prion protein (McKenzie *et al.*, 1998). The use of copper chelators during disease progression has been shown to lengthen the disease process (Sigurdsson *et al.*, 2003). These findings imply that the relation between metals and amyloidogenesis is not a simple one. Interactions occur between proteins and copper and need to be regulated. Therefore, cells have numerous regulatory and chaperoning proteins especially for copper. If this regulation breaks down, then there are consequences that could potentially trigger neurodegenerative disease.

In the case of Alzheimer's disease the picture is further complicated in that copper binds both to the normal precursor and the disease specific protein (Hesse *et al.*, 1994; Huang *et al.*, 2004). Copper binds to APP and prevents dimerisation of the protein which is protective against cleavage of the protein by beta-secretase, the enzyme complex known to cause the formation of Aβ. Once Aβ is formed it can also bind copper. The presence of copper in preparations of Aβ is associated with its ability to aggregate. Alpha-synuclein is a protein associated with a number of diseases including Parkinson's disease and is the main component within Lewy bodies. Aggregation of alpha-synuclein can be triggered simply by high concentration of the protein. Binding of copper to this protein can increase its rate of aggregation (Uversky *et al.*, 2001).

Neurodegenerative diseases associated with metalloproteins often show disturbances in the metabolism of the metals that the proteins bind (Kralovicova *et al.*, 2009). The consequence of this is that neurodegenerative diseases can alter the behaviour of many other proteins. Thus indirectly, other metalloproteins can show disturbances in neurodegenerative conditions even if those proteins are not critically related to the cause. Metallothioneins are a family of protein important in the regulation of levels of copper and zinc in cells. They also play a role in the regulation of redox balance. In other words they can influence the ability of copper to cause oxidative stress. Redox active metal such as iron, copper and manganese can exist in either oxidized or reduced forms and the balance between these forms can be very important to cellular metabolism as well as to the viability of neurons. This potentially relates to why metalloproteins can be causal to disease processes as controlling metals in cells is clearly important to their viability. While metals are vital for a variety of activities in the cell they can also cause the generation of destructive reactive oxygen species by reactions such as the Fenton reaction. Metallothioneins are therefore important in a number of processes that could be disrupted in neurodegeneration. Chapter 4 provides a thorough analysis of these proteins and their role in the cell.

This volume brings together the work of a number of distinguished researchers. The field of neurodegeneration and metalloproteins has existed for almost 20 years now and continues to expand. The chapters here therefore represent a small subset of researchers presenting a broad overview of a diverse area but also illuminates each researcher's own distinctive specialization within this field. The two opening chapters present general overviewed of a number of diseases. Chapter 2 from the group of David Allsop looks at the formation of oligomeric species of proteins, their ability to interact with metals and the consequence in terms of oxidative stress. Chapter 3 from the group of Henryk Kozlowski looks at the use of peptides to model binding of metals to proteins associated with neurodegeneration and the interesting insights such studies can provide.

The next chapters deal with specific proteins. Chapter 4 from Duncan Sutherland and Martin Stillman provides an extensive overview of metallothioneins while chapter 5 from Sharon La Fontaine, James Camakaris and Julian Mercer looks at the P type ATPases. The following chapters concentrate on specific diseases or the proteins associated with them. Chapters 6 and 7 are related to Alzheimer's disease. The chapter by Robert Cappai, Gerd Multhaup and their colleagues deals with APP, interaction with copper and the relevance of this to Alzheimer's disease. The next chapter by Silvia Bolognin and Paolo Zatta focuses more on beta-amyloid and its interaction copper, aluminium and other metals.

Chapter 8 is my own review of the prion disease field and focuses on the positive role of copper binding in PrP's role as an antioxidant and the negative role of manganese as a catalyst for aggregation and a stabilizer of protein survival. Chapters 9 and 10 deal with alpha-synuclein. Heather Lucas and Jennifer Lee discuss copper binding and the potential consequences of this for cellular metabolism and Parkinson's disease with emphasis on the potential oxidation of proteins. Andrés Binolfi and Claudio Fernández focus more specifically on metal binding to alpha-synuclein and the structural consequences of this. As copper bound protein is more rapidly converted to amyloid fibrils, then copper binding may have important consequences for Parkinson's disease. The last chapter by Shin-Ichi Ono and colleagues deals with the use of animal models of ALS to attempt to generate potential treatments.

While this book does not include every approach and is far from exhaustive in describing the studies of neurodegeneration and metal binding proteins, it does provide a very important overview of the current trends in the research. I hope it provides an important starting point for further reading into this dynamic and exciting area of research.

References

Bocharova, O. V., Breydo, L., Salnikov, V. V. and Baskakov, I. V. (2005). Copper(II) inhibits *in vitro* conversion of prion protein into amyloid fibrils. *Biochemistry*, 44, pp. 6776–6787.

Chiti, F. and Dobson, C. M. (2006). Protein misfolding, functional amyloid, and human disease. *Annu Rev Biochem*, 75, pp. 333–366.

Hesse, L., Beher, D., Masters, C. L. and Multhaup, G. (1994). The beta A4 amyloid precursor protein binding to copper. *FEBS Lett*, 349, pp. 109–116.

Honti, V. and Vecsei, L. (2005). Genetic and molecular aspects of spinocerebellar ataxias. *Neuropsychiatr Dis Treat*, 1, pp. 125–133.

Huang, X., Atwood, C. S., Moir, R. D., Hartshorn, M. A., Tanzi, R. E. and Bush, A. I. (2004). Trace metal contamination initiates the apparent auto-aggregation, amyloidosis, and oligomerization of Alzheimer's Abeta peptides. *J Biol Inorg Chem*, 9, pp. 954–960.

Kertesz, A. (2008). Frontotemporal dementia: a topical review. *Cogn Behav Neurol*, 21, pp. 127–133.

Kralovicova, S., Fontaine, S. N., Alderton, A., Alderman, J., Ragnarsdottir, K. V., Collins, S. J. and Brown, D. R. (2009). The effects of prion protein expression on metal metabolism. *Mol Cell Neurosci*, 41, pp. 135–147.

Lubarsky, M. and Juncos, J. L. (2008). Progressive supranuclear palsy: Aa current review. *Neurologist*, 14, pp. 79–88.

Madsen, E. and Gitlin, J. D. (2007). Copper and iron disorders of the brain. *Annu Rev Neurosci*, 30, pp. 317–337.

McKenzie, D., Bartz, J., Mirwald, J., Olander, D., Marsh, R. and Aiken, J. (1998). Reversibility of scrapie inactivation is enhanced by copper. *J Biol Chem*, 273, pp. 25545–25547.

Prusiner, S. B. (1998). Prions. *Proc Natl Acad Sci USA*, 95, pp. 13363–13383.

Ross, C. A. and Poirier, M. A. (2004). Protein aggregation and neurodegenerative disease. *Nat Med*, 10 Suppl, pp. S10–S17.

Sigurdsson, E. M., Brown, D. R., Alim, M. A., Scholtzova, H., Carp, R., Meeker, H. C., Prelli, F., Frangione, B. and Wisniewski, T. (2003). Copper chelation delays the onset of prion disease. *J Biol Chem*, 278, pp. 46199–46202.

Strong, M. and Rosenfeld, J. (2003). Amyotrophic lateral sclerosis: a review of current concepts. *Amyotroph Lateral Scler Other Motor Neuron Disord*, 4, pp. 136–143.

Uversky, V. N., Li, J. and Fink, A. L. (2001). Metal-triggered structural transformations, aggregation, and fibrillation of human alpha-synuclein. A possible molecular NK between Parkinson's disease and heavy metal exposure. *J Biol Chem*, 276, pp. 44284–44296.

Will, R. G. (2003). Acquired prion disease: iatrogenic CJD, variant CJD, kuru. *Br Med Bull*, 66, pp. 255–265.

Chapter 2

A Possible Key Role for Redox-Active Metal Ions and Soluble Oligomers in Neurodegenerative Disease

Brian J. Tabner, Susan Moore, Jennifer Mayes, and David Allsop

Division of Biomedical and Life Sciences, School of Health and Medicine, Lancaster University, Lancaster LA1 4YQ, UK

2.1 Summary

The deposition of abnormal protein fibrils is the defining pathological feature of the "protein misfolding" disorders. Early-stage protein aggregates ("soluble oligomers") may be responsible for some of the reported toxic effects of these fibrillogenic proteins. In the case of Aβ, which accumulates in the brain in Alzheimer's disease (AD), these oligomers also have potent effects on long-term potentiation (LTP) and memory. A significant number of the aggregating proteins found in neurodegenerative disease can bind to redox-active metal ions and this interaction could explain some of their toxic properties. There is good evidence that Aβ can generate hydrogen peroxide (H_2O_2), a key reactive oxygen species (ROS). We have confirmed this by employing the electron spin resonance spin-trapping technique, and have obtained similar results with α-synuclein, toxic fragments of the prion protein (PrP), the British dementia peptide (ABri) and amylin. Our data suggest that H_2O_2 accumulates during the early stages

Brain Diseases and Metalloproteins
Edited by David R. Brown

ISBN 978-981-4316-01-9 (Hardcover), 978-981-4364-07-2 (eBook)
www.panstanford.com

of aggregation and that it can be readily converted into hydroxyl radicals, by Fenton's reaction. We hypothesize that a low-n oligomer could be the optimal size for ROS generation, and that a fundamental molecular mechanism underlying the pathogenesis of some of the protein misfolding disorders could be the direct production of ROS by soluble, early stage protein oligomers.

2.2 Introduction to the Proteopathies

The formation of extracellular deposits of amyloid and/or intracellular inclusions consisting of abnormal protein fibrils is a common feature of numerous different "protein misfolding disorders" (Dobson, 1999; Soto, 2001) or "proteopathies" (Walker *et al.*, 2006). The underlying protein component of the pathological fibrils found in these diseases almost invariably adopts, predominantly, a cross-β-pleated sheet configuration (Sipe and Cohen, 2000). These protein deposits can be local (e.g., restricted to the brain, pancreas, thyroid) or systemic (widespread throughout the body); they can also be primary (with no obvious underlying cause), secondary (to another underlying disease process, such as inflammation) or due to inherited disease (Westermark, 2005). In some forms of amyloidosis (e.g., in cardiac amyloidosis), the total mass of the accumulated protein fibrils can exceed that of the unaffected organ, and the presence of such a vast quantity of this material can cause extensive structural damage, leading to organ dysfunction, and eventually to organ failure.

In those diseases where the proteins concerned accumulate locally in the central nervous system (CNS) in the form of a large number of small, focal lesions (Soto, 2003), their contribution to organ dysfunction (in this case related to brain neurodegeneration) is much less clear. Prominent examples of these diseases are presented in Table 2.1. The protein deposits associated with these CNS diseases can take the form of extracellular amyloid fibrils (e.g., senile plaques, cerebrovascular amyloid), cytoplasmic inclusions (e.g., neurofibrillary tangles, Lewy bodies (LBs), glial cell inclusions) or fibrils found elsewhere within affected nerve cells (e.g., Lewy neurites, intranuclear inclusions). Numerous studies have concluded that these various protein deposits play an important role in the early stages of disease pathogenesis, but, in this situation, it may not be the

Table 2.1 Aggregating proteins associated with various neurodegenerative diseases

Protein(s)	Disease(s)	Lesion(s)
Aβ tau	Alzheimer's disease Down's syndrome	Senile plaques and neurofibrillary tangles (NFTs)
α-synuclein	Parkinson's disease Dementia with Lewy bodies	Lewy bodies, Lewy Neurites
Tau protein	Frontotemporal dementia (FTD)	Tau inclusions
Prion protein	Transmissible prion disease	Amyloid plaques, Prion rods
Huntingtin	Huntington's disease	Intranuclear inclusions
ABri	British familial dementia	Amyloid plaques, NFTs
ADan	Danish familial dementia	Amyloid plaques, NFTs
SOD-1	Motor neuron disease (MND)	SOD-1 inclusions
TDP-43	FTD/MND	Intracellular inclusions

sheer physical bulk of the protein fibrils that is the main problem, but much more subtle effects of early-stage protein aggregates (soluble oligomers) on the properties of cell membranes and the functioning of neuronal synpases (Kayed *et al.*, 2004). A significant number of the aggregating proteins found in the neurodegenerative diseases presented in Table 2.1 have been shown to bind to redox-active metal ions (Brown, 2009; Allsop *et al.*, 2008) and this interaction could explain some of their toxic properties, as discussed further below.

2.3 Mutations in Genes Associated with Protein Processing and Aggregation Cause Inherited Forms of Proteopathy

As a general rule, inherited forms of the proteopathies are almost invariably caused by a mutation in the gene coding for either the fibrillar protein itself, or another protein involved in the production or clearance (i.e., metabolism) of this protein. This provides clear evidence for the importance of these proteins in disease pathogenesis.

In line with this general principle, mutations in the genes for *all* of the proteins mentioned in Table 2.1 (or their precursors) have been found in inherited forms of an associated neurodegenerative disease (Allsop *et al.*, 2008). In rare familial forms of AD and Parkinson's disease (PD), for example, mutations have been found in the genes encoding the β-amyloid precursor protein (APP) and α-synuclein, respectively (St. George-Hyslop, 2000; Hardy *et al.*, 2009). APP is the precursor for β-amyloid (Aβ), the peptide that accumulates at the centre of senile plaques and in the walls of cerebral blood vessels in AD, whereas α-synuclein is the fibrillar component of LBs and Lewy neurites, the characteristic histopathological features of PD and dementia with Lewy bodies (DLB). Mutations responsible for other forms of familial AD have been found in the genes encoding the presenilins (presenilin 1 and presenilin 2) which are an essential component of γ-secretase, a multi-protein complex with aspartyl proteinase activity (Kaether *et al.*, 2006). The latter is involved in proteolytic cleavage of APP at the C-terminal end of Aβ, which is the final step in generation of the Aβ peptide. It can be argued that a similar situation exists in PD, because many of the other causative genes for inherited forms of this disease are also connected with protein processing pathways, in this case involving either the ubiquitin-proteasome or lysosomal systems, which could interact with α-synuclein (Hardy *et al.*, 2009). The ubiquitin-proteasome system is responsible for regulating intracellular homeostasis, including the removal of misfolded proteins from the cell (Layfield *et al.*, 2005). Failure of this system could, therefore, lead to the accumulation of aggregated forms of α-synuclein within cells. Although some familial forms of PD do not develop overt LBs, there is growing evidence that soluble oligomers of α-synuclein could be one of the causes of neurodegeneration in this disease (see below), and, at present, these cannot be revealed by histopathological methods in post-mortem human brain, and so their presence could remain undetected. Alternatively, the death of nigral cells in some forms of inherited parkinsonism (as opposed to PD itself) could be due to other mechanisms, such as mitochondrial dysfunction. The situation for most of the other inherited diseases mentioned in Table 2.1 is more straightforward, with mutations in the genes *MAPT, PRNP, IT15* (and other trinucleotide repeat genes), *BRI2, SOD1* and *TARDBP*, which all encode an aggregating protein, being responsible for an inherited neurodegenerative disease

(Allsop *et al.*, 2008). Mutations in the *PGRN* and *FUS* genes have recently been identified in familial motor neuron disease (alternatively called amyotrophic lateral sclerosis) or frontotemporal dementia (see, for example, Baker *et al.*, 2006; Vance *et al.*, 2009) and must link in somehow with TDP-43 protein aggregation, although the pathways involved are not presently understood.

A common experimental observation regarding pathogenetic amyloid-forming mutant proteins is that, when incubated in solution *in vitro*, they often show an increased propensity, compared with the wild-type protein, to fold into β-pleated sheets and to aggregate into oligomers or fibrils (Chiti and Dobson, 2006). The effects of these various pathogenetic mutations therefore provides clear evidence that the aggregation of the protein concerned is likely to be the initiating factor in the pathogenesis of the disease in question. Another common theme to emerge from genetic studies is that overexpression of the wild-type protein, due to gene multiplication events, can lead to inherited disease. Examples of this are Down's syndrome (trisomy 21), with Alzheimer-type pathology being due to overexpression of APP (Beyreuther *et al.*, 1993), and inherited forms of PD or DLB, due to duplication or triplication of the α-synuclein gene (*SNCA*) locus (Hofer *et al.*, 2005). This is presumably because protein aggregation is a concentration-dependent phenomenon and is stimulated by increased expression due to a simple effect of gene dosage.

In many instances, the importance of protein aggregation in disease pathogenesis is supported by transgenic animal studies, where expression of the mutant or wild-type human fibril-forming protein leads to a pathological picture in the animal model with many similarities to the equivalent human disease.

2.4 The Toxic Effects of Protein Oligomers

Further evidence of an important role for these aggregating proteins in the initiation of disease is the fact that many of the fibril-forming proteins or peptides concerned, or smaller fragments thereof, have been shown to be toxic to cells in culture. However, it is not clear how these studies relate to cell death *in vivo*. Nor is it clear if the mechanism of cell death is similar in all cases where toxicity has been observed.

The cytotoxic effects of Aβ, which accumulates in the brain in AD, have been studied particularly extensively. Earlier studies by Pike *et al.* (1993) suggested that the aggregation state of Aβ is a key factor in determining its toxicity. However, the precise molecular mechanism by which Aβ mediates cell death has remained a matter of considerable dispute. There is now general agreement that the influx of calcium ions into cells and the production of ROS, resulting in oxidative damage, are both involved, but it is still unclear how these are linked with exposure to Aβ. Behl *et al.* (1994) reported that exposure to Aβ and other amyloid proteins elevated H_2O_2 levels in cultured cells, suggesting that the resulting oxidative damage could be a common toxic mechanism. One fascinating possibility arising from this is that the aggregating peptide itself may be able to generate ROS directly, and this is discussed further below. Whereas earlier studies emphasised a role for fully formed amyloid fibrils in the toxic mechanism, more recent attention has focussed on the possibility that earlier protein assemblies (referred to as oligomers, protofibrils, annular protofibrils, globular neurotoxins or, in the case of Aβ, ADDLs) could be involved (Lambert, *et al.*, 1998; Walsh, *et al.*, 1999; Walsh, *et al.*, 2002; Lashuel, *et al.*, 2002; Wang, *et al.*, 2002; Kim, *et al.*, 2003; Chromy, *et al.*, 2003; Kayed, *et al.*, 2003; Klein *et al.*, 2004; Kayed *et al.*, 2004; Demuro, *et al.*, 2005; Cleary, *et al.*, 2005; Walsh and Selkoe, 2007; Haass and Selkoe, 2007). These types of small protein assemblies can be derived from most (if not all) of the various proteins implicated in the proteopathies and they are now thought to be at least partially responsible for their toxic effects on cells. This could involve a common (probably conformation-dependent) mechanism, based on insertion of small oligomers into cell membranes, which causes local damage and leads to changes in membrane permeability (Kayed *et al.*, 2004; Demuro *et al.*, 2005). However, the detailed molecular mechanisms responsible for this toxic effect are not clear. One possibility is that annular protofibrils (ring-shaped structures) can form pores/ion channels in the membrane (Lashuel *et al.*, 2002). Another possibility is that toxic oligomers can insert into membranes and generate ROS *in situ*, which would inflict local oxidative damage to lipids and proteins, and this might explain reported changes in membrane permeability (Allsop *et al.*, 2008). In the case of Aβ, there is still no clear consensus on the precise nature of the toxic form of the peptide, with oligomers (Walsh *et al.*, 2002; Wang *et al.*, 2002;

Kim *et al.*, 2003; Kayed *et al.*, 2003; Kayed *et al.*, 2004; Demuro *et al.*, 2005; Cleary, *et al.*, 2005; Walsh and Selkoe, 2007; Haass and Selkoe, 2007), protofibrils (Walsh *et al.*, 1999), annular protofibrils (Lashuel *et al.*, 2002), ADDLs (Lambert *et al.*, 1998; Chromy *et al.*, 2003; Klein *et al.*, 2004), dimers (Shankar *et al.*, 2008; Hung *et al.*, 2008), trimers (Hung *et al.*, 2008) and Aβ*56 (9-12mer) (Lesné *et al.*, 2006) all being implicated. Other researchers, however, have emphasized the fact that it may be the aggregation process itself that is responsible for the damaging effects of amyloid, rather than any specific form of protein aggregation intermediate (Wogulis *et al.*, 2005). Aβ oligomers have also been reported to inhibit long-term potentiation (LTP) and impair the memory of complex learned behaviours when injected at very low (nM) concentrations into the lateral ventricles of rats, suggesting that they are potent neurotoxins with selective effects on synaptic function and memory (Walsh *et al.*, 2002; Cleary *et al.*, 2005). The potential relevance of this to the symptoms and pathogenesis of AD is striking. However, it is still not clear how these observations relate to the neurotoxic properties of protein aggregates towards cultured cells, which are usually carried out at much higher concentrations. Most importantly, it is not yet clear how these observations relate to neurodegeneration, synaptic changes and memory loss in the human brain.

2.5 The Generation of Hydrogen Peroxide by Aβ

We now focus on the possibility that Aβ (and other amyloidogenic proteins) might be involved directly in the generation of two important ROS, namely H_2O_2 and the hydroxyl radical (•OH). Behl *et al.* (1994) reported that exposure of cultured cells to Aβ resulted in the elevation of H_2O_2 levels, but it was not clear whether this was a direct or indirect effect of the peptide. This was followed by two key publications by Ashley Bush and colleagues, who noted the generation of H_2O_2 during the incubation of solutions of both Aβ (1-40) and Aβ (1-42) (Huang *et al.*, 1999a; Huang *et al.*, 1999b) and suggested that the Aβ peptide may be able to generate this ROS directly. These authors were also able to show that Aβ had the ability to reduce Fe(III) to Fe(II) and Cu(II) to Cu(I) and, also, that these metal ions were bound to the peptide. These latter observations have great significance since these metal ion levels are found to be elevated in the amyloid deposits found in individuals with AD (Lovell *et al.*, 1998; Smith *et al.*, 1997). Furthermore, not only has H_2O_2 (an important oxidizing species) shown to be generated, but in the presence of Fe(II) (and also Cu(I)) it readily releases •OH *via* the Fenton reaction:

$$\text{Fe(II)} + H_2O_2 \longrightarrow \text{Fe(III)} + {}^{\bullet}\text{OH} + {}^{-}\text{OH}$$

Importantly, this radical is released in the immediate vicinity of the peptide. The reduction of oxygen to form H_2O_2 requires the transfer of electrons from the bound Aβ/Cu(I), (or Aβ/Fe(II)), complex with the subsequent formation of •OH also requiring an electron. In order for the generation of these two ROS to be maintained over a significant period of time the complex needs to be redox active (i.e., both metal ion oxidation states need to be readily available). The release of the hydroxyl radical is highly significant. It is extremely reactive, with a reaction rate with a very wide range of molecules very close to the diffusion limit and, largely as a consequence of this, is very unselective in its site of attack. Consequently, extensive attack by this radical on the peptide itself, and other nearby molecules, would be expected. Recent data indicate that oxidative damage is an early-stage event in AD (see, for example, Nunomura *et al.*, 2001) and this feature will be considered later.

Bush and colleagues employed a fluorescent dye-based technique for the spectrophotometric detection of H_2O_2 (Huang *et al.*, 1999a). In our laboratories, we were more familiar with an alternative and well-established technique based on electron spin resonance (ESR) spectroscopy. In this latter method, H_2O_2 is converted into •OH radicals upon the addition of a solution of Fe(II) ions. These radicals are then trapped, employing 5,5-dimethyl-1-pyrroline *N*-oxide (DMPO) as a spin trap, to produce the corresponding (ESR active) DMPO-OH adduct, see Scheme 2.1 below:

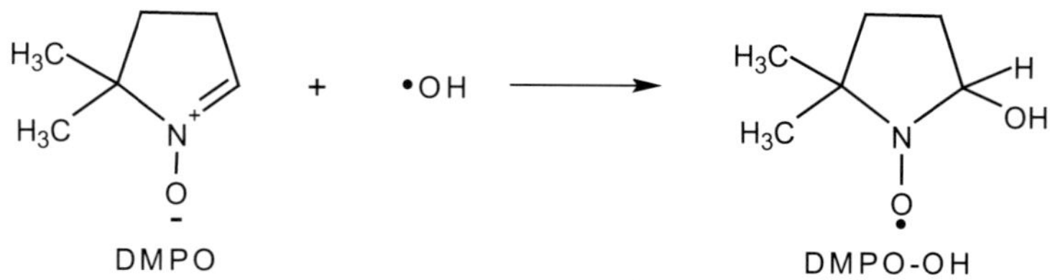

Scheme 2.1

This adduct has a uniquely characteristic ESR spectrum and provides an alternative and very different, approach to the detection of H_2O_2 (see Fig. 2.1). Employing the ESR technique, (the full experimental details of this method have been reported elsewhere (Turnbull *et al.*, 2001; Turnbull *et al.*, 2003a; Turnbull *et al.*, 2003b)), H_2O_2 was detected during the incubation of both Aβ (1–40) and Aβ (1–42) and also during incubation of the smaller, toxic Aβ (25–35) fragment (Tabner *et al.*, 2001). In contrast, no H_2O_2 was generated during the

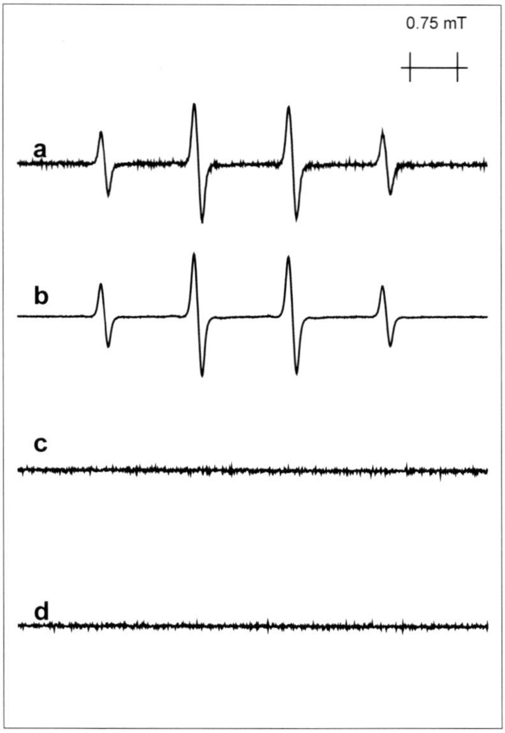

Figure 2.1 ESR spectra obtained from Aβ-related peptidesESR spectra recorded when solutions of DMPO, diethylenetriaminepentaacetic acid (DETAPAC) and Fe(II) were added to individual solutions of (a) Aβ (1–40), (b) Aβ (1–42) (spectrum intensity × 0.25), (c) Aβ (40–1) and (d) Aβ (1–40) Met35Nle, all of which had been incubated at 100 μM in phosphate buffered saline (PBS) for 1 h at 37°C. The intensity of the DMPO-OH adduct spectrum in both (a) and (b), after different incubation times of up to 48 h, followed the characteristic profile described in the text, whilst no adduct was discernible over the same time period in the cases of both (c) and (d).

incubation of various controls, such as the non-toxic reverse Aβ (40–1) and Aβ (1–40)Met35Nle peptides (see Fig. 2.1 and Turnbull *et al.*, 2003b; Tabner *et al.*, 2001; Tabner *et al.*, 2002; Tabner *et al.*, 2003). We note three significant differences between our experiments and those of Bush and colleagues. Firstly, in our experiments metal ion concentrations were at trace levels, whereas Bush and colleagues used metal ion levels of 0.1 of those of the peptide. Secondly, we observed that the concentration of H_2O_2 generated during incubation was significantly less than that reported by Bush and colleagues. Thirdly, in our experiments in which H_2O_2 was generated, its concentration always followed the same pattern, i.e., there was a time-lag before H_2O_2 was detected, and then the DMPO-OH concentration would rise to a maximum before slowly declining away towards zero as the peptide incubation time was further increased. It is interesting that Behl *et al.* (1994) had noted that H_2O_2 levels in cells treated with Aβ also reached a maximum before declining.

2.6 Other Aggregating Protein Systems

Confirmation of the generation of H_2O_2 by the Aβ (1–40) and Aβ (1–42) peptides led us to query whether this property was unique to these peptides, or whether various other peptides and proteins implicated in neurodegenerative diseases (and possibly in some peripheral amyloidoss) might share this same feature. If so, this might help to explain the origins of the oxidative brain damage (to proteins, lipids and nucleic acids) which has been described in numerous publications related to AD and the other neurodegenerative diseases presented in Table 2.1. We found that α-synuclein (Turnbull *et al.*, 2001; Tabner *et al.*, 2002; Tabner *et al.*, 2003), the British dementia peptide (ABri) (Tabner *et al.*, 2005) and certain toxic fragments of the prion protein (PrP) (Turnbull *et al.*, 2003a; Turnbull *et al.*, 2003b; Tabner *et al.*, 2003) all had the ability to generate H_2O_2. In addition, we also established that the human form of the amylin peptide (alternatively called islet amyloid polypeptide (IAPP)) found in the islets of Langerhans of the pancreas in late-onset type-2 diabetes mellitus also generated H_2O_2 (Masad *et al.*, 2007). In these latter experiments, the highly sensitive Amplex Red fluorescence dye method was supplemented by the ESR technique. The generation of H_2O_2 from Aβ was also confirmed using the Amplex Red method (Masad *et al.*, 2007).

Table 2.2 summarises the results of these experiments. In all of these experiments, there is a clear positive correlation between the ability of the various proteins and peptides to generate H_2O_2 and their toxic effects on cultured cells. Other key features also emerge. For example, several of those peptides and proteins which generate H_2O_2

Table 2.2 ESR observations during the incubation of various proteins and peptides (all at 100 μM in PBS at 37°C). The spectra were monitored for selected time periods from 1 h up to 48 h.* In these studies, the production of H_2O_2 was demonstrated by both the ESR and Amplex red techniques. We also noted that the concentration of H_2O_2 generated in the experiments with human amylin was significantly enhanced when incubation was in the presence of Cu(II)

	DMPO-OH observed	DMPO-OH not observed
Aβ peptides	Aβ (1–40)*	Aβ (1–40)Met35Nle
	Aβ (1–42)*	Aβ (40–1)
	Aβ (25–35)	
Synucleins	α-synuclein	β-synuclein
		γ-synuclein
	NAC (1–35)	NAC (19–35)
	NAC (1–18)	NAC (35–1)
		NAC (18–1)
PrP-related		full length PrPc
		Doppel
	PrP (106–126) + 2 mM Cu(II)	PrP (106–126) (without Cu(II))
	PrP (106–126) + 0.2 mM Cu(II)	PrP (106–126) scrambled + Cu(II)
	PrP (121–231) D178N	PrP (121–231)
	PrP (121–231) F198S	
	PrP (121–231) E200K	
ABri	ABri (oxidised form)	ABri (reduced form)
		ABri (wild-type)
Amylin	Human amylin*	Rodent amylin

(see Table 2.2) also bind certain metal ions, often with high affinity. The evidence for metal ion binding to key fibril-forming peptides and proteins has been summarised (Brown, 2009). In addition to the high affinity in the cases of Aβ (1–40) and Aβ (1–42) (Huang *et al.*, 1999a; Huang *et al.*, 1999b), the N-terminal domain of PrP can bind up to six Cu(II) ions (Millhauser, 2007). The toxic peptide PrP (106–126) also binds Cu(II) ions (Jobling *et al.*, 2001), as does α-synuclein, where it has been reported that this binding facilitates its aggregation (Paik *et al.*, 2000). Preliminary results indicate binding of human amylin to Cu(II) (Masad *et al.*, 2007). Although binding has not yet been established, there is strong evidence that both Fe(III) and Cu(II) influence fibril formation in ABri (Khan *et al.*, 2004) and accelerate aggregation in the 'NAC' fragment (amino acid residues 61–95) of α-synuclein (Khan *et al.*, 2005).

2.7 Mechanism of Hydrogen Peroxide Formation and Aβ Oxidation

In the case of Aβ, it has been proposed (Opazo *et al.*, 2002) that the first step in the generation of H_2O_2 (in the absence of any external source of electrons) is the transfer of an electron from Aβ to the bound redox-active metal ion, illustrated here for Cu(I)/Cu(II) binding, leading to the formation of the corresponding peptide radical cation. The transfer of electrons from the Cu(I) complex to molecular oxygen then leads to H_2O_2 formation:

$$2A\beta/Cu(I) + 2H^+ + O_2 \longrightarrow 2A\beta/Cu(II) + H_2O_2$$

This process requires two electrons and so, *in vitro*, and in the absence of an external electron source, the second electron must be provided by a second Aβ/Cu(I) complex. However, it is now appreciated that, when present, a reductant can also act as an external source of electrons. In the original research undertaken by Bush and colleagues (Huang *et al.*, 1999a; Huang *et al.*, 1999b) one of the reagents required in the spectrophotometric technique employed, (tris(2-carboxyethyl)phosphine hydrochloride), was present during the incubation of the peptide, and it was subsequently established by Opazo *et al.* (2002) that this reagent could act as such a reductant. These experiments were undertaken with a Cu(II):Aβ ratio of 1:1 or, sometimes even 2:1. These workers also noted that the concentration of H_2O_2 was significantly enhanced in the presence of this and other suitable reductants, such as cholesterol, dopamine (DA) and vitamin C. In the case of DA, H_2O_2 generation was monitored every 30 minutes over a 2-h period in the presence of Cu(II)/Aβ. These results revealed that there was a linear increase in generated H_2O_2 with time and that the time-lag observed when Cu(II) levels were at trace concentrations was completely eliminated. H_2O_2 concentrations, measured after 1 h incubation, were also found to increase as the concentration of the Aβ/Cu complex increased. In addition, Himes *et al.* (2008) nicely illustrated that, in the presence of equimolar Cu(I), the H_2O_2 generation from various Aβ fragments exceeds those generated in its absence by a factor close to 10 after the same incubation period (i.e., 1 h). These various results indicate that a large number of factors are important in determining H_2O_2 levels *in vitro* which often hinder a simple comparison between the various published data. Clearly the length of the incubation period before H_2O_2 measurement and the sensitivity of the assay technique employed are of paramount importance.

Since H_2O_2 is generated in the immediate vicinity of the Aβ peptide, it could also be responsible for much of the oxidation of the peptide itself. Possible reactions are (i) the loss of an electron from the peptide/Cu(I) complex to form the Aβ radical cation, which then undergoes rapid rearrangement to give side chain oxidation products, (ii) a bimolecular reaction with H_2O_2 leading directly to peptide oxidation, (iii) the liberation of the •OH radical from H_2O_2 via the Fenton reaction and subsequent direct attack of this radical on the peptide, and (iv) the further attack of peptidyl radicals, formed following •OH attack. Any combination of these reactions could explain how Aβ becomes oxidatively modified in the brain in AD (Dong *et al.*, 2003). In support of mechanism (iii), we have already shown that sufficient H_2O_2 is self-generated during the early stages of aggregation of Aβ (1–40) to form detectable peptidyl radicals, on addition of Fe(II) (Tabner *et al.*, 2006).

2.8 Concluding Comments

Protein aggregation plays a crucial role in a whole range of different diseases, including some important neurodegenerative diseases.

In the case of AD, the 'amyloid cascade hypothesis' originally proposed that the β-amyloid (Aβ) fibrils found in senile plaques are the neurotoxic agent responsible for initiating the events leading to neurofibrillary tangle pathology and neurodegenerative brain damage in this disease (Hardy and Allsop, 1991). However, support for this 'end-stage' view has waned in recent years, due in part to the fact that the correlation between the severity of the disease and the density of amyloid plaques is poor (Terry *et al.*, 1991), and also due to increasing evidence for the importance of earlier-stage Aβ assemblies ("soluble oligomers") in AD and other proteopathies, which are more closely linked to disease severity (see, for example, Fukumoto *et al.*, 2010). Another key finding is that oxidative damage is a very early event in AD, preceding the formation of amyloid plaques and neurofibrillary tangles, and that this damage decreases as the disease progresses (Nunomura *et al.*, 2001; Markesbery *et al.*, 2005). During the early stages of disease, soluble oligomers are likely to be the major type of Aβ aggregate present. Clearly, if these oligomers are involved in the generation of ROS, they could be responsible for much of this early oxidative damage. As Aβ aggregation proceeds, ROS levels and associated oxidative damage would then decline, mirroring the time-course of H_2O_2 generation during the aggregation of synthetic Aβ *in vitro*.

There is also the question as to why toxicity appears to decrease as the size of the protein aggregate increases beyond the oligomer stage. It is undoubtedly the case that one reason for this is that the small, soluble oligomers are much more mobile than larger aggregates, and so are able to diffuse from their site of origin and insert into surrounding cell membranes, causing local damage, including changes in membrane permeability. However, there is another possibility that warrants serious consideration, namely that the ability to generate ROS could be a particular property of these small oligomers. In our own experiments, we have already shown that mature amyloid fibres derived from Aβ have lost the ability to generate H_2O_2 (Tabner *et al.*, 2005). This could be because, during aggregation, the amyloidogenic protein and/or any suitable external reductant reach a point where the supply of available electrons becomes virtually exhausted, or that the metal ion bound to the fibrillar protein loses its redox activity. A more likely possibility, however, is that the decrease in ability to generate ROS with increasing size of the aggregate could be the result

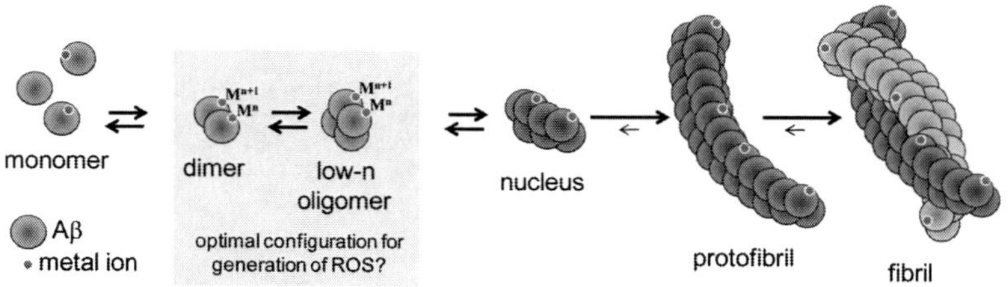

Figure 2.2 The potential role of soluble oligomers and redox-active metal ions in ROS generation in this diagram M^n and M^{n+1} represent Cu(I) and Cu(II) when Cu is the bound metal ion and Fe(II) and Fe(III) when it is Fe. The box encloses those species which appear to have an optimal configuration for ROS generation and where the bound metal ions must be redox-active. Recent experimental evidence shows that once bound, Cu ions remain bound to Aβ throughout aggregation (Karr and Szalai, 2008) and, therefore, the lack of toxicity of mature Aβ fibrils cannot be due to a reduction in the number of bound Cu ions. See also Colour Insert.

of a steric problem. With increasing size and complexity, the metal ion and the oxygen molecule required to form H_2O_2 could become sterically removed from one another, resulting in a lack of ability of the required electrons to transfer. This could become the case as the aggregate becomes more complex and eventually reaches its solubility limit. Furthermore, a monomer bound to a single metal ion may not be particularly active in the formation of H_2O_2 as this reaction requires two electrons. Rapid diffusion could prevent the close proximity of two such monomers to oxygen, thus impeding this reaction. This could explain the time-delay before detection of H_2O_2 in our own aggregation experiments. We hypothesize that a very small (low-n) oligomer could be the optimal size for ROS generation (Fig. 2.2) and that this, in addition to their ability to insert into membranes, could explain why these oligomers are a potent toxic entity.

It is becoming increasingly recognised that ROS, including H_2O_2, act as diffusible signaling molecules in the CNS. Not only this, but they have also been shown to be involved in the molecular mechanisms underlying synaptic plasticity and memory formation (Knapp and Klann, 2002; Kishida and Klann, 2007). Thus, the generation of ROS by soluble oligomers of Aβ could also, potentially, explain their potent effects on LTP, memory and learning. This would seem to an important area for future research that seems to have been largely overlooked by other investigators.

Acknowledgements

We are grateful to the Alzheimer's Society (PhD Studentship to JM) and the Wellcome Trust (Project Grant) for supporting our work in this area of research.

References

Allsop, D., Mayes, J., Moore, S., Masad, A. and Tabner, B. J. (2008) Metal-dependent generation of reactive oxygen species from amyloid proteins implicated in neurodegenerative disease. *Biochem. Soc. Trans.* 36, pp. 1293–1298.

Baker, M., Mackenzie, I. R., Pickering-Brown, S. M., Gass, J., Rademakers, R., Lindholm, C., Snowden, J., Adamson, J., Sadovnick, A. D., Rollinson, S., Cannon, A., Dwosh, E., Neary, D., Melquist, S., Richardson, A., Dickson, D., Berger, Z., Eriksen, J., Robinson, T., Zehr, C., Dickey, C. A., Crook, R., McGowan, E., Mann, D., Boeve, B., Feldman, H. and Hutton, M. (2006) Mutations in progranulin cause tau-negative frontotemporal dementia linked to chromosome 17. *Nature* 442, pp. 916–919.

Behl, C., Davis, J. B., Lesley, R. and Schubert, D. (1994) Hydrogen peroxide mediates amyloid β protein toxicity. *Cell* 77, pp. 817–827.

Beyreuther, K., Pollwein, P., Multhaup, G., Mönning, U., König, G., Dyrks, T., Schubert, W. and Masters, C. L. (1993) Regulation and expression of the Alzheimer's β/A4 amyloid protein precursor in health, disease, and Down's syndrome. *Annals N. Y. Acad. Sci.* 695, pp. 91–102.

Brown, D. R. (2009) Brain proteins that mind metals: a neurodegenerative perspective. *Dalton Trans.* 7, pp. 4069–4076.

Chiti, F. and Dobson, C. M. (2006) Protein misfolding, functional amyloid, and human disease. *Ann. Rev. Biochem.* 75, pp. 333–366.

Chromy, B. A., Nowak, R. J., Lambert, M. P., Viola, K. L., Chang, L., Velasco, P. T., Jones, B. W., Fernandez, S. J., Lacor, P. N., Horowitz, P., Finch, C. E., Krafft, G. A. and Klein, W. L. (2003) Self-assembly of Aβ (1–42) into globular neurotoxins. *Biochemistry* 42, pp. 12749–12760.

Cleary, J. P., Walsh, D. M., Hofmeister, J. J., Shankar, G. M., Kuskowski, M. A., Selkoe, D. J. and Ashe, K. H. (2005) Natural oligomers of the amyloid-β protein specifically disrupt cognitive function. *Nature Neurosci.* 8, pp. 79–84.

Demuro, A., Mina, E., Kayed, R., Milton, S. C., Parker, I. and Glabe, C. G. (2005) Calcium dysregulation and membrane disruption as a ubiquitous neurotoxic mechanism of soluble amyloid oligomers. *J. Biol. Chem.* 280, pp. 17294–17300.

Dobson, C. M. (1999) Protein misfolding, evolution and disease. *Trends Biochem. Sci.* 24, pp. 329–332.

Dong, J., Atwood, C. S., Anderson, V. E., Siedlak, S. L., Smith, M. A., Perry, G. and Carey, P. R. (2003) Metal binding and oxidation of amyloid-β within isolated senile plaque cores: Raman microscopic evidence. *Biochemistry* 42, pp. 2768–2773.

Fukumoto, H., Tokuda, T., Kasai, T., Ishigami, N., Hidaka, H., Kondo, M., Allsop, D. and Nakagawa, M. (2010) High-molecular weight β-amyloid oligomers are elevated in cerebrospinal fluid of Alzheimer patients. *FASEB J.* 24, pp. 2716–1726.

Haass, C. and Selkoe, D. J. (2007) Soluble protein oligomers in neurodegeneration: Lessons from the Alzheimer's amyloid β-peptide. *Nature Rev. Mol. Cell Biol.* 8, pp. 101–112.

Hardy J. and Allsop D. (1991) Amyloid deposition as the central event in the aetiology of Alzheimer's disease. *Trends Pharmacol. Sci.* 12, pp. 383–388.

Hardy, J., Lewis, P., Revesz, T., Lees, A. and Paisan-Ruiz, C. (2009) The genetics of Parkinson's disease: A critical review. *Current Opinion Gen. Devel.* 19, pp. 254–265.

Himes, R. A., Park, G. Y., Siluvai, G. S., Blackburn, N. J. and Karlin, K. D. (2008) Structural studies of copper(I) complexes of amyloid-β peptide fragments: formation of two-coordinate bis(histidine) complexes. *Angew. Chem. Int. Ed. Engl.* 47, pp. 9084–9087.

Huang, X. D., Atwood, C. S., Hartshorn, M. A., Multhaup, G., Goldstein, L. E., Scarpa, R. C., Cuajungco, M. P., Gray, D. N., Lim, J., Moir, R. D., Tanzi, R. E. and Bush, A. I. (1999a) The Aβ peptide of Alzheimer's disease directly produces hydrogen peroxide through metal ion reduction. *Biochemistry* 38, pp. 7609–7616.

Huang, X. D., Cuajungco, M. P., Atwood, C. S., Hartshorn, M. A., Tyndall, J. D. A., Hanson, G. R., Stokes, K. C., Leopold, M., Multhaup, G., Goldstein, L. E., Scarpa, R. C., Saunders, A. J., Lim, J., Moir, R. D., Glabe, C., Bowden, E. F., Masters, C. L., Fairlie, D. P., Tanzi, R. E. and Bush, A. I. (1999b) Cu(II) potentiation of Alzheimer Aβ neurotoxicity - correlation with cell-free hydrogen peroxide production and metal reduction. *J. Biol. Chem.* 274, pp. 37111–37116.

Hofer, A., Berg, D., Asmus, F., Niwar, M., Ransmayr, G., Riemenschneider, M., Bonelli, S.-B., Steffelbauer, M., Ceballos-Baumann, A., Haussermann, P., Behnke, S., Krüger, R., Prestel, J., Sharma, M., Zimprich, A., Riess, O. and Gasser, T. (2005) The role of α-synuclein gene multiplications in early-onset Parkinson's disease and dementia with Lewy bodies. *J. Neural Trans.* 112, pp. 1249–1254.

Hung, L. W., Ciccotosto, G. D., Giannakis, E., Tew, D. J., Perez, K., Masters, C. L., Cappai, R., Wade, J. D. and Barnham, K. J. (2008) Amyloid-β peptide (Aβ) neurotoxicity is modulated by the rate of peptide aggregation: Aβ dimers and trimers correlate with neurotoxicity. *J. Neurosci.* 28, pp. 11950–11958.

Jobling, M. F., Huang, X., Stewart, L. R., Barnham, K. J., Curtain, C., Volitakis, I., Perugini, M., White, A. R., Cherny, R. A., Masters, C. L., Barrow, C. J., Collins, S. J., Bush, A. I. and Cappai, R. (2001) Copper and zinc binding modulates the aggregation and neurotoxic properties of the prion peptide PrP106–26. *Biochemistry* 40, pp. 8073–8084.

Kaether, C., Haass, C. and Steiner, H. (2006) Assembly, trafficking and function of γ secretase. *Neurodegen. Dis.* 3, pp. 275–283.

Karr, J. W. and Szalai, V. A. (2008) Cu(II) binding to monomeric, oligomeric, and fibrillar forms of the Alzheimer's diseases amyloid-β peptide. *Biochemistry* 47, pp. 5006–5016.

Kayed, R., Head, E., Thompson, J. L., McIntire, T. M., Milton, S. C., Cotman, C. W. and Glabe, C. G. (2003) Common structure of soluble amyloid oligomers implies common mechanism of pathogenesis. *Science* 300, 486–489.

Kayed, R., Sokolov, Y., Edmonds, B., McIntire, T. M., Milton, S. C., Hall, J. E. and Glabe, C. G. (2004) Permeabilization of lipid bilayers is a common conformation-dependent activity of soluble amyloid oligomers in protein misfolding diseases. *J. Biol. Chem.* 279, pp. 46363–46366.

Khan, A., Ashcroft, A. E., Korchazhkina, O. V. and Exley, C. (2004) Metal-mediated formation of fibrillar ABri amyloid. *J. Inorg. Biochem.* 98, pp. 2006–2010.

Khan, A., Ashcroft, A. E., Higenell, V., Korchazhkina, O. V. and Exley, C. (2005) Metals accelerate the formation and direct the structure of amyloid fibrils of NAC. *J. Inorg. Biochem.h* 99, pp. 1920–1927.

Kim, H. J., Chae, S. C., Lee, D. K., Chromy, B., Lee, S. C., Park, Y. C., Klein, W. L., Krafft, G. A., and Hong, S. T. (2003) Selective neuronal degeneration induced by soluble oligomeric amyloid β protein. *FASEB J.* 17, pp. 118–120.

Kishida, K. T. and Klann, E. (2007) Sources and targets of reactive oxygen species in synaptic plasticity and memory. *Antioxid. Redox Signal.* 9, pp. 233–244.

Klein, W. L., Stine, W. B. and Teplow, D. B. (2004) Small assemblies of unmodified amyloid β protein are the proximate neurotoxin in Alzheimer's disease. *Neurobiol. Aging* 25, pp. 569–580.

Knapp, L. T. and Klann, E. (2002) Role of reactive oxygen species in hippocampal long-term potentiation: contributory or inhibitory? *J. Neurosci Res.* 70, pp. 1–7.

Lambert, M. P., Barlow, A. K., Chromy, B. A., Edwards, C., Freed, R., Liosatos, M., Morgan, T. E., Rozovsky, I., Trommer, B., Viola, K. L., Wals, P., Zhang, C., Finch, C. E., Krafft, G.A., and Klein, W. L. (1998) Diffusible, nonfibrillar ligands derived from Aβ 1–42 are potent central nervous system neurotoxins. *Proc. Natl. Acad. Sci. USA* 95, pp. 6448–6453.

Lashuel, H. A., Petre, B. M., Wall, J., Simon, M., Nowak, R. J., Walz, T. and Lansbury, P. T. (2002) α-Synuclein, especially the Parkinson's disease-associated mutants, forms pore-like annular and tubular protofibrils. *J. Mol. Biol.* 322, pp. 1089–1102.

Layfield, R., Lowe, J. and Bedford, L. (2005) The ubiquitin-proteasome system and neurodegenerative disorders. *Essays Biochem.* 41, pp. 157–171.

Lesné, S., Koh, M. T., Kotilinek, L., Kayed, R., Glabe, C. G., Yang, A., Gallagher, M. and Ashe, K. H. (2006) A specific amyloid-β protein assembly in the brain impairs memory. *Nature* 440, pp. 352–357.

Lovell, M. A., Robertson, J. D., Teesdale, W. J., Campbell, J. L. and Markesbery, W. R. (1998) Copper, iron and zinc in Alzheimer's disease senile plaques. *J. Neurol. Sci.* 158, pp. 47–52.

Markesbery, W. R., Kryscio, R. J. Lovell, M. A. and Morrow, J. D. (2005) Lipid peroxidation is an early event in the brain in amnestic mild cognitive impairment. *Ann. Neurol.* 58, pp. 730–735.

Masad, A., Hayes, L., Tabner, B. J., Turnbull, S., Cooper, L. J., Fullwood, N. J., German, M. J., Kametani, F., El-Agnaf, O. M. A. and Allsop, D. (2007)

Copper-mediated formation of hydrogen peroxide from the amylin peptide: A novel mechanism for degeneration of islet cells in type-2 diabetes mellitus? *FEBS Lett.* 581, pp. 3489–3493.

Millhauser, G. L. (2007) Copper and the prion protein: Methods, structures, function, and disease. *Annu. Rev. Phys. Chem.* 58, pp. 299–320.

Nunomura, A., Perry, G., Aliev, G., Hirai, K., Takeda, A., Balraj, E. K., Jones, P. K., Ghanbari, H., Wataya, T., Shimohama, S., Chiba, S., Atwood, C. S., Petersen, R. B. and Smith, M. A. (2001) Oxidative damage is the earliest event in Alzheimer disease. *J. Neuropathol. Exp. Neurol.* 60, pp. 759–767.

Opazo, C., Huang, X., Cherny, R. A., Moir, R. D., Roher, A. E., White, A. R., Cappai, R., Masters, C. L., Tanzi, R. E., Inestrosa, N. C. and Bush, A. I. (2002) Metalloenzyme-like activity of Alzheimer's disease β-amyloid. Cu-dependent catalytic conversion of dopamine, cholesterol, and biological reducing agents to neurotoxic H_2O_2. *J. Biol. Chem.* 277, pp. 40302–40308.

Paik, S. R., Shin, H. J. and Lee, J. H. (2000) Metal-catalyzed oxidation of α-synuclein in the presence of copper(II) and hydrogen peroxide. *Arch. Biochem. Biophys.* 378, pp. 269–277.

Pike, C. J., Burdick, D., Walencewicz, A. J., Glabe, C. G. and Cotman, C. W. (1993) Neurodegeneration induced by β-amyloid peptides *in vitro*: The role of peptide assembly state. *J. Neurosci.* 13, pp. 1676–1687.

Shankar, G. M., Li, S., Mehta, T. H., Garcia-Munoz, A., Shepardson, N. E., Smith, I., Brett, F. M., Farrell, M. A., Rowan, M. J., Lemere, C. A., Regan, C. M., Walsh, D. M., Sabatini, B. L. and Selkoe, D. J. (2008) Amyloid-β protein dimers isolated directly from Alzheimer's brains impair synaptic plasticity and memory. *Nature Med.* 14, pp. 837–842.

Sipe, J. D. and Cohen, A. S. (2000) Review: History of the amyloid fibril. *J. Struct. Biol.* 130, pp. 88–98.

Smith, M. A., Harris, P. L. R., Sayre, L. M. and Perry, G. (1997) Iron accumulation in Alzheimer's disease as a source of redox-generated free radicals. *Proc. Natl. Acad. Sci. USA* 94, pp. 9866–9868.

Soto, C. (2001) Protein misfolding and disease; protein refolding and therapy. *FEBS Lett.* 498, pp. 204–207.

Soto, C. (2003) Unfolding the role of protein misfolding in neurodegenerative diseases. *Nature Rev. Neurosci.* 4, pp. 49–60.

St. George-Hyslop, P. H. (2000) Molecular genetics of Alzheimer's disease. *Biol. Psychiat.* 47, pp. 183–199.

Tabner, B. J., Turnbull, S., El-Agnaf, O. M. A. and Allsop, D. (2001) Production of reactive oxygen species from aggregating proteins implicated in Alzheimer's disease, Parkinson's disease and other neurodegenerative diseases. *Curr. Top. Med. Chem.* 1, pp. 507–517.

Tabner, B. J., Turnbull, S., El-Agnaf, O. M. A. and Allsop, D. (2002) Formation of hydrogen peroxide and hydroxyl radicals from Aβ and α-synuclein as a possible mechanism of cell death in Alzheimer's disease and Parkinson's disease. *Free Radic. Biol. Med.* 32, pp. 1076–1083.

Tabner, B. J., Turnbull, S., El-Agnaf, O. M. A. and Allsop, D. (2003) Direct production of reactive oxygen species from aggregating proteins and peptides implicated in the pathogenesis of neurodegenerative diseases. *Curr. Med. Chem. – Immun., Endoc. & Metab. Agents* 3, pp. 299–308.

Tabner, B. J., El-Agnaf, O. M. A., Turnbull, S., German, M. J., Paleologou, K. E., Hayashi, Y., Cooper, L. J., Fullwood, N. J. and Allsop, D. (2005) Hydrogen peroxide is generated during the very early stages of aggregation of the amyloid peptides implicated in Alzheimer's disease and familial British dementia. *J. Biol. Chem.* 280, pp. 35789–35792.

Tabner, B. J., Turnbull, S., King, J. E., Benson, F. E., El-Agnaf, O. M. A. and Allsop, D. (2006) A spectroscopic study of some of the peptidyl radicals formed following hydroxyl radical attack on β-amyloid and α-synuclein. *Free Radic. Res.* 40, pp. 731–739.

Terry, R. D., Masliah, E., Salmon, D. P., Butters, N., DeTeresa, R. Hill, R., Hansen, L. A. and Katzman, R. (1991) Physical basis of cognitive alterations in Alzheimer's disease: Synapse loss is the major correlate of cognitive impairment. *Ann. Neurol.* 30, pp. 572–580.

Turnbull, S., Tabner, B. J., El-Agnaf, O. M. A. and Allsop, D. (2001) α-Synuclein implicated in Parkinson's disease catalyses the formation of hydrogen peroxide *in vitro*. *Free Radic. Biol. Med.* 30, pp. 1163–1170.

Turnbull, S., Tabner, B. J., Brown, D. R. and Allsop, D. (2003a) Copper-dependent generation of hydrogen peroxide from the toxic prion protein fragment PrP106-126. *Neurosci. Lett.* 336, pp. 159–162.

Turnbull, S., Tabner, B. J., Brown, D. R. and Allsop, D. (2003b) Generation of hydrogen peroxide from mutant forms of the prion protein fragment PrP121-231. *Biochemistry* 42, pp. 7675–7681.

Vance, C., Rogelj, B., Hortobágyi, T., De Vos, K. J., Nishimura, A. L., Sreedharan, J., Hu, X., Smith, B., Ruddy, D., Wright, P., Ganesalingam, J., Williams, K. L., Tripathi, V., Al-Saraj, S., Al-Chalabi, A., Leigh, P. N., Blair, I. P., Nicholson, G., de Belleroche, J., Gallo, J. M., Miller, C. C. and Shaw, C. E. (2009) Mutations in FUS, an RNA processing protein, cause Familial Amyotrophic Lateral Sclerosis Type 6. *Science* 323, pp. 1208–1211.

Walker, L. C., Levine, H. 3rd, Mattson, M. P. and Jucker, M. (2006) Inducible proteopathies. *Trends Neurosci.* 29, pp. 438–443.

Walsh, D. M., Hartley, D. M., Kusumoto, Y., Fezoui, Y., Condron, M. M., Lomakin, A., Benedek, G. B., Selkoe, D. J. and Teplow, D. B. (1999) Amyloid β-protein fibrillogenesis. Structure and biological activity of protofibrillar intermediates. *J. Biol. Chem.* 274, pp. 25945–25952

Walsh, D. M., Klyubin, I., Fadeeva, J. V., Cullen, W. K., Anwyl, R., Wolfe, M. S., Rowan, M. J., and Selkoe, D. J. (2002) Naturally secreted oligomers of amyloid β protein potently inhibit hippocampal long-term potentiation *in vivo*. *Nature* 416, pp. 535–539.

Walsh, D. M. and Selkoe, D. J. (2007) Aβ oligomers – a decade of discovery. *J. Neurochem.* 101, pp. 1172–1184.

Wang, H. W., Pasternak, J. F., Kuo, H., Ristic, H., Lambert, M. P., Chromy, B., Viola, K. L., Klein, W. L., Stine, W. B., Krafft, G. A., and Trommer, B. L. (2002) Soluble oligomers of β amyloid (1–42) inhibit long-term potentiation but not long-term depression in rat dentate gyrus. *Brain Res.* 924, pp. 133–140.

Westermark, P. (2005) Aspects on human amyloid forms and their fibril polypeptides. *FEBS J.* 272, 5942–5949.

Wogulis, M., Wright, S., Cunningham, D., Chilcote, T., Powell, K. and Rydel, R. E. (2005) Nucleation-dependent polymerization is an essential component of amyloid-mediated neuronal cell death. *J. Neurosci.* 25, pp. 1071–1080.

Chapter 3

Modelling of the Metal Binding Sites in Proteins Involved in Neurodegeneration

Ewa Gralka, Daniela Valensin, Maurizio Remelli, and Henryk Kozlowski

Faculty of Chemistry, University of Wroclaw, F. Joliot-Curie 14, 50383 Wroclaw, Poland

Dipartimento di Chimica, Università di Siena, Siena, Italy

Dipartimento di Chimica, Università di Ferrara, Ferrara, Italy

3.1 Introduction

Many proteins involved in the neurodegeneration processes are potential metalloproteins. In some of them the metal binding domain is flexible or unstructured (e.g., prion proteins, β-amyloid peptide, α-synuclein) resembling oligo-peptide chains. The regular protein structure usually has a critical impact on the binding ability of metal ion to protein scaffold, while in peptides with random structures the individual binding ability of the particular amino acid residue may decide about the peptide binding to metal ion (Kozlowski *et al.*, 1999; Kozlowski *et al.*, 2005; Kozlowski *et al.*, 2008). Thus, it seems to be reasonable to use short peptide fragments of the unstructured protein to establish the detailed characteristics of the metal-protein interactions. In the peptide or protein systems with no distinct structural arrangements the major anchoring binding sites for metal ions like Cu^{2+} are the N-terminal amino

Brain Diseases and Metalloproteins
Edited by David R. Brown

ISBN 978-981-4316-01-9 (Hardcover), 978-981-4364-07-2 (eBook)
www.panstanford.com

nitrogen and nitrogen donor of imidazole side chain of histydyl residue. When N-terminal function is protected like it happens for peptide fragments in the protein sequence, the major anchoring sites are imidazoles of His residues. Cu^{2+} ions anchored on the His side chain can then coordinate to the successive deprotonated nitrogen donors of adjacent amide bonds (Kozlowski *et al.*, 1999; Kozlowski *et al.*, 2005). The side chains of the other amino acid residues can affect distinctly the peptide/protein binding ability towards metal ions but it happens usually *via* indirect interactions e.g., with involvement of hydrogen bond network or stacking between the aromatic side chains. The metal ion interaction with whole protein within its unstructured domain may differ from that observed in metal-peptide system e.g. in thermodynamic stability or kinetics, but the binding sites and the structure around the binding sites should be very similar to each other. In any case the precise information obtained from the studies on the metal-peptide models are very useful in understanding of the interactions with whole proteins showing e.g., the binding preferences and most characteristic features resulting from the metal ion binding to protein. Thus, proteins containing His-rich domains within their unstructured sequences represent strong potential to bind metal ions with possible biological consequences. In this chapter the selected examples for prion proteins, α-synuclein, and β-amyloid peptide will be discussed both from thermodynamic and structural point of view. Some oxidant and anti-oxidant properties of these metallopeptide systems will also be discussed.

3.2 Peptides as Models for Unstructured Protein Interactions with Metal Ions

The most frequently used approach to study the stability of a complex formed between a protein and metal ion is the same generally employed in biochemistry to study the binding between a low-molecular-weight ligand and a macromolecule: during titration of a known substrate amount with the ligand, the free and bound substrate and/or free ligand concentrations are determined and the dissociation constant (K_d) is computed. If the ligand is a metal ion (M^{n+}) and the substrate is a protein (P), the dissociation equilibrium is the following:

$$\mathrm{MP} \rightleftharpoons \mathrm{M} + \mathrm{P} \qquad K_\mathrm{d} = \frac{[\mathrm{M}][\mathrm{P}]}{[\mathrm{MP}]} \tag{3.1}$$

where charges are omitted for the sake of simplicity and square brackets represent molar concentrations. K_d value corresponds to $[M]_{50}$, the concentration of free metal ion when [P] = [MP], i.e., when the protein is 50% bound. K_d can correctly describe the affinity between M^{n+} and P only if the stoichiometry of the adduct is 1:1 and side acid-base reactions are not involved.

When the protein bears *n* independent and equivalent binding sites, K_d can be taken as the affinity of M for a single site, and it can be estimated by using the *Scatchard* equation:

$$\frac{r}{[\mathrm{M}]} = \frac{n}{K_\mathrm{d}} - r\frac{1}{K_\mathrm{d}} \tag{3.2}$$

where *r* is the *saturation fraction*, i.e., the mean number of occupied sites per protein unit. A plot of $r/[M]$ vs. *r* would give a straight line with negative slope corresponding to $-1/K_d$; from the intercept it is possible to determine *n*. If the sites are not equivalent, the plot will show a curvature and Scatchard equation should be modified by introducing more parameters. Moreover, if the substrate contains *n* non-independent sites, *Hill* equation can be employed:

$$\log\left(\frac{r}{1-r}\right) = n\log[\mathrm{M}] - \log K'_\mathrm{d} \tag{3.3}$$

where K'_d is the apparent and overall dissociation constant:

$$\mathrm{M}_n\mathrm{P} \rightleftharpoons n\mathrm{M} + \mathrm{P} \qquad K'_\mathrm{d} = \frac{[\mathrm{M}]^n[\mathrm{P}]}{[\mathrm{M}_n\mathrm{P}]} \tag{3.4}$$

and it will be equal to $[\mathrm{M}]_{50}^n$, where $[\mathrm{M}]_{50}$ can be considered an average estimation of the microscopic dissociation constant of each single site. A plot of $\log\left(\frac{r}{1-r}\right)$ vs. log[M] would give a straight line with slope *n*, the *Hill constant*, interpreted to describe the binding cooperativity: $n = 1$ means no cooperativity (independent sites); $n > 1$ means positive cooperativity and $n < 1$ means negative cooperativity.

The above procedure, relatively simple if [M] and [P] or [MP] can be measured during the titration, is not free from drawbacks. First of all, the experimental K_d value depends on pH (since protonation equilibria are not explicitly considered) and on the possible presence of both a buffer and a competing metal chelator. As a result, K_d values are hardly comparable if determined under different experimental conditions. Second, the binding site is often depicted as a template where the metal ion can accommodate: on the contrary, the geometry and also the donor-atom set of a given binding site can change with experimental conditions and a mixture of different species is most often present in solution, each one characterized by a different binding constant. Third, when two or more binding sites are present, the binding pathway can be not simply sequential: different complexes can be simultaneously present in solution, possessing the same metal/protein stoichiometry but characterized by different combinations of occupied sites. In this case it is not correct to attribute the first K_d value to the species 1:1, the second one to the species 2:1, and so on. Finally, the most realistic representation of the system is often that considering different *binding modes*, corresponding to species with different complexation geometries, rather than different *binding sites* representing distinct domains of the protein. The last considerations make it evident that K_d can give only an unfocused picture of the metal-protein interactions since it is unable to account for all the species formed in the system. As a consequence, the species really active *in vivo* could be unrecognized if they are formed in low fraction. On the other hand, this procedure is valuable to give a general description of the behaviour of high-molecular-weight substrates containing a high number of acidic or basic groups and of potentially donor atoms.

The method of choice to completely describe the complex-formation equilibria involving a protein and a metal ion is the investigation on model systems containing relatively short and soluble peptides, corresponding to only one active site (or a few of them): their thermodynamic characterization can lead to the full knowledge of the solution composition, in a wide range of experimental conditions. Moreover, the peptides models are suitable for spectroscopic investigations better than wild type proteins often scarcely soluble at physiological pH, thus allowing the proposal of structural hypotheses for the main complexes, in solution. Finally, short models are better handled for theoretical calculations.

Most of thermodynamic information is contained in the *speciation model*, including the stoichiometry and the thermodynamic stability (independent of pH and reagent concentrations) of all the complex species formed in solution, provided they reach a percentage higher than ca. 5% with respect to the total metal ion or peptide. Evidently, the use of models implies a simplification of the system, neglecting long-range effects. Moreover, short peptides can effectively represent only unstructured protein domains. In other words, if the binding site is not hindered by constraints deriving from the secondary or tertiary protein structure and it can adjust its shape to accommodate at the best the metal ion, then a fragment containing all the residues involved in complex-formation is able to reproduce the behavior of the protein. This has been well documented in the case of prion protein (see below).

Complex-formation constants contained in the speciation model only depend on temperature and ionic strength. Considering a metal ion, a ligand (L) and the proton (H), a complex-formation reaction can be described as follows:

$$pM + qL + rH \rightleftarrows M_pL_qH_r \qquad \beta_{pqr} = \frac{[M_pL_qH_r]}{[M]^p[L]^q[H]^r} \tag{3.5}$$

where charges and coordinated solvent-molecules are omitted for clarity; β_{pqr} is the "overall" complex-formation constant and p, q and r are the stoichiometric-coefficients of M, L and H respectively. The proton stoichiometric-coefficient r can be negative if the number of dissociated protons exceeds the maximum number of protons that the ligand can release in the absence of metal ion or the dissociated protons derive from water molecules in the first coordination sphere. The speciation model can be determined by using an analytical technique able to measure the free concentration of one species involved in complex-formation equilibria, along a titration. Potentiometry is the best choice, especially if acid-base equilibria are present and the glass electrode can be employed; other used techniques are spectrophotometry, NMR or solution calorimetry. Suitable computer programs are available to determine the speciation model that fits at the best the experimental titration curve (Gans *et al.*, 1996; Gans *et al.*, 2008). The solution composition, under each given experimental condition, can be computed and distribution or competition diagrams can be plotted (Alderighi *et al.*, 1999). In addition, the knowledge of speciation models relative to different metal/ligand systems, allows the computation of molar concentrations of every species forming in a multi-metal and/or multi-ligand system under the unique hypothesis (often accomplished) that no mixed species is formed. Finally, speciation models also allow to calculate the pM value (= $-\log[M]_{free}$, under given experimental conditions) a parameter often employed to compare the binding ability of different competing ligands (Crisponi and Remelli, 2008).

It should be clear at this point that K_d and β_{pqr} can have very different meanings, especially when many complex species, having different metal/ligand stoichiometric ratio and/or protonation degree, are simultaneously present in solution. In the simplest cases it is possible to calculate *apparent* (pH dependent) constants from the *absolute* constant, described by eq. (5) (Dawson *et al.*, 1989) or *vice versa*. Otherwise, the knowledge of the complete speciation model allows one to "simulate" an experiment of the same type of those used to determine K_d, calculating at each titration point the values of [M], [P] and [MP] requested to built up the Scatchard or Hill plots. The same procedure also allows to compute directly the corresponding $[M]_{50}$ value (Kozlowski *et al.*, 2010).

3.3 Structural Approach to Metal-Peptide Interactions (Role of His as the Metal Ion Binding Site)

The prion, the β-amyloid and the α-synuclein proteins have the peculiarities to bind several transition metal ions, such as Cu^{2+}, Zn^{2+}, Mn^{2+} and Ni^{2+}. It is worthy to emphasize that Cu^{2+} binding is a general feature for prion, β-amyloid and α-synuclein proteins suggesting that copper homeostasis might be related to most of the neurodegenerative disorders. In this section the metal binding abilities and the metal complex structural features of these proteins, determined by means of peptide models will be discussed.

Before talking about the structural characterization of the metal binding sites some general considerations on Cu^{2+} and Zn^{2+}

inorganic chemistry are outlined. Firstly, according to the principle of hard/soft acids and bases Cu^{2+} has a preference towards N over O or S donors. Secondly, Cu binding is in competition with protonation, which means that Cu^{2+} binds preferentially to N or O ligands with a low pK_a. Considering copper binding to amino acids the possible nitrogen donor atoms derive from side-chains of His ($pK_a \sim 6.5$), N-terminal amino group ($pK_a \sim 8$) and to much less extent from side-chains of Lys ($pK_a \sim 10.5$) or Arg ($pK_a \sim 12.5$), eventually from backbone peptide bonds ($pK_a > 15$). So His residue and then N-terminus are the preferred Cu^{2+} anchoring sites. Functions with high pK_a are only available when Cu^{2+} ion is brought in their proximity (e.g., to form a chelate). Indeed, the close proximity of Cu^{2+} to a donor system induces a decrease in its pK_a value. Cu^{2+} anchored to His imidazole or N-terminus may induce deprotonation of amide function close to copper ion (amide of residue(s) preceding or following the anchoring site). This results into the formation of a stable five-, six- or even seven-membered chelate rings. On the contrary Zn^{2+} is not able to coordinate to amide nitrogens, and imidazole(s), N-terminus and carboxyl moieties are the only possible donor groups.

3.3.1 *Peptide Models for Prion Proteins*

The long flexible N-terminal region of the human prion protein, $hPrP_{23-120}$, contains six His residues (His61, His69, His77, His85, His96 and His111) which can anchor up to six Cu^{2+} ions (for reviews see Davies *et al.*, 2008; Walter *et al.*, 2009; Viles *et al.*, 2008; Kozlowski *et al.*, 2006; Kozlowski *et al.*, 2008). The $hPrP_{60-91}$ region, usually refereed as the octarepeat domain (Octa4), is highly conserved across all mammalian PrP and it is composed of four tandem repeats of the fundamental octapeptide sequence PHGGGWQG (Octa1). A huge number of investigations indicate that Octa4 region can bind up to four copper ions and that there are two additional Cu^{2+} binding sites outside this region. These two independent binding sites are located at His96 and His111, within the so called amyloidogenic region, $hPrP_{90-126}$, of the flexible N-terminal PrP^C tail. As regards the copper preference between the octarepeat and the amyloidogenic region no general consensus is reached and a detailed discussion on this topic will be tackled in the next section.

The determination of the copper binding donors atoms and the structural features of Cu^{2+} complexes has been obtained by using different peptide models encompassing the N-terminal region from mammals and non mammals prion proteins. The major findings and the derived biological relevance will be reported below on the basis of the investigated peptide model.

3.3.1.1 The single octapeptide Ac-PHGGGWQG-NH_2 (Octa1) model

The copper coordination sphere of the single octapeptide unit (Octa1) has been investigated by using many different spectroscopic techniques. It is widely accepted that copper coordination is pH dependent and that the His imidazole is the copper anchoring site. Additionally, the amide coordination from the next Gly residues occurs at physiological pH. However slightly different conclusions have been reported on the exact copper coordination sphere. Raman and absorption spectroscopy showed that at neutral and basic pH, the single Octa1 forms a 1:1 complex with the metal ion bound to imidazole and to two deprotonated main-chain amide nitrogens (Miura *et al.*, 1999). A study based on circular dichroism (CD) and electron paramagnetic resonance (EPR) measurements, performed at pH>6, have detected the presence of a square pyramidal 4N Cu^{2+} geometry with three amide backbone and the imidazole nitrogens as the copper donor atoms (Bonomo *et al.*, 2000). An extensive EPR and electron spin-echo envelope modulation (ESEEM) investigation on Octa1 and its related fragments in *N*-ethylmorpholine (NEM) buffer, showed that Ac-HGGGW-NH_2, detected as the minimal copper binding unit, has a planar 3N1O copper coordination sphere (Aronoff-Spencer *et al.*, 2000). At physiological pH, the involvement of three nitrogen atoms in copper coordination has been later on confirmed by further potentiometric and spectroscopic analysis where the detection of Cu^{2+} induced ^{13}C and ^{1}H nuclear magnetic resonance (NMR) line broadening has given the first evidence of copper binding to His2 imidazole Nπ and Gly3, Gly4 deprotonated backbone amide nitrogen atoms in solution (Luczkowski *et al.*, 2002) (see below). Furthermore, the measurements of paramagnetic relaxation enhancements have also indicated that Trp indole is approaching the copper center, anticipating a Trp stabilizing role in copper binding (Luczkowski *et al.*, 2002). The proximity of Trp has been also

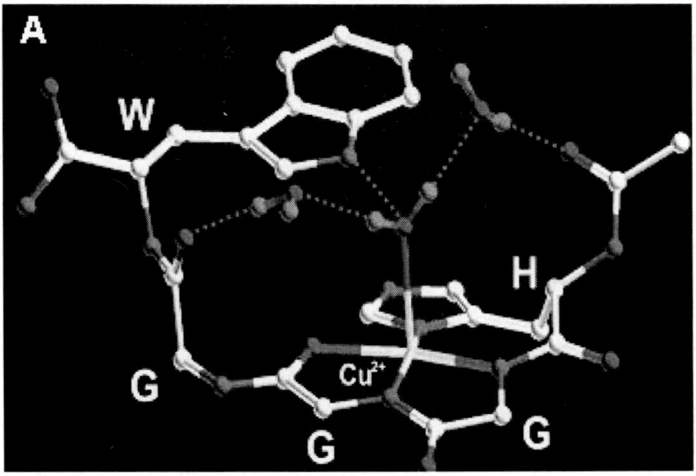

Figure 3.1 X-Ray Structure of Cu^{2+}-Ac-HGGGW-NH_2 complex (from Burns *et al.*, 2002).

probed by three-pulse ESEEM and hyperfine sub-level correlation (HYSCORE) analysis, which additionally have indicated the nature of copper bound O donor atom, from carbonyl oxygen rather than from a water molecule (Hureau *et al.*, 2008). The first structural features of the copper binding sites to PHGGGWGQ sequence was provided by the X-ray structure of the Cu^{2+}-Ac-HGGGW-NH_2 complex (Burns *et al.*, 2002). The metal ion is penta-coordinated with four donors in a square planar arrangement and an axial water molecule. In the equatorial plane the His imidazole Nπ, the deprotonated amides from the next two Gly residues and the carbonyl of the second Gly are the copper donor atoms. The NH of Trp indole forms an hydrogen bond with the copper bound water molecule (Fig. 3.1).

Successively, numerous similar structures have been obtained from theoretical calculations on Cu^{2+} binding to different Octa1 model systems (Ac-HGGG-NH_2, Ac-HGGGW-NH_2 and Ac-PHGGGWGQ-NH_2). Extensive X-Ray absorption fine structure (EXAFS), EPR, electron nuclear double resonance (ENDOR), ESEEM together with computer calculation in iterative way, have revealed a different orientation of the Trp indole and the relative position of the copper bound axial water, which has been found on the other side of the metal center respect to the structure obtained by X-Ray (Mentler *et al.*, 2005).

Ab initio molecular dynamics (MD) simulations of the Car-Parrinello (CP) type and B3LYP hybrid density functional methods show that (i) deprotonation and coordination of the backbone NH atoms of the two following Gly is strongly favored by their trans position in respect to His imidazole ring (Pushie and Rauk, 2003); (ii) the Cu^{2+}-N (Gly) bonds are stronger than the Cu^{2+}-Nπ one (Furlan *et al.*, 2007); (iii) while the square planar coordination (3N1O) is maintained in all the CPMD trajectory, the Cu^{2+} bound axial water moves away from the metal center (Furlan *et al.*, 2007) and (iv) the Cu^{2+} bond to the carbonyl oxygen of the second Gly is weaker than the other three N-Cu^{2+} bonds (Furlan *et al.*, 2007).

Other recent theoretical calculations, performed on Cu^{2+}-Octa1 model systems, point out the lability (Riihimaki *et al.*, 2007), or even exclude the presence (Marino *et al.*, 2007) of the axial Cu^{2+} bound water molecule. The final structures indicate that Trp indole ring is located at the top of the equatorial Cu^{2+} coordination plane without the presence of any hydrogen bonding interaction with the axial water molecule (Riihimaki *et al.*, 2008) suggesting electrostatic interaction between the Trp and His side-chains.

The strong structural rearrangements of Octa1 upon Cu^{2+} coordination, have also been evaluated by analyzing the far UV CD spectra, which are distinctly different from the typical random coil peptide spectra (Garnett and Viles, 2003). The CD analysis of analogues peptides (Ac-HAAAW-NH_2 and Ac-GQAHGGGWG-NH_2) revealed that beyond His and Trp directly involved in the Cu^{2+} complexes, (i) also the Gly residues are essential for main chain coordination; (ii) only when Pro is present Cu^{2+} coordinates the two Gly main chain amides, indicating a key role of Pro in forcing the amide backbone deprotonation towards the C-terminus. The effects of the Gly/Ala substitution have been additionally investigated by NMR and potentiometric analysis of Cu^{2+}-Ac-HGGGW-NH_2 and Ac-HAAAW-NH_2 complexes (Luczkowski *et al.*, 2003; Gralka *et al.*, 2008). Both peptides bind Cu^{2+} similarly: at neutral pH, the Cu^{2+} ion is bound to His Nπ, two adjacent amide nitrogen and one oxygen (carbonyl or water) atoms, resulting in a 3N1O binding mode. The NMR behavior of the two systems is also similar, indicating the involvement of the imidazole ring and the proximity of Trp indole to the metal center. For both peptides, the copper induced paramagnetic relaxation effects (much more pronounced on the imidazole than on the rest of the protons) support Cu^{2+} binding as a two step process. Firstly Cu^{2+} anchors at the imidazole nitrogen and secondly deprotonation and binding to adjacent amide nitrogens occurs. The NMR derived 3D structures of the Cu^{2+}-Ac-HAAAW- NH_2 system show two major Trp orientations with respect to the Cu^{2+} coordination plane. The found Cu^{2+}–Trp indole NH distances (0.49–0.55 nm), are consistent with the occurrence of a hydrogen bond between the indole NH and the oxygen of an axial Cu^{2+}-bound water, as detected in the X-ray structure (Gralka *et al.*, 2008).

In conclusion, the most interesting features in the Cu^{2+} interactions with the mammalian Octa1 are: the formation of a seven-membered ring involving His imidazole and the next amide group in the C-terminus direction and the close proximity of Trp side chain to the metal ion both in solution and in solid state. The Cu^{2+} indole interaction (mediated by water hydrogen bond or by electrostatic interaction) is believed to play a key role for the stability of the Cu^{2+}-Octa1 complex.

3.3.1.2 The four octapeptide $(PHGGGWQG)_4$ (Octa4) model

The first evidence on copper binding to Octa4 was obtained by using mass spectrometry (Hornshaw *et al.*, 1995a; Hornshaw *et al.*, 1995b). The primary studies aimed to determine the copper coordinating atoms have revealed that Cu^{2+} coordination is strongly dependent on pH, and on copper concentration. Copper binding to only the four His residues (multi-His coordination) is found at pH lower than 7.4 (Miura *et al.*, 1999) and in presence of equimolar Cu^{2+} concentration (Valensin *et al.*, 2004). The schematic representation of copper multi His binding is shown in Fig. 3.2 (Kozlowski *et al.*, 2009).

Additionally, CD, NMR spectroscopy (Viles *et al.*, 1999) and electrospray ionization mass spectrometry (ESI-MS) (Whittal *et al.*, 2000) performed at physiological pH, showed that Octa4 peptide binds up to four copper ions. The speciation profile of Cu^{2+}-Octa4 complexes, obtained in presence of 4.0 Cu^{2+} equivalents, shows that the species, dominating at pH around 7, is characterized by spectroscopic parameters consistent with each copper anchored to His imidazole and further coordinated by two or three amide nitrogens (Valensin *et al.*, 2004). A recent detailed EPR study also shows that at pH 7.4 three different types of copper coordination can occur according to

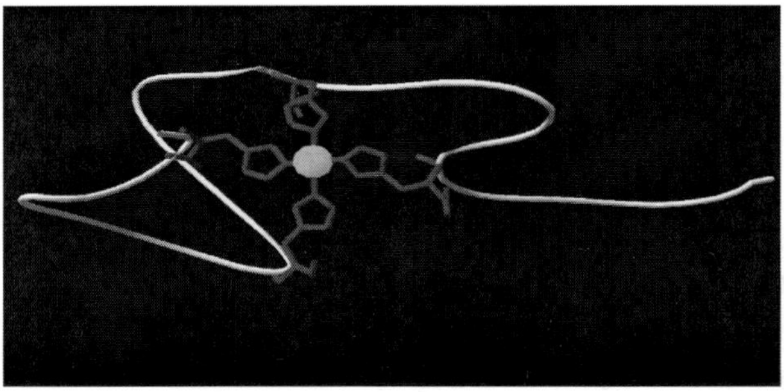

Figure 3.2 Schematic representation of the multi-His coordination of the Cu^{2+}-Octa4 complex (from Kozlowski *et al.*, 2009 with permission from Elsevier). See also Colour Insert.

metal concentration (Chattopadhyay *et al.*, 2005). These three copper binding modes, classified as component 1, 2 and 3, have been widely characterized and are now well accepted. Component 1 is the main copper binding mode at high Cu^{2+} occupancy (2–4 Cu^{2+} equivalents) and it is characterized by copper coordination to each octapeptide unit (PHGGGWGQ) in a mode exactly coincident to Fig. 3.1. Component 2, dominating at intermediate Cu^{2+} occupancy (1–2 Cu^{2+} equivalents) is due to copper binding to one or two His imidazoles and to a deprotonated amide nitrogen. Component 3, present at low Cu^{2+} occupancy (1 and less Cu^{2+} equivalents) is represented by a single Cu^{2+} ion bound to four His imidazoles (multiple His coordination, Fig. 3.2). The components 1 and 3 are also referred to intra and inter-repeat copper binding modes respectively. No experimental structures of the above mentioned Cu^{2+}-Octa4 binding modes have been determined so far due to the apparent intrinsic flexibility of this domain. The only available structure is that one of Cu^{2+}-Octa2 complex, determined by NMR relaxation rate enhancements calculated with trace amount of metal, showing copper binding to the two His $N\tau/N\pi$ (Valensin *et al.*, 2004).

Theoretical calculations, CD spectroscopy and other techniques have been applied to gain insight on the structural conformation of all the three Cu^{2+}-Octa4 complexes. Dipolar couplings arising from proximal copper centers have been determined for component 1, suggesting the occurrence of packing interaction between the four Cu^{2+}-HGGGW domain. The suggested Cu-Cu distances are consistent with van-der Waals contact between the metal centers and with hydrophobic collapse of the four octapeptide units upon copper binding (Chattopadhyay *et al.*, 2005). The CD spectra of Octa2, Octa3 or Octa4 fully loaded copper systems are also indicative of interactions between multiple Cu^{2+} centers and support that the -GQP- motifs linking the copper binding sites, are highly structured upon the coordination of the four Cu^{2+} ions (Garnett and Viles, 2003). In line with the experimental observations, molecular dynamics (MD) simulations performed on peptide sequences modeling the Octa2 bound to Cu^{2+} at full occupancy, show that the linker region between each –HGGG- sequence adopts a bent or turn structure (Pushie and Vogel, 2007). MD simulations performed on the whole N-terminal domain ($hPrP_{30-120}$) indicate that the coordination of four copper ions results in compact structural rearrangements which allow the metal coordinating sites to come close to each other (Valensin *et al.*, 2009).

Component 2 and 3 structures have been recently obtained by theoretical calculations. DFT modelling and the calculation of the aqueous free energy change on the component 3 binding mode (multi His coordination) show that Cu^{2+} coordination to Nτ is preferred to Nπ (Pushie and Vogel, 2008). MD simulation of component 3 indicates that Trp80 and Trp89 are relatively close to the metal ions (Pushie and Vogel, 2008). In line with these reports, a recent developed hybrid DFT/DFT method applied to the multiple His component 3, supports copper binding to four His Nτ in a planar geometry, as shown in Fig. 3.3A (Hodak *et al.*, 2009). The calculations show that the binding of four His creates three intermediate segments of identical sequence GGGWGQP, which, although flexible, form turns. The formed turns are strongly dependent on the distance between the copper bound His and they become regular β-turns when the His Cα-Cu^{2+} distances are less than 0.8 nm (Fig. 3.3B). Similar calculations have been performed on component 2 binding mode showing that 2 His imidazoles, one of them in axial position, one deprotonated amide and two water molecules constitute the copper coordination sphere (Fig. 3.4) (Hodak *et al.*, 2009).

3.3.1.3 The $hPrP_{90-126}$ model

His96 and His111 are the Cu^{2+} anchoring sites within the amyloidogenic region of hPrP ($hPrP_{90-126}$). At physiological pH,

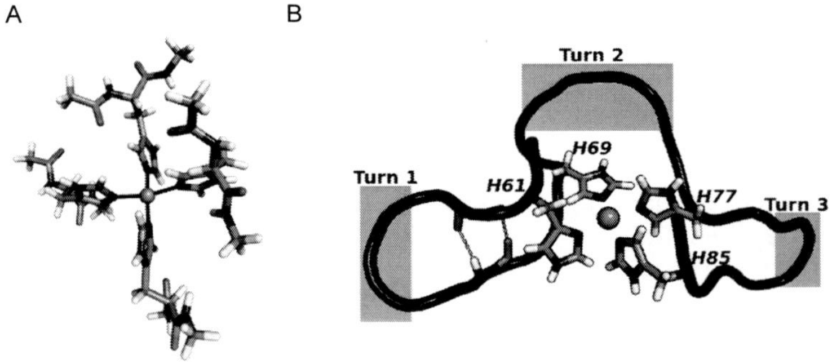

Figure 3.3 (A) Binding geometry of the multi His coordination of Cu^{2+}-Octa4 complex. (B) MD snapshot of the component 3 binding mode of Cu^{2+}-Octa4 complex (from Hodak *et al.*, 2009, reproduced by permission of PNAS).

the existence of two independent binding sites located at His96 and His111 is also found at low copper concentration, excluding the occurrence of multiple His binding modes (Remelli *et al.*, 2009). The two metal complexes (centered at His96 and His111), dominating at physiological pH, are a mixture of a 3N1O and 4N species in a tetragonal/square planar copper geometry. After copper anchoring to His imidazole (Nπ) metal coordination proceeds by step-wise deprotonation of backbone amides. A study on fragments $hPrP_{92-100}$ and $hPrP_{106-113}$ showed that copper amide deprotonation can proceed either towards the N-terminal or C-terminal direction forming a six-membered or seven membered chelate ring respectively (Berti *et al.*,

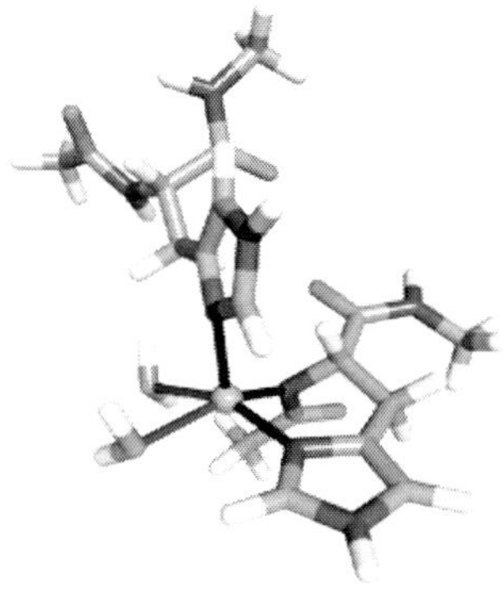

Figure 3.4 Binding geometry of the component 2 binding mode of Cu^{2+}-Octa4 complex (from Hodak *et al.*, 2009, reproduced by permission of PNAS).

2007), even though the former is preferred (Gralka *et al.*, 2008). The NMR structures of the four copper complexes involving the fragments $hPrP_{91-96}$, $hPrP_{96-100}$, $hPrP_{106-111}$ and $hPrP_{111-115}$ have been obtained by using NMR relaxations enhancements constraints (Fig. 3.5).

CD experiments on copper complexes with model peptides containing His96 and/or His111 are notably different form those observed for the apo system indicating that Cu^{2+} coordination forces the backbone to a fixed conformation dependent on geometrical arrangement of the metal ion and the stability of the chelate ring (Di Natale *et al.*, 2005; Di Natale *et al.*, 2009). In particular the CD difference spectra of the copper bound $hPrP_{91-115}$ show positive ellipticity at 195 nm and negative ellipticity at 217 nm, suggesting Cu^{2+} induced β-sheet conformations (Jones *et al.*, 2004).

MD simulations, performed on the $hPrP_{23-130}$ fragment binding four Cu^{2+} at Octa4 and one Cu^{2+} ion at His111 evidence a highly compact structure of the N-terminal domain upon Cu^{2+} coordination.

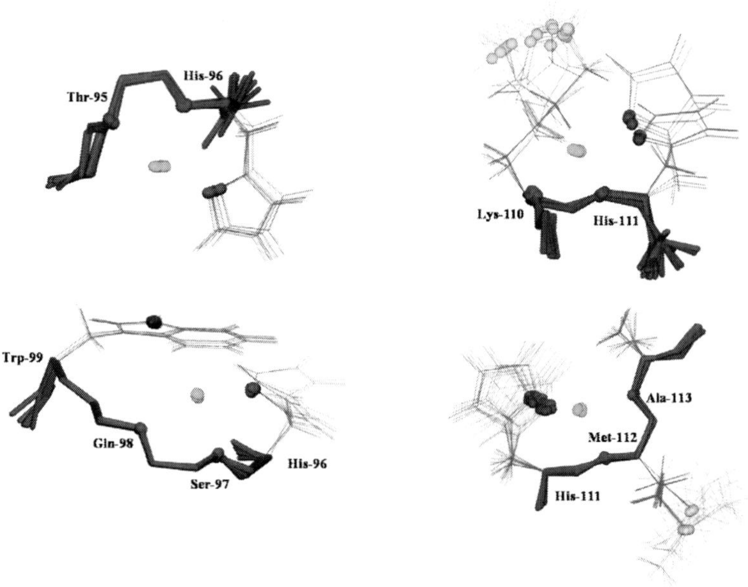

Figure 3.5 NMR structures of the following complexes: (A) Cu^{2+}-$hPrP_{91-96}$; (B) Cu^{2+}-$hPrP_{96-100}$; (C) Cu^{2+}-$hPrP_{106-111}$; (D) Cu^{2+}-$hPrP_{111-115}$. All the structures were obtained through restrained simulated annealing. Figure was created with MOLMOL 2K.1.0 (from Gralka *et al.*, 2008, reproduced by permission of the Royal Society of Chemistry). See also Colour Insert.

The presence of five different Cu^{2+} sites have yielded the detection of a relatively huge number of inter-residues contacts indicating a large changes in tertiary structure with all the five metal coordination sites close to each other (Valensin *et al.*, 2009b).

Although most of the experimental results on the His111 copper site are consistent with 3N1O copper coordination, few reports indicate the involvement of thioether from either Met109 or Met112 resulting in 2N1O1S square planar coordination environment (Shearer *et al.*, 2008). However the absence of any UV-Vis band characteristic of Cu-S charge transfer transition and the absence of large NMR effects on both Met methyl protons lead us to exclude copper coordination by Met sulphur atoms. On the contrary, as recently reported, NMR investigations on Cu^{+} binding to PrP_{91-124} fragments support the coordination of Met sulfur atoms to the diamagnetic copper ion. The proposed Cu^{+} coordination sphere consists of His96 and His111 imidazole nitrogen and Met109 and Met112 sulphur atoms (Badrick and Jones, 2009).

3.3.1.4 The non-mammalian PrP models

Other species different from mammals (e.g., birds and fishes) have also prion proteins. The investigations on non-mammalian PrP is at very early stages and there is no strong evidence that binding of biological metal ions (Cu^{2+} and Zn^{2+}) is biologically relevant. It is worthy to stress that these proteins also contain tandem repeat region rich in His and Gly, which can bind Cu^{2+} with relatively high affinity (Kozlowski *et al.*, 2008). Similarly to mammalian PrP, the N-terminal region of chicken PrP (chPrP) contains tandem repetitions of a hexapeptide sequence (PHNPGY). Copper coordination to the repeat chPrP region strongly depends on the number of repetitions. A single unit displays a 2N coordination mode, with the involvement of His Nπ and deprotonated His amide nitrogen (Fig. 3.6) (Stanczak *et al.*, 2005b).

On the contrary fragments containing more than one PHNPGY sequence coordinate Cu^{2+} by multiple His imidazoles, similarly to the component 3 binding mode detected for mammalian octarepeat region (Stanczak *et al.*, 2005a). A similar multiple His binding mode is also detected for Cu^{2+} coordination to the tandem repeat region of fish PrP-like proteins (Valensin D. *et al.*, 2009a; Gaggelli *et al.*, 2008a; Stanczak *et al.*, 2006).

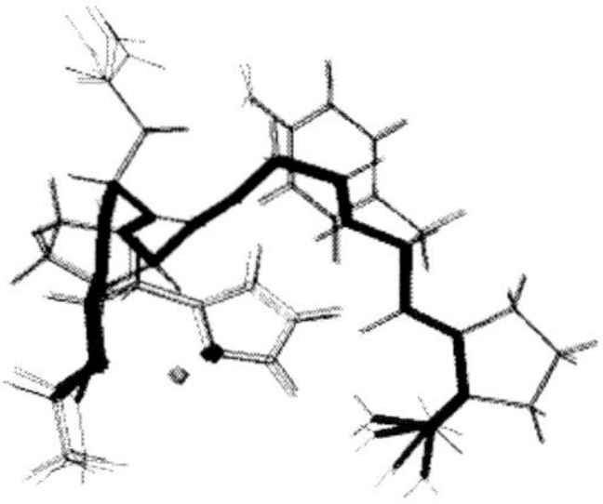

Figure 3.6 The structure families of Cu^{2+}-Ac-HNPGYP-NH_2 complex. The structures were obtained through restrained simulated annealing. Figure was created with MOLMOL 2K.1. (from Stanczak *et al.*, 2005b, reproduced by permission of the Royal Society of Chemistry).

Moreover, chPrP and hPrP posses very similar amyloidogenic regions (scheme 3.1): (i) both contains two His (His96 and His111 for $hPrP_{91-126}$ and His110 and His124 for $chPrP_{105-140}$); (ii) the hydrophobic tail, rich in Ala and Gly is exactly conserved; (iii) the few amino acid substitutions are mainly located close to the two His.

	96 / 111
$hPrP_{91-126}$	QGGGTHSQWNKPSKP--KTNMKHMAGAAAAGAVVGGLGG
	110 / 124
$chPrP_{105-140}$	SGGSYHNQ-- --KPWKPPKTNFKHVAGAAAAGAVVGGLGG

Scheme 3.1 Amino acid sequence of $hPrP_{91-126}$ and $chPrP_{105-140}$.

It has been recently found that also the amyloidogenic chPrP region binds Cu^{2+} ions very efficiently, two independent Cu^{2+} sites being located at His110 and His124, respectively (Gralka *et al.*, 2009). At physiological pH, both copper sites posses a 3N1O and/or 4N donor set in square planar geometry similarly to His96 and His111 copper binding domains in hPrP. The structural and the kinetic properties of the His110 and His124 Cu^{2+} complexes has been obtained by using Ni^{2+} as a diamagnetic probe (Valensin *et al.*, 2010). The comparison of the human and chicken metal sites shows large structural changes between the His96 (hPrP) and His110 (chPrP) copper domain with a strong stabilizing role played by Tyr109 aromatic ring in the latter case. On the contrary the His sites close to the hydrophobic tail are very similar in both human and chicken PrP. Such differences result in different copper preferences between two His sites in human and chicken amyloidogenic regions.

3.3.1.5 The biological relevance of the model copper-PrP peptide systems

The structures of the copper binding sites obtained through model studies yield detailed information on the metal binding sites, especially on (i) anchoring site(s), (ii) the direction taken by successive copper binding to deprotonated amide nitrogens and (iii) the role of side-chains, close to the binding site, in stabilizing metal-binding. All the collected data suggest that the copper-peptide interaction is modulated by subtle changes in pH, which is reminiscent of an uptake-release mechanism for metal-transport. The binding mode shifts from inter- to intra-repeat as Cu^{2+} concentration is raised. The inter-repeat multi-imidazole coordination is found in mammalian, avian, Japanese pufferfish, and zebrafish PrP, with a strength apparently related to the number of His residues coordinating the metal. At relatively high Cu^{2+} concentration (and relatively high pH) the prevailing intra-repeat binding mode implies binding to amide nitrogen donors. Moreover the transition between the intra- and inter-repeat binding modes, characterized by different structural rearrangements of the flexible PrP N-terminal tail, suggests a copper dependent switch that facilitates specific cellular processes. Finally, the different behaviors of chicken and human amyloidogenic regions towards copper and the fact that chicken are not affected by prion disease induce to speculate that Cu^{2+} binding to His111 site may be responsible for particular structural rearrangements of the hydrophobic $hPrP_{113-126}$ tail, favoring the generation of β-sheet structures.

3.3.2 *The Amyloid-β (Aβ) Peptide Models*

Aβ strongly interacts with Cu^{2+} and Zn^{2+} ions (for reviews see Adlard and Bush 2006; Faller and Hureau 2009). The metal binding sites have been localized in the N-terminal part of the peptide encompassing the first 16 amino acid residues, so that most of the structural investigations have been performed with Aβ peptide models missing the highly hydrophobic C-terminal part. Most of the reports on Cu^{2+} binding to Aβ agree with the formation of monomeric Cu^{2+}- Aβ complex. EPR and potentiometric experiments are consistent with the presence of two different copper species around physiological pH (Kowalik-Jankowska *et al.*, 2003; Guilloreau *et al.*, 2006). It's well accepted that the first species, dominating at pH 6, is characterized by a 3N1O (4N or 2N2O) donor atom set. Concerning the nature of the nitrogen donors no total consensus has been reached so far. EPR results indicate two His residues and the N-terminus as the residues providing the nitrogen Cu^{2+} coordinating atoms (Kowalik-Jankowska *et al.*, 2003; Karr *et al.*, 2005). On the contrary NMR copper induced line broadening, and CD experiments performed on His/Ala substituted peptides support the involvement of all the three His and the N-terminus nitrogens (Guilloreau *et al.*, 2006; Syme *et al.*, 2004; Danielsson *et al.*, 2007; Hou and Zagorski, 2006). Such different conclusions have been recently discussed and it has been suggested the presence of different Cu^{2+}-Aβ complexes in equilibrium at room temperature, being the two discussed models the most populated states (Faller and Hureau 2009). The presence of multiple conformations in equilibrium at room temperature makes the structural determination of copper bound Aβ tricky. On the other hand, the NMR structure of the rat Aβ complex in water micelle environment has been obtained (Gaggelli *et al.*, 2008b). The metal ion is bound to two His Nπ imidazole, to the N-terminal amino and to an oxygen atom from a water molecule. (Fig. 3.7). Also for the possible copper O-ligand there are several hypotheses: the Asp1, Glu11 carboxylates, the Tyr10 phenolate and a water molecule. However, EPR studies of the ^{17}O-labelled H_2O (Karr *et al.*, 2005) and the absence of intense Cu^{2+}-phenolate charge-transfer

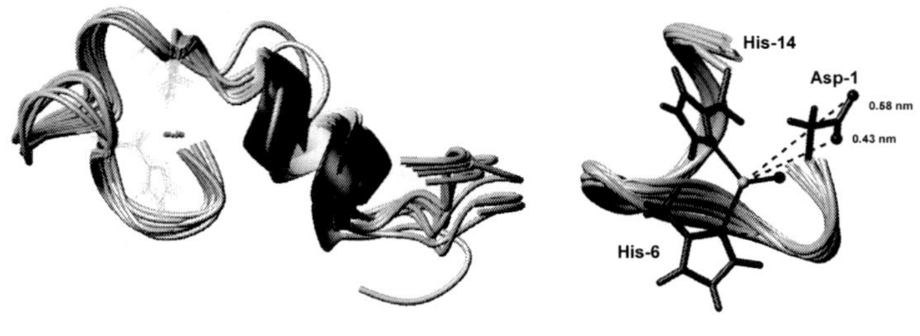

Figure 3.7 Structure of the Cu^{2+}- $A\beta_{1-28}$ complex in water-micelle (SDS) environment (from Gaggelli *et al.*, 2008b, reproduced by permission of the Royal Society of Chemistry).

UV-Vis band (Kowalik-Jankowska *et al.*, 2003; Guilloreau *et al.*, 2006) strongly suggest Asp1 or Glu11 carboxilate as the main candidate for copper coordination.

As already mentioned, at pH above 8 a different Cu^{2+}- Aβ complex is formed, which is characterized by additional involvement of deprotonated amide nitrogen(s). In particular, the similarity between the potentiometric and the spectroscopic results on $A\beta_{1-16}$ and $A\beta_{1-10}$ metal complexes excluded the involvement of His13 and His14 in copper binding (Kowalik-Jankowska *et al.*, 2001; Kowalik-Jankowska *et al.*, 2003). Recent NMR and pulse EPR investigations point out the importance of Asp1-Ala-2 amide bond in Cu^{2+} binding to $A\beta_{1-16}$ even at physiological pH (Dorlet *et al.*, 2009; Hureau *et al.*, 2009b). A model with a pentacoordinated Cu^{2+} is suggested: the equatorial donor atoms derived from the Asp1 N-terminus amine, the Ala2 main chain amide, the imidazole nitrogen atoms from one of the three His and the Ala2 carbonyl oxygen. The additional involvement of Asp1 carboxylate in axial position is proposed. A slightly different Cu^{2+} chemical environments have been reported using the data obtained by X-ray absorption spectroscopy (XAS) and low temperature density functional theory (DFT) analysis (Streltsov *et al.*, 2008). A distorted six-coordinated (3N3O) copper geometry has been found with three His and a carboxylate oxygen (Asp1 or Glu11) in the equatorial plane, a water molecule and a carboxylate oxygen (Glu11 or Asp1) as the axial donor atoms (Streltsov *et al.*, 2008).

The Cu^{2+}-Aβ complexes can readily be reduced to Cu^{+}-Aβ in presence of reductant species such as ascorbate or dithionite (Streltsov and Varghese 2008). XANES and EXFAS analysis on Cu^{+} interacting with $A\beta_{1-16}$, $A\beta_{1-40}$ and other Aβ peptide models indicated the information of linear diimidazole-Cu^{+} adduct, where the His13-Cu^{+}-His14 moiety is found to be responsible for Cu^{+}-Aβ redox activity (Himes *et al.*, 2008; Shearer and Szalai, 2008). On the other hand, NMR chemical shift variations of all three His aromatic protons suggest the involvement of all His in Cu^{+} binding, indicating the existence of a dynamic exchange between the His ligands in solution (Hureau *et al.*, 2009a).

The N-terminal hydrophilic region of Aβ can also bind Zn^{2+}. Most of the available results on the metal binding site relied on NMR and EXAFS spectroscopy (Curtain *et al.*, 2001; Mekmouche *et al.*, 2005;

Syme and Viles, 2006; Zirah *et al.*, 2006; Danielsson *et al.*, 2007; Minicozzi *et al.*, 2008). Zinc binds Aβ and forms a monomeric metal complex. All the collected results are in line with the simultaneous Zn^{2+} coordination to all three His imidazoles. There is not much agreement for the other ligand(s) involved in metal binding. The participation of Asp1 N-terminus and/or carboxylate (Mekmouche *et al.*, 2005; Danielsson *et al.*, 2007; Gaggelli *et al.*, 2008c), of Tyr10 phenolate (Curtain *et al.*, 2001; Minicozzi *et al.*, 2008) and of Glu11 carboxilate (Zirah *et al.*, 2006; Gaggelli *et al.*, 2008c) have been proposed as well. Among all these hypothesis the binding of Tyr10 seems the least probable by considering UV-Vis studies (Talmard *et al.*, 2007). NMR line broadenings detected on ^{1}H-^{13}C HSQC experiments strongly suggest the involvement of Asp1 (Danielsson *et al.*, 2007). The NMR structure of the Zn^{2+} complex with the N-acetylated Aβ16 peptide shows Zn^{2+} coordination of Glu11 carboxyl group (Zirah *et al.*, 2006). The chemical shifts variations and the NMR structures determined for Zn^{2+}-Aβ28 systems in presence of SDS give evidence of simultaneous binding of Glu11 and Asp1 (Gaggelli *et al.*, 2008c).

In conclusion, all the studies on Cu^{2+} and Zn^{2+} interactions with Aβ peptide show that both metal ions specifically bind the N-terminal region of Aβ. Copper and zinc binding is very specific, each metal ion has its own donor set and this could reflect the different behaviors of the two metal ions on Aβ aggregation.

3.3.3 *Metal Ion Binding to α-Synuclein Peptide Fragments*

α-Synuclein (α-Syn), a 140 amino acid unstructured protein, is found in cytosol and its aggregation could contribute to pathologies like Parkinson disease (Hsu *et al.*, 1998). Although its biological function is not understood it seems to be a metalloprotein (Rasia *et al.*, 2005).

α-Syn contains three potential binding sites, which can effectively bind Cu^{2+} ions, the N-terminal amino group of Met with successive amide nitrogens, His50 imidazole with adjacent peptide nitrogens and much less effective C-terminal set of carboxylates of Asp and Glu residues. The recent studies on cytotoxicity on α-Syn peptide fragments have shown that used peptides are toxic and their toxicity could be copper dependent or independent depending on the used sequences. The studies suggested that copper dependent toxicity could be related to His50 and C-terminal sites. The combined impact on the cell toxicity of more than one metal ion binding site is likely (Brown, 2009). The conclusions of the chemical studies on Cu^{2+} binding to α-Syn and its fragments are quite controversial. Earlier works suggested the negatively charged C-terminal region as the main binding site for Cu^{2+} ions (Paik *et al.*, 1999; Uversky *et al.*, 2001). This rather unusual hypothesis (Cu^{2+} is not in favor for carboxylates) was later challenge by other works suggesting the N-terminal as the main coordination site for Cu^{2+} (Rasia *et al.,* 2005; Binolfi *et al.*, 2008; Jackson and Lee, 2009; Kowalik-Jankowska *et al.*, 2006). The possible involvement of the His50 residue in the interactions with metal ion is also likely. The latter findings agree with concepts of coordination chemistry although exact relation between binding site and aggregation process leading to fibrillization is not yet quite well understood (Rasia *et al.*, 2005; Brown, 2009).

3.4 Thermodynamic and Speciation Studies

3.4.1 *Prion Protein*

A huge amount of work on the thermodynamic characterization of a metal/protein system is that referring to the prion protein (PrP^C). Several reviews have been recently published on this topic (Brown and Kozlowski, 2004; Gaggelli *et al.*, 2006; Kozlowski *et al.*, 2006; Kozlowski *et al.*, 2008; Millhauser, 2007). Large consensus can be found in literature on the fact that PrP^C can bind several Cu^{2+} ions *in vivo* with high affinity and selectivity with respect to other metal ions like Zn^{2+}, Ni^{2+} or Mn^{2+}. The Cu^{2+} binding-sites are located in the N-terminal, unstructured region of the protein. Therefore, many different peptide fragments have been synthesized and separately investigated as model systems; they are characterized by different chain lengths and contain one or more binding sites. The first paper on this topic dates 2002 (Luczkowski *et al.*, 2002) and opens a series of studies on the PrP^C octarepeat region (Hureau *et al.*, 2006; Luczkowski *et al.*, 2003; Valensin *et al.*, 2004; Valensin *et al.*, 2009b). The investigation has successively been extended to the fifth binding-site (Belosi *et al.*, 2004; Berti *et al.*, 2007; Di Natale *et al.*, 2005; Di Natale *et al.*, 2009; Gaggelli *et al.*, 2005; Gralka *et al.*, 2008; Joszai *et al.*, 2006; Osz *et al.*, 2007; Osz 2008; Remelli *et al.*, 2005; Remelli *et al.*, 2009). Analogous studies have been performed on avian (Gralka *et al.*, 2009; La Mendola *et al.*, 2009; Stanczak *et al.*, 2005a; Stanczak *et al.*, 2005b) and fish (Camponeschi *et al.*, 2009; Gaggelli *et al.*, 2008a; Kozlowski *et al.*, 2008; Stanczak *et al.*, 2006; Szyrwiel *et al.*, 2008; Valensin *et al.*, 2009a) prion proteins.

It is widely accepted that the octarepeat domain of PrP^C can form Cu^{2+} complexes with different stoichiometries and geometries mainly depending on pH and the metal/ligand concentration ratio. The first "contact" between the metal ion and octarepeat is realized through the side imidazoles of the four His residues, most likely by means of the $N\tau$ of the hetero-aromatic rings, the most exposed to the environment. At acidic pH, where Cu^{2+} is not strong enough to displace amide protons, or at sub-stoichiometric levels of copper, a 1:1 species is preferentially formed where the metal is bound to two or more N_{Im} atoms. This result was confirmed by comparing the binding ability of peptides Octa1, Octa2 and Octa4, respectively containing 1, 2 or 4 octarepeatitions (Valensin *et al.*, 2004), and it is in good agreement with spectroscopic studies (Millhauser, 2004) (see Figs. 3.1 and 3.2). The stabilities of 1:1 complexes, not influenced by protonation equilibria, are in the order Octa4 > Octa2 > Octa1, due to the involvement of more and more imidazolic units. The poly-histidine complexes are the only that Zn^{2+} can form with PrP^C (Walter *et al.*, 2007). Increasing pH and Cu^{2+} concentration, copper can displace amidic protons and poly-nuclear species are formed, with a metal/peptide stoichiometry up to 4:1 in the case of Octa4. Here, each octarepeat section behaves as an independent binding site with a binding mode identical to that followed by Octa1 at neutral pH. Due to the presence of a Pro residue preceding His, the amidic nitrogens involved in coordination are those belonging to the two Gly residue following His (in the C-terminus direction), thus forming a rather unusual 7-membered ring. It is worth noting that the structure proposed through solution equilibria studies (Fig. 3.8) is exactly the same independently found at the solid state by X-ray analysis (Fig. 3.1). The gradual change from the multi-His to the mono-His binding mode gives a slope lower than one in the corresponding Hill plot calculated both from spectroscopic (Walter *et al.*, 2006; Walter *et al.*, 2009) and from potentiometric (Kozlowski *et al.*, 2010) data. However, it is important to underline that the classical concept of cooperativity looks not very appropriate to be applied here, where

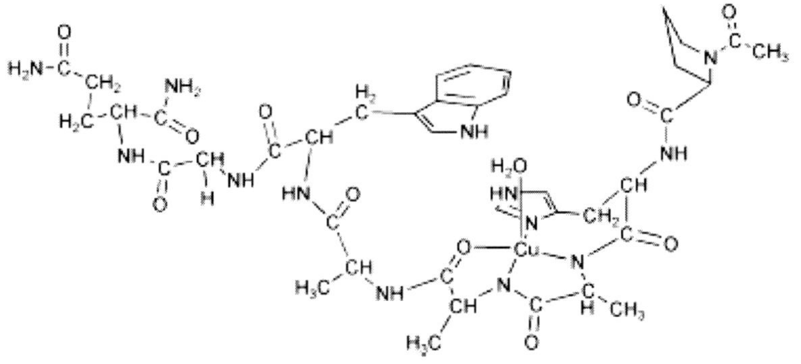

Figure 3.8 Solution structure of the main Cu^{2+}/Octa1 complex, at neutral pH (from Luczkowski *et al.*, 2003, reproduced by permission of the Royal Society of Chemistry).

different *binding modes* rather than *binding sites* are present. As a matter of fact, the mono-histidine coordination of Octa1 at neutral pH is weaker than the multi-histidine mode of Octa4 as demonstrated by K_d data calculated from the corresponding speciation models: 1.2 μM for Octa1 (Kozlowski *et al.*, 2010) and 1.9 nM for Octa4. The copper binding ability of Octa1 fragment was also compared with that of analogues respectively having three Ala or Lys residues in the place of glycines following His (Luczkowski *et al.*, 2003). The major complex formed in the pH range 6–8 is again the $\{N_{Im}, 2N^-, O\}$ species of Fig. 3.8, but the substitution of Gly with bulkier Ala or Lys residues weakens the complex to some extent. It has been suggested that this glycine-rich region in the octarepeat fragment plays a role in assuring both a high peptide flexibility and the requested selectivity towards the Cu^{2+} ions, in the biological pH range.

PrP^C po ssesses two additional binding sites outside the octarepeat domain; they are located around residues His96 and His111. The investigations performed using peptidic fragments containing both the binding sites, established they are independent (Berti *et al.*, 2007; Osz *et al.*, 2007). At neutral pH, the preferred coordination mode is again $\{N_{Im}, 2N^-, O\}$ for each site, imidazole acting as anchoring site and equatorial oxygen belonging to either a neighbouring carbonyl group or a water molecule; also at low metal-concentration levels neither intramolecular bis-histidine species nor intermolecular bis-complexes are formed. Although the coordination geometry and the donor-atom set is the same for the two binding sites, their binding strength is slightly different, in favour to His111. This is clearly shown by the competition diagram built up for a solution containing equimolar amounts of Cu^{2+}, PrP_{92-100} and $PrP_{106-113}$ (Berti *et al.*, 2007), and it is also reflected by the calculated K_d values for the two fragments, 160 nM for the former and 58 nM for the latter (Kozlowski *et al.*, 2010). The difference is not negligible and reveals that not only geometry or donor groups play a role in complex stability, but also the side chains of amino acidic residues surrounding the anchoring unit, which can interact each other and/or with the metal ion. Moreover, for both His96 and His111 sites it has been suggested that amide coordination can take place both in the N-terminal and in C-terminal direction (Berti *et al.*, 2007); potentiometric and calorimetric results on short fragments (Gralka *et al.*, 2008) supported that hypothesis and demonstrated that thermodynamic parameters for the two coordination modes are very close to each other and the preference can simply depend on weak interactions due to uncoordinated side chains.

In the fifth-site domain (PrP_{91-120}) further residues are present bearing potentially metal-binding groups: Lys at positions 101, 104, 106 and 110; Met at positions 109 and 112. The participation of their side chains to metal-ion coordination has been extensively debated: the most widely accepted conclusions are that amino groups of Lys residues do not participate to complexation (Di Natale *et al.*, 2005; Remelli *et al.*, 2005) while thioether sulphur atoms of Met residues are suggested to interact with Cu^{2+}, especially at acidic pH values where amidic nitrogens of polypeptidic backbone are not available (Berti *et al.*, 2007; Di Natale *et al.*, 2005; Osz *et al.*, 2007). Since Met residues are located around His111, this can also explain the higher binding affinity of this site, with respect to His96 in hPrP. The opposite affinity order was instead found for mouse PrP (mPrP), where Met109 and Met112 are not present (Klewpatinond *et al.*, 2008). Interestingly, the preference between the His96 and His111 in hPrP is reversed also when the metal ion is Ni^{2+} (Klewpatinond and Viles, 2007). Comprehensive thermodynamic data involving mPrP or Ni^{2+} are not available to date.

The last controversial subject concerning Cu^{2+} binding to hPrP refers to the selectivity between the octarepeat and the fifth-site domains. Solution-equilibria studies suggest that the former is

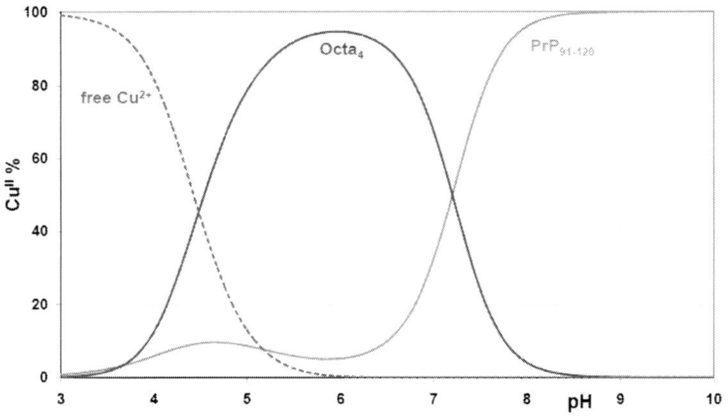

Figure 3.9 Competition between $Octa_4$ (red) and $hPrP_{91-120}$ (green) in forming Cu^{2+} complexes. $[Cu^{2+}] = [Octa_4] = [hPrP_{91-120}] = 1$ mM. (Data from Berti *et al.*, 2007; Valensin *et al.*, 2004). See also Colour Insert.

stronger at acidic and neutral pH, while the latter is preferred in the alkaline pH range (Fig. 3.9). This can be rationalized bearing in mind that octarepeat can take advantage from the multi-histidine binding mode, already operational at pH 4, while the fifth-site can efficiently bind Cu^{2+} only when the metal ion can displace amide protons, i.e. above pH 6–7. Computed K_d values (1.9 nm for Octa4 and 27 nM for $hPrP_{91-115}$) confirm this conclusion, in agreement with most spectroscopic investigations.

3.4.2 *β-Amyloid Peptide*

The investigation method based on model peptides has been fruitfully employed also in the case of interaction between metal (copper and zinc) ions with the slightly soluble amyloid peptide (Aβ), involved in Alzheimer's disease. NMR studies showed that $A\beta_{1-40}$ is mostly unstructured in solution (Gaggelli *et al.*, 2006); it can bind Cu^{2+} and Zn^{2+} ions (Atwood *et al.*, 2000) and these metals, along with iron, accumulate in amyloid deposits (Lovell *et al.*, 1998). There is a general literature agreement (Faller and Hureau, 2009; Kozlowski *et al.*, 2009) on the fact that the Cu^{2+} and Zn^{2+} binding sites are located at the N-terminal section of Aβ, where the main donor groups are present: terminal amine and side carboxylate of Asp1,

Table 3.1 Solution-equilibria studies on Aβ model-peptides

Model peptide(s)	**Metal ion**	**Metal/ peptide ratio**	**Ref.**
$A\beta_{11-16}$ human and mouse	Cu^{2+}	1:1	(Kowalik-Jankowska *et al.*, 2000)
$A\beta_{1-6}$, $A\beta_{1-9}$, $A\beta_{1-10}$ human and mouse	Cu^{2+}	1:1	(Kowalik-Jankowska *et al.*, 2001)
$A\beta_{11-20}$, $A\beta_{11-28}$ human and mouse	Cu^{2+}	1:1	(Kowalik-Jankowska *et al.*, 2002)
$A\beta_{1-16}$, $A\beta_{1-28}$ human and mouse	Cu^{2+}	1:1	(Kowalik-Jankowska *et al.*, 2003)
$A\beta_{1-4}$, $A\beta_{1-6}$, Ac-$A\beta_{1-6}$, $A\beta_{1-16}$, $A\beta_{1-16}$Y10A, Ac-$A\beta_{8-16}$Y10A, $A\beta_{1-16}$PEG	Cu^{2+}	4:1 ÷ 1:2	(Damante *et al.*, 2008)
$A\beta_{1-4}$, $A\beta_{1-6}$, Ac-$A\beta_{1-6}$, $A\beta_{1-16}$, $A\beta_{1-16}$Y10A, Ac-$A\beta_{8-16}$Y10A, $A\beta_{1-16}$PEG	Zn^{2+}	4:1 ÷ 1:2	(Damante *et al.*, 2009)

side imidazole rings of His6, His13 and His14. In the same region other residues bearing potential donor atoms are present, i.e., Glu3, Glu11, Tyr10 and Lys16; their involvement in complexation has been widely discussed and not completely solved, yet. Only few papers deal with solution-equilibria studies on Aβ model-peptides: they are listed in Table 3.1.

When Cu^{2+} ions and the considered peptide are present in equimolar amounts, the formation of only mononuclear 1:1 species have been detected. Complex-formation starts at pH close to 4: the first anchoring point could in principle be either the terminal amine or one of His residues. There is general agreement on the presence, in the acidic pH range, of macrochelate species where both N-terminal amine and one or more His imidazoles are bound to Cu^{2+}, depending on fragment length. Side carboxylate group of Asp1 participates in metal ion coordination (Faller and Hureau, 2009). At neutral pH, the accepted coordination mode is {3N, 1O}, with the contribution of one or more deprotonated peptide nitrogens, located after Asp1 or around His residues. Participation of all His residues was confirmed using peptides where His residues were ^{15}N labelled (Shin and Saxena, 2008) or substituted with alanines (Syme *et al.*, 2004). The most likely situation is a mixture of species with the same stoichiometry but different donor-atom sets. The different length of model peptides $A\beta_{1-16}$ and $A\beta_{1-28}$ does not affect the binding geometry, even though the K_d values, computed from stability constant values determined at 1:1 concentration ratio, are not identical (0.21 and 0.024 nM, respectively) (Faller and Hureau, 2009).

Two very recent papers describe the synthesis and use, as a Aβ model peptide, of $A\beta_{1-16}$PEG, where a polyethylene glycol moiety has been conjugated with $A\beta_{1-16}$, thus considerably increasing the peptide solubility without significantly affecting its metal-binding capabilities. This modification made it possible to study complex-formation equilibria in the presence of a large excess of Cu^{2+} or Zn^{2+} (Damante *et al.*, 2008; Damante *et al.*, 2009), since the metal is kept in solution also at alkaline pH values. In the case of Cu^{2+}, the formation of mono-, di-, tri- and tetra-nuclear species was detected. In the latter, the four anchoring sites (Asp1, His6, His13 and His14) behave as independent sites, involving in coordination their neighbouring amide nitrogens. It is worth noting that, in the case of His14, amide coordination is suggested to proceed towards the C-terminus direction, (with the formation of a seven-membered ring) as already found for PrP (see above). Finally, the study on the Zn^{2+}/ $A\beta_{1-16}$PEG complex-formation equilibria (Damante *et al.*, 2009) showed that Aβ can accommodate up to three metal ions, independently bound to the three His residues.

3.4.3 *α-Synuclein*

The solution equilibria involving some α-Synuclein (α-Syn) fragments or analogues and the Cu^{2+} ion have been recently investigated (Kowalik-Jankowska *et al.* 2005, 2006, 2007). The role played by the N-terminus in Cu^{2+} complexation was first studied by means of the following fragments:

α-Syn_{1-17} (MDVFMKGLSKAKEGVVA-NH_2),
α-Syn_{1-28} (MDVFMKGLSKAKEGVVAAAEKTKQGVAENH_2) and
α-Syn_{1-39} (MDVFMKGLSKAKEGVVAAAEKTKQGVAEAPGKTKEGVLY-NH_2).

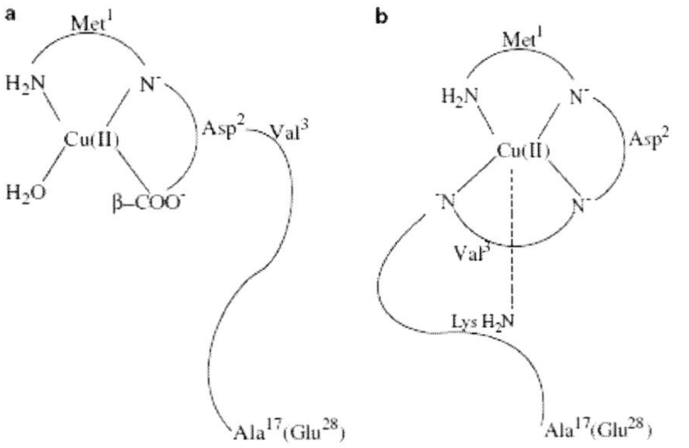

Figure 3.10 The 2N (a) and 4N (b) species for the 1–17 and 1–28 fragments of α-synuclein (from Kowalik-Jankowska *et al.*, 2005, with permission from Elsevier).

The main results are the following (see Fig. 3.10): (i) thioether sulfur of Met1 does not directly participate in metal-ion binding; (ii) 2N species, involving terminal NH_2, amide N^- and β-carboxylate of Asp2, are formed in the acidic-neutral pH range while 3N and 4N complexes are formed at alkaline pH, with the involvement of the side amino group of Lys residue, in apical position; (iii) Tyr39 does not take part in the copper coordination; (iv) the C- terminal tail (18–39) does not influence the coordination mode but slightly affects the complex stabilities. The behaviour of His50 towards the Cu^{2+} ion was then examined and the following α-Syn fragments were employed:

α-Syn_{29-56} (AAGKTKEGVLYVGSKTKEGVVHGVATVA-NH_2),
α-Syn_{29-56} (A29M, A30D), Ac-α-Syn_{29-56} (A29M, A30D),
α-Syn_{26-56} (V26M, A27D, A30P, A53T),
Ac-α-Syn_{26-56} (V26M, A27D, A30P, A53T);

the use of wild-type Ac-α-Syn_{26-56} or Ac-α-Syn_{29-56} was prevented by their low solubility. When the amino terminal group is acetylated, the imidazole nitrogen of His50 can act as a Cu^{2+} anchoring site, at acidic pH, leading to sequential amide-nitrogen binding at higher pH. The substitution of Ala30 and Ala53 respectively with Pro and Thr residues did not significantly affect the binding behaviour of these peptides towards the Cu^{2+} ion. It was confirmed that Tyr39 is not involved in complexation.

3.5 Impact of Metal Ions on Prion Protein Fibril Formation

The studies with the wild type and modified sequences of proteins and their peptide fragments under different solution conditions have been performed to understand the mechanisms by which polypeptides/proteins misfold and aggregate. The local factors such as pH, metal ions, small molecules which may act as ligands for proteins or lipid membranes are capable to induce aggregation and fibril formation. Analysis of proteins found in pathological deposits e.g., α-synuclein, β-amyloid peptide or prion protein *in vitro* experiments provided evidences that transition metal ions are able to accelerate the aggregation process as well as conformational changes. The most studied metal ions are copper, iron, zinc, nickel and manganese.

3.5.1 *$PrP_{106-126}$, Model Neurotoxic Peptide*

Prion protein contains two Cu^{2+} binding domains, one within the octarepeat region and the other one in the neurotoxic peptide region. The latter region in human protein comprises two histidine residues, His96 and His111. The neurotoxic peptide with $PrP_{106-126}$ sequence is characterized by two regions with hydrophilic (KTNMKHM) and hydrophobic (AGAAAAGAVVGGLG) sequences. $PrP_{106-126}$ is highly fibrilogenic and toxic to neurons in vitro, it shares many physicochemical and biological properties with PrP^{Sc} e.g., resistance to protease digestion, activation of astroglial and microglial cells, neurotoxic activity toward primary cultures of hippocampal neurons, it also can mediate the conversion of native cellular PrP^{C} to the scrapie form PrP^{Sc}. Polymers encompassing residues 106–126 show a secondary structure composed largely of β-sheets suggesting that this region might be a good model for studying the aggregation process (Brown *et al.*, 1996). Cu^{2+} ions have the capacity to convert PrP^{C} into protease resistant forms in vitro conditions and promote self association process (Quaglio *et al.,* 2001). The unstructured region PrP_{90-126} seems to be essential in the amyloid formation. Jones *et al.* using UV CD technique indicated loss of irregular architecture in this region in the presence of Cu^{2+} ions and the increase of β-sheet conformation (Jones *et al.*, 2004).

The addition of Cu^{2+} or Zn^{2+} ions to $PrP_{106-126}$, which are responsible for the β-sheet fibril formation, is crucial for aggregation process (Brown *et al.*, 1996). Presence of Cu^{2+} and Zn^{2+} ions is necessary for proper fibril polimeryzation although whether metals are involved in seed or profofibril formation is not yet clear. Similar investigation for Ni^{2+} ions have not indicated any direct contribution to aggregation process (Jobling *et al.*, 2001). His111 was shown to be crucial for Cu^{2+} binding and its replacement abolish aggregation. Hydrophobic tail of a peptide chain seems to be crucial for the formation of β-sheet conformation.

Brown *et al.* suggested that the conformation of PrP and β-sheet formation might be modified by manganese ions. Binding of Mn^{2+} to PrP stimulates the fibrils formation and plaque generation with simultaneously enhanced protease resistance. Manganese binding to PrP generate protein species, which can act as a seed and may catalyze the protein aggregation (Brown *et al.*, 2000).

In whole prion protein not only neurotoxic region is implicated in misfolding process. The octarepeat region have a distinct impact on the process as well. Under incubation with Cu^{2+} ions (concentration 50 μM), full length PrP became proteinase resistant (Kuczius *et al.*, 2004). The insertion of extra octarepeat into the PrP can trigger spontaneous disease in humans (Pushie *et al.*, 2009). Transgenic mice that lack octarepeat region are still susceptible to disease but the progression is slowed down (Flechsig *et al.*, 2000). In the absence of Octa4 region His96 and His111 are the only possible anchoring sites for copper ions. In the low occupancy of metal ions CD spectra and MD simulation suggest that Cu^{2+} binding to His96 and His111 generate the local β-sheet conformation (Brown *et al.*, 1996; Pushie *et al.*, 2009).

3.5.2 *Amyloid-β Peptide Aggregation*

Aβ is derived from Amyloid Precursor Protein (APP) after β and γ-secretase cleavages. $A\beta_{1-40}$ is the most common and probably non toxic species, the second species $A\beta_{1-42}$, with increased ability to aggregation, is mainly associated with AD (Gaeta and Hider, 2005). The sequence of the amyloid peptide has two hydrophilic (1–16 and 22–28) and two hydrophobic (17–21 and 29–42) regions. The first 16 amino acid peptide has been proposed to play a main role in metal ion binding and it shows no tendency to aggregate or fibril formation (Kowalik-Jankowska *et al.*, 2002). The residues 25–35 are similar to the hydrophobic sequence core of $PrP_{106-126}$, and have a critical role in aggregation process. There is a lot of evidence that transition metal ions have important influence on the Aβ aggregation and toxicity. Copper, zinc and iron ions have been found in extracellular amyloid plaques in AD patients brains (Hung *et al.*, 2010).

Effect of Cu^{2+} on the aggregation of Aβ was studied for model-peptides $A\beta_{14-23}$, $A\beta_{11-23}$ and $A\beta_{11-28}$. It was suggested that Cu^{2+} may have significant impact on the formation and structure of the amyloid fibrils. Copper accelerates fibril formation for $A\beta_{14-23}$, but it is inhibiting fibril formation for $A\beta_{11-28}$. Cu^{2+} binds to the N-terminal albumin-like sequence Xxx-Xxx-His in $A\beta_{11-23}$ and $A\beta_{11-28}$ thus decreasing the β-sheet assembly, perhaps by excluding the participation of the first three amino-acids in the β-sheet structure. In the case of $A\beta_{14-23}$, Cu^{2+} binds to N-terminal His14 via histamine-like binding mode and may enhance the amyloid fibrils formation by bridging two peptide moieties and stabilization of the peptide aggregation (Pradines *et al.*, 2008).

Different levels of zinc ions in the brain may contribute to precipitation and aggregation of Zn-Aβ which is more amorphous and less fibrilar than Aβ. Zn^{2+} promotes aggregation of $A\beta_{1-28}$ by changing partially α-helical structured intermediate into β-sheet structured aggregates (Faller *et al.*, 2009). The coordination of Zn^{2+} to His13 is crucial to the zinc ion induced aggregation of Aβ, the substitution of His13 residues changing the aggregation behavior (Liu *et al.*, 1999).

3.5.3 α-Synuclein Aggregation

α-Synuclein, the main component of inclusions commonly known as Lewy bodies in PD, is a small natively unfolded protein possessing a random coil secondary structure. The central hydrophobic region of 35 amino acid peptide, known as NAC (residues 61–95), is responsible for conformational change from random coil to β-sheet structure, where individual β-strand run perpendicularly to the fiber axis, followed by an acidic C-terminal tail (amino acids 96–140) that blocks synuclein filament assembly (Binolfi *et al.*, 2008). The study concerning metals and α-synuclein are mainly focused on the oxidative stress; however, it was found that metal ions are able to affect α-synuclein conformation. First information about interactions of Cu^{2+} ions and α-synuclein concerned the C-terminal part of protein and involved its aggregation. Cu^{2+}, at slightly acidic pH (6.5), is a potent inducer of synuclein oligomerization mediated by the C-terminal acidic region of the protein (Davies *et al.*, 2008).

More investigation suggested that the high affinity binding sites for Cu^{2+} ions is located at the N-terminal domain in the region ^{1}MDVFMKGLS^{9} and ^{48}VVHGV^{52}. Studies on the long fragment α-Syn$_{1-108}$ and α-Syn$_{95-137}$ confirm that the binding of Cu^{2+} ions to the N-terminal region of α-Synuclein is independent of the presence of the C-terminus. The presence of Cu^{2+} ions in the fibrils suggested that the protein aggregates in its copper-bound form (Rasia *et al.*, 2005).

Iron was found in increased concentration in PD brain in Lewy bodies. Contribution of Fe in pathology of PD may be linked with its ability to catalyze of oxidative reaction (Bolognin *et al.*, 2009). Other studies suggested that iron ions promote the kinetic of α-Syn aggregation (Uversky *et al.*, 2001).

3.6 Impact of Metal-Peptide Interaction on Oxidative Stress

Oxidative stress plays a critical role in pathology of neurodegenerative disorders. Oxidative stress appears as a result of the imbalance between production of ROS (reactive oxygen species) and the ability of the cells to defend against them. It has been widely demonstrated that oxidative stress play main role in the dysfunction and degeneration during AD. The antioxidant and pro-oxidant activity of Aβ is linked to its ability to bind metal ions (Cu^{2+}, Fe^{3+}) and as consequence to mediate redox reactions. It has been reported that Aβ generates hydrogen peroxide from molecular oxygen, through electron transfer interaction involving redox active metal ions (Huang *et al.*, 1999). The ROS production by aggregation intermediates of Cu-Aβ$_{1-42}$ produces five-times more $OH^{\bullet}$ than monomeric Aβ (Guilloreau *et al.*, 2007). A variety of oxidized Aβ species are present in amyloid plaques especially methionine sulfur atom (Met35) is susceptible to oxidation and peptide lacking methionine was not able to reduce Cu^{2+} to Cu^{+} (Rauk, 2009; Hung *et al.*, 2010). A few models of ROS production during development of AD were proposed. It is likely, however, that in the first step the $2e^-/2H^+$ pathway for the O_2 activation occurs (Faller and Hureau, 2009).

Both avian and mammalian prion proteins were shown to possess SOD-like activity and it is likely that one of the biological functions of prion protein when copper is bound concerns the anti-oxidant activity. Model studies have clearly shown that at low metal concentrations with multi-imidazole binding of Cu^{2+} to octarepeat region metal ion is distinctly more effective as SOD center than that formed at high copper amounts when metal ion binds to imidazole and two or three amide nitrogens. The SOD activity of model Cu^{2+}-peptide systems are much lower than that of real SOD1 enzyme, however, compare to other metallopeptides it is relatively high (Stanczak and Kozlowski, 2007; Stanczak *et al.*, 2007). The reason for different SOD activity of Cu^{2+} bound to four imidazoles and to one imidazole and two or three amide nitrogens is rather clear. The binding of Cu^{2+} to amide nitrogens strongly stabilizes Cu^{2+} compared to Cu^+, while in SOD activity the redox cycle of $Cu^{2+/+}$ couple should be rather easy. The latter is favored with multi-imidazole binding. Comparing different coordination modes with possible biological functioning it seems, at least on the chemistry level of understanding that at low copper concentrations SOD-activity prevails while prion protein loaded with high amount of metal will work as copper transporter during internalization of copper-protein system.

References

Adlard, P. A. and Bush, A. I., (2006) Metals and Alzheimer's disease *J Alzheimers Dis.*, 10, pp. 145–63.

Alderighi, L., Gans, P., Ienco, A., Peters, D., Sabatini, A. and Vacca, A. (1999). Hyperquad simulation and speciation (HySS): a utility program for the investigation of equilibria involving soluble and partially soluble species, *Coord. Chem. Rev.* 184, pp. 311–18.

Aronoff-Spencer, E., Burns, C. S., Avdievich, N. I., Gerfen, G. J., Peisach, J., Antholine, W. E., Ball, H. L., Cohen, F. E., Prusiner, S. B. and Millhauser, G. L. (2000). Identification of the Cu^{2+} binding sites in the N-terminal domain of the prion protein by EPR and CD spectroscopy, *Biochemistry*, 39, pp. 13760–71.

Atwood, C. S., Scarpa, R. C., Huang, X. D., Moir, R. D., Jones, W. D., Fairlie, D. P., Tanzi, R. E. and Bush, A. I. (2000). Characterization of copper interactions with Alzheimer amyloid beta peptides: Identification of an attomolar-affinity copper binding site on amyloid beta 1–42, *J. Neurochem.* 75, pp. 1219–33.

Badrick, A. C. and Jones, C. E. (2009) The amyloidogenic region of the human prion protein contains a high affinity (Met)(2)(His)(2) Cu(I) binding site, *J. Inorg. Biochem.*, 103, pp. 1169–75.

Belosi, B., Gaggelli, E., Guerrini, R., Kozlowski, H., Luczkowski, M., Mancini, F. M., Remelli, M., Valensin, D. and Valensin, G. (2004). Copper binding to the neurotoxic peptide PrP106–126: Thermodynamic and structural studies, *Chembiochem* 5, pp. 349–59.

Berti, F., Gaggelli, E., Guerrini, R., Janicka, A, Kozlowski, H., Legowska, A., Miecznikowska, H., Migliorini, C., Pogni, R., Remelli, M., Rolka, K., Valensin, D.and Valensin, G. (2007). Structural and dynamic characterization of copper(II) binding of the human prion protein outside the octarepeat region, *Chem. Eur. J.*, 13, pp. 1991–2001.

Binolfi, A., Lamberto, G. R., Duran, R., Quintanar, L., Bertoncini, C. W., Souza, J. M., Cervenansky, C., Zweckstetter, M., Griesinger, C. and Fernandez, C. O. (2008) Site-Specific Interactions of Cu(II) with α and β-Synuclein: Bridging the Molecular Gap between Metal Binding and Aggregation, *J. Am. Chem. Soc.* 130, pp. 11801–12.

Bonomo, R. P., Impellizzeri, G., Pappalardo, G., Rizzarelli, E. and Tabbì, G. (2000). Copper(II) binding modes in the prion octapeptide PHGGGWGQ: a spectroscopic and voltammetric study, *Chemistry*, 6, pp. 4195–202.

Bolognin, S., Messori, L. and Zatta, P. (2009). Metal Ion Physiopathology in Neurodegenerative Disorders. *Neuromol. Med.*, 11, pp. 223–38.

Brown, D. R., (2009), Metal binding to alpha-synuclein peptides and its contribution to toxicity. *Biochem. Biophys. Res. Commun.*, 380, pp. 377–81

Brown, D. R., Hafiz, F., Glasssmith, L. L., Boon-Seng Wong, B.-S., Jones, I. M., C. Clive, C., and Haswell, S. J. (2000) Consequences of manganese replacement of copper for prion protein function and proteinase resistance. *EMBO J.* 19, pp. 1180–86.

Brown, D. R. and Kozlowski, H. (2004). Biological inorganic and bioinorganic chemistry of neurodegeneration based on prion and Alzheimer diseases, *Dalton Trans.*, pp. 1907–17.

Brown, D. R., Schmidt, B. and Kretzschmar, H. A. (1996) Role of microglia and host prion protein in neurotoxicity of a prion protein fragment. *Nature*, 380, pp. 345–47.

Burns, C. S., Aronoff-Spencer, E., Dunham, C. M., Lario, P., Avdievich, N. I., Antholine, W. E., Olmstead, M. M., Vrielink, A., Gerfen, G. J., Peisach, J., Scott, W. G. and Millhauser, G. L. (2002). Molecular features of the copper binding sites in the octarepeat domain of the prion protein, *Biochemistry*, 41, pp. 3991–4001.

Camponeschi, F., Gaggelli, E., Kozlowski, H., Valensin, D. and Valensin, G. (2009). Structural features of the Zn^{2+} complex with the single repeat region of "prion related protein" (PrP-rel-2) of zebrafish zPrP63-70 fragment, *Dalton Trans.*, pp. 4643–45.

Chattopadhyay, M., Walter, E. D., Newell, D. J., Jackson, P. J., Aronoff-Spencer, E., Peisach, J., Gerfen, G. J., Bennett, B., Antholine, W. E. and Millhauser, G. L. (2005). The octarepeat domain of the prion protein binds Cu(II) with three distinct coordination modes at pH 7.4, *J. Am. Chem. Soc.*, 127, pp. 12647–56.

Crisponi, G. and Remelli, M. (2008). Iron chelating agents for the treatment of iron overload, *Coord. Chem. Rev.* 252, pp. 1225–40.

Curtain, C. C., Ali, F., Volitakis, I., Cherny, R. A, Norton, R. S, Beyreuther, K., Barrow, C. J., Masters, C. L, Bush, A. I. and Barnham, K. J. (2001). Alzheimer's disease amyloid-beta binds copper and zinc to generate an allosterically ordered membrane-penetrating structure containing superoxide dismutase-like subunits. *J Biol Chem.*, 276. pp 20466–73.

Damante, C. A., Osz, K., Nagy, Z., Pappalardo, G., Grasso, G., Impellizzeri, G., Rizzarelli, E. and Sovago, I. (2008). The metal loading ability of beta-amyloid N-terminus: a combined potentiometric and spectroscopic study of copper(ii) complexes with beta-amyloid(1–16), its short or mutated peptide fragments, and its polyethylene glycol (PEG)-ylated Analogue, *Inorg. Chem.* 47, pp. 9669–83.

Damante, C. A., Osz, K., Nagy, Z., Pappalardo, G., Grasso, G., Impellizzeri, G., Rizzarelli, E. and Sovago, I. (2009). Metal loading capacity of A beta N-terminus: a combined potentiometric and spectroscopic study of zinc(II) complexes with A beta(1–16), its short or mutated peptide fragments and its polyethylene glycol-ylated analogue, *Inorg. Chem.* 48, pp. 10405–15.

Danielsson, J., Pierattelli, R., Banci, L. and Gräslund, A. (2007). High-resolution NMR studies of the zinc-binding site of the Alzheimer's amyloid beta-peptide. *FEBS J.*, 274, pp. 46–59.

Dawson, R. M. C., Elliott, D. C., Elliott, W. H. and Jones, K. M. (1989). *Data for Biochemical Research*, Oxford University Press.

Davies, P. and Brown, D. R. (2008). The chemistry of copper binding to PrP: is there sufficient evidence to elucidate a role for copper in protein function? *Biochem J.*, 410, pp. 237–44.

Davies, P., Fontanie, S. N., Moualla, D., Wang, X., Wright, J. A. and Brown, D. R., (2008). Amyloidogenic metal-binding proteins: new investigative pathways. *Biochem. Soc. Trans.*, 36, pp. 1299–303.

Di Natale, G., Grasso, G., Impellizzeri, G., La Mendola, D., Micera, G., Mihala, N., Nagy, Z., Osz, K., Pappalardo, G., Rigó, V., Rizzarelli, E., Sanna, D. and Sóvágó, I. (2005). Copper(II) interaction with unstructured prion domain outside the octarepeat region: speciation, stability, and binding details of copper(II) complexes with PrP106–126 peptides, *Inorg. Chem.*, 44, pp. 7214–25.

Di Natale, G., Osz, K., Nagy, Z., Sanna, D., Micera, G., Pappalardo, G., Sóvágó, I. and Rizzarelli, E. (2009). Interaction of copper(II) with the prion peptide fragment HuPrP(76–114) encompassing four histidyl residues within and outside the octarepeat domain, *Inorg. Chem.*, 48, pp. 4239–50.

Dorlet, P., Gambarelli, S., Faller, P. and Hureau, C. (2009). Pulse EPR spectroscopy reveals the coordination sphere of copper(II) ions in the 1–16 amyloid-beta peptide: a key role of the first two N-terminus residues. *Angew Chem Int Ed Engl.*, 48, pp. 9273–76.

Faller, P. and Hureau, C. (2009). Bioinorganic chemistry of copper and zinc ions coordinated to amyloid-beta peptide, *Dalton Trans.*, pp. 1080–94.

Faller, P. (2009) Copper and zinc binding to amyloid-beta: coordination, dynamics, aggregation, reactivity and metal-ion transfer. *Chembiochem*, 10, pp. 2837–45.

Flechsig E., Shmerling D., Hegyi I., Raeber A. J., Fischer, M., Cozzio, A., von Mering C., Aguzzi A. and Weissmann C. (2000) Prion protein devoid of the octapeptide repeat region restores susceptibility to scrapie in PrP knockout mice. *Neuron,* 27, pp. 399–408.

Furlan, S., La Penna, G., Guerrieri, F., Morante, S. and Rossi, G. C. (2007). Ab initio simulations of Cu binding sites on the N-terminal region of prion protein. *J. Biol. Inorg. Chem.*, 12, pp. 571–83.

Gaeta, A. and Hider, R. C. (2005). The crucial role of metal ions in neurodegeneration: The basis for a promising therapeutic strategy. *British J. Pharmacol.* 146, pp. 1041–59.

Gaggelli, E., Bernardi, F., Molteni, E., Pogni, R., Valensin, D., Valensin, G., Remelli, M., Luczkowski, M. and Kozlowski, H. (2005). Interaction of the human prion PrP(106–26) sequence with copper(II), manganese(II), and zinc(II): NMR and EPR studies, *J. Am. Chem. Soc.* 127, pp. 996–1006.

Gaggelli, E., Jankowska, E., Kozlowski, H., Marcinkowska, A., Migliorini, C., Stanczak, P., Valensin, D. and Valensin, G. (2008a). Structural Characterization of the Intra- and Inter-Repeat Copper Binding Modes within the N-Terminal Region of "Prion Related Protein" (PrP-rel-2) of Zebrafish, *J. Phys. Chem. B* 112, pp. 15140–50.

Gaggelli, E., Kozlowski, H., Valensin, D. and Valensin, G. (2006). Copper homeostasis and neurodegenerative disorders (Alzheimer's, prion, and Parkinson's diseases and amyotrophic lateral sclerosis), *Chem. Rev.* 106, pp. 1995–2044.

Gaggelli, E., Grzonka, Z., Kozłowski, H., Migliorini, C., Molteni, E., Valensin, D. and Valensin, G. (2008b). Structural features of the Cu(II) complex with the rat Abeta(1–28) fragment. *Chem. Commun. (Camb).*, pp. 341–43.

Gaggelli, E., Janicka-Klos, A., Jankowska, E., Kozlowski, H., Migliorini, C., Molteni, E., Valensin, D., Valensin, G., and Wieczerzak, E. (2008c). NMR studies of the Zn^{2+} interactions with rat and human beta-amyloid (1–28) peptides in water-micelle environment. *J. Phys. Chem. B.*, 112, pp. 100–109.

Gans, P., Sabatini, A. and Vacca, A. (1996). Investigation of equilibria in solution. Determination of equilibrium constants with the HYPERQUAD suite of programs, *Talanta* 43, pp. 1739–53.

Gans, P., Sabatini, A. and Vacca, A. (2008). Simultaneous calculation of equilibrium constants and standard formation enthalpies from calorimetric data for systems with multiple equilibria in solution, *J. Solut. Chem.* 37, pp. 467–76.

Garnett, A. P. and Viles, J. H. (2003). Copper binding to the octarepeats of the prion protein. Affinity, specificity, folding, and cooperativity: insights from circular dichroism, *J. Biol. Chem.*, 278, pp. 6795–802.

Gralka, E., Valensin, D., Gajda, K., Bacco, D., Szyrwiel, L., Remelli, M., Valensin, G., Kamasz, W., Baranska-Rybak, W. and Kozłowski, H. (2009). Copper(II) coordination outside the tandem repeat region of an unstructured domain of chicken prion protein. *Mol. Biosyst.*, 5, pp. 497–510.

Gralka, E., Valensin, D., Porciatti, E., Gajda, C., Gaggelli, E., Valensin, G., Kamysz, W., Nadolny, R., Guerrini, R., Bacco, D., Remelli, M. and Kozlowski, H. (2008). Cu^{II} binding sites located at His-96 and His-111 of the human prion protein: thermodynamic and spectroscopic studies on model peptides, *Dalton Trans.*, 14, pp. 5207–219.

Guilloreau, L., Combalbert, S., Sournia-Saquet, A., Mazarguil, H. and Faller, P. (2007). Redox chemistry of copper-amyloid-β: the generation of hydroxyl radical in the presence of ascorbate is linked to redox-potentials and aggregation state, *ChemBioChem*, 8, pp. 1317–25.

Guilloreau, L., Damian. L., Coppel, Y., Mazarguil, H., Winterhalter, M. and Faller, P. (2006). Structural and thermodynamical properties of CuII amyloid-beta16/28 complexes associated with Alzheimer's disease. *J. Biol. Inorg. Chem.*, 11, pp. 1024–38.

Himes, R. A., Park, G. Y., Siluvai, G. S., Blackburn, N. J. and Karlin, K. D. (2008). Structural studies of copper(I) complexes of amyloid-beta peptide fragments: formation of two-coordinate bis(histidine) complexes. *Angew. Chem. Int. Ed Engl.*, 47, pp. 9084–87.

Hodak, M., Chisnell, R., Lu, W. and Bernholc, J. (2009). Functional implications of multistage copper binding to the prion protein, *Proc. Natl. Acad. Sci. USA*, 106, pp. 11576–81.

Hou, L. and Zagorski, M. G. (2006). NMR reveals anomalous copper(II) binding to the amyloid Abeta peptide of Alzheimer's disease. *J. Am. Chem Soc.*, 128, pp. 9260–61.

Hornshaw, M. P., McDermott, J. R. and Candy, J. M. (1995a). Copper binding to the N-terminal tandem repeat regions of mammalian and avian prion protein, *Biochem. Biophys. Res. Commun.*, 207, pp. 621–29.

Hornshaw, M. P., McDermott, J. R., Candy, J. M. and Lakey, J. H. (1995b). Copper binding to the N-terminal tandem repeat region of mammalian and avian prion protein: structural studies using synthetic peptides, *Biochem. Biophys. Res. Commun.*, 214, pp. 993–99.

Hsu, L. J., Mallory, M., Xia, Y., Veinbergs, I., Hashimoto, M., Yoshimoto, M., Thal, L. J., Saitoh, T. and Masliah, E. (1998). Expression pattern of synucleins (non Aβ component of Alzheimer disease amyloid precursor protein-α-synuclein) during murine brain development. *J. Neurochem.,* 71, pp. 338–44.

Huang, X., Atwood, C. S., Hartshorn, M. A., Multhaup, G., Goldstein, L. E., Scarpa, R. C., Cuajungco, M. P., Gray, D. N., Lim, J., Moir, R. D., Tanzi, R. D. and Bush, A. I. (1999). The Aβ peptide of Alzheimer's disease directly produces hydrogen peroxide through metal ion reduction, *Biochemistry*, Vol. 38, No. 24, pp. 7609–16.

Hung Y. H., Bush A. I. and Cherny, R. A. (2010) Copper in the brain and Alzheimer's disease. *J Biol. Inorg. Chem.*, 15, pp. 61–76.

Hureau, C., Balland, V., Coppel, Y., Solari, P. L., Fonda, E. and Faller, P. (2009a). Importance of dynamical processes in the coordination chemistry and redox conversion of copper amyloid-beta complexes. *J. Biol. Inorg. Chem.*, 14, pp. 995–1000.

Hureau, C., L. Charlet, P. Dorlet, F. Gonnet, L. Spadini, E. Anxolabehere-Mallart and J. J. Girerd (2006). A spectroscopic and voltammetric study of the pH-dependent Cu(II) coordination to the peptide GGGTH: relevance to the fifth Cu(II) site in the prion protein, *J. Biol. Inorg. Chem.* 11, pp. 735–44.

Hureau, C., Coppel, Y., Dorlet, P., Solari, P. L., Sayen, S., Guillon, E., Sabater, L. and Faller, P. (2009b). Deprotonation of the Asp1-Ala2 peptide bond induces modification of the dynamic copper(II) environment in the amyloid-beta peptide near physiological pH. *Angew. Chem. Int. Ed Engl.*, 48, pp. 9522–25.

Hureau, C., Mathé, C., Faller, P., Mattioli, T. A. and Dorlet, P. (2008) Folding of the prion peptide GGGTHSQW around the copper(II) ion: identifying the oxygen donor ligand at neutral pH and probing the proximity of the tryptophan residue to the copper ion. *J. Biol. Inorg. Chem.*, 13, pp. 1055–64.

Jackson and Lee, (2009), Identyfication of the minimal copper(II)-binding α-synuclein. *Inorg. Chem.*, 48, pp. 9303–07

Jobling, M. F., Huang, X., Stewart, L. R., Barnham, K. J., Curtain, C. C., Volitakis, I., Perugini. M., White, A. R., Cherny, R. A., Masters, C. L., Barrow, C. J., Collins, S. J., Bush, A. I. and Cappai, R. (2001). Copper and zinc binding modulates the aggregation and neurotoxic properties of the prion peptide PrP106–126. *Biochemistry*, 40, pp. 8073–84.

Jones, C. E., Abdelraheim, S. R., Brown, D. R. and Viles, J. H. (2004). Preferential Cu^{2+} coordination by His96 and His111 induces beta-sheet formation in the unstructured amyloidogenic region of the prion protein, *J. Biol. Chem.*, 279, pp. 32018–27.

Joszai, V., Nagy, Z., Osz, K., Sanna, D., Di Natale, G., La Mendola, D., Pappalardo, G., Rizzarelli, E. and Sovago, I. (2006). Transition metal complexes of terminally protected peptides containing histidyl residues, *J. Inorg. Biochem.* 100, pp. 1399–409.

Karr, J. W., Akintoye, H., Kaupp, L. J. and Szalai, V. A. (2005). N-Terminal deletions modify the Cu^{2+} binding site in amyloid-beta. *Biochemistry*, 44, pp. 5478–87.

Klewpatinond, M., Davies, P., Bowen, S., Brown, D. R. and Viles, J. H. (2008). Deconvoluting the Cu^{2+} binding modes of full-length prion protein, *J. Biol. Chem.* 283, pp. 1870–81.

Klewpatinond, M. and Viles, J. H. (2007). Fragment length influences affinity for Cu^{2+} and Ni^{2+} binding to His(96) or His(111) of the prion protein and spectroscopic evidence for a multiple histidine binding only at low pH, *Biochem. J.* 404, pp. 393–402.

Kowalik-Jankowska, T., Ruta-Dolejsz, M., Wisniewska, K., Lankiewicz, L. and Kozlowski, H. (2000). Copper(II) complexation by human and mouse fragments (11–16) of beta-amyloid peptide, *J. Chem. Soc. Dalton Trans.*, pp. 4511–19.

Kowalik-Jankowska, T., Ruta-Dolejsz, M., Wisniewska, K. and Lankiewicz, L. (2001). Cu(II) interaction with N-terminal fragments of human and mouse beta-amyloid peptide, *J. Inorg. Biochem.* 86, pp. 535–45.

Kowalik-Jankowska, T., Ruta-Dolejsz, M., Wisniewska, K., and Lankiewicz, L. (2002). Coordination of copper(II) ions by the 11–20 and 11–28 fragments of human and mouse beta-amyloid peptide, *J. Inorg. Biochem.* 92, pp. 1–10.

Kowalik-Jankowska, T., Ruta, M., Wisniewska, K. and Lankiewicz, L. (2003). Coordination abilities of the 1–16 and 1–28 fragments of beta-amyloid peptide towards copper(II) ions: a combined potentiometric and spectroscopic study, *J. Inorg. Biochem.* 95, pp. 270–82.

Kowalik-Jankowska, T., Rajewska, A., Wisniewska, K., Grzonka, Z. and Jezierska, J. (2005). Coordination abilities of N-terminal fragments of alpha-synuclein towards copper(II) ions: A combined potentiometric and spectroscopic study, *J. Inorg. Biochem.* 99, pp. 2282–91. With permission from Elsevier.

Kowalik-Jankowska, T., Rajewska, A., Jankowska, E. and Grzonka, Z. (2006). Copper(II) binding by fragments of alpha-synuclein containing M-1-D-2- and -H-50-residues; a combined potentiometric and spectroscopic study, *Dalton Trans.*, pp. 5068–76.

Kowalik-Jankowska, T., Rajewska, A., Jankowska, E. and Grzonka, Z. (2007). Coordination abilities of alpha-synuclein fragments modified in the 30th (A30P) and 53rd (A53T) positions and products of metal-catalyzed oxidation, *Dalton Trans.*, pp. 4197–206.

Kozlowski, H., Bal, W., Dyba, M., Kowalik-Jankowska, T. (1999). Specific Structure-Stability Relations in Metallopeptides. *Coord. Chem. Rev.* 184, pp. 319–46.

Kozlowski, H., Janicka-Klos, A., Brasun, J., Gaggelli, E., Valensin, D. and Valensin, G. (2009). Copper, iron, and zinc ions homeostasis and their role in neurodegenerative disorders (metal uptake, transport, distribution and regulation), *Coord. Chem. Rev.* 253, pp. 2665–85.

Kozlowski, H., Janicka-Klos, A., Stanczak, P., Valensin, D., Valensin, G., and Kulon, K., (2008). Specificity in the Cu^{2+} interactions with prion protein fragments and related His-rich peptides from mammals to fishes, *Coord. Chem. Rev.* 252, pp. 1069–78.

Kozlowski, H., Kowalik-Jankowska, T., Jeżowska-Bojczuk, M., (2005). Chemical and biological aspects of Cu^{2+} interactions with peptides and aminoglycosides. *Coord. Chem. Rev.*, 249, pp. 2323–34.

Kozlowski, H., Luczkowski, M. and Remelli, M. (submitted). Prion proteins and copper ions. Biochemical and chemical controversies, *Dalton Trans.* 2010, pp. 6371–85.

Kozlowski, H., Luczkowski, M., Valensin D. and Valensin, G. (2006). Metal ion binding properties of proteins related to neurodegeneration. *Neurodegenerative Diseases and Metal Ions.* A. Sigel, H. Sigel and R. K. O. Sigel. Chichester, Wiley.

Kuczius, T., Buschmann, A., Zhang, W., Karch, H., Becker, K., Peters, G. and Groschup, M. H. (2004). Cellular prion protein acquires resistance to proteolytic degradation following copper ion binding. *Biol. Chem.* 385, pp. 739–47.

La Mendola, D., Bonomo, R. P., Caminati, S., Di Natale, G., Emmi, S. S., Hansson, O., Maccarrone, G., Pappalardo, G., Pietropaolo, A. and Rizzarelli, E. (2009). Copper(II) complexes with an avian prion N-terminal region and their potential SOD-like activity, *J. Inorg. Biochem.* 103, pp. 195–204.

Liu, S. T., Howlett, G. and Barrow, C. J. (1999). Histidine-13 is a crucial residue in the zinc ion-induced aggregation of the A beta peptide of Alzheimer's disease. *Biochemistry*, 38 pp. 9373–78.

Lovell, M. A., Robertson, J. D., Teesdale, W. J., Campbell, J. L. and Markesbery, W. R. (1998). Copper, iron and zinc in Alzheimer's disease senile plaques, *J. Neurol. Sci.* 158, pp. 47–52.

Luczkowski, M., Kozlowski, H. Legowska, A., Rolka, K. and Remelli, M. (2003). The possible role of Gly residues in the prion octarepeat region in the coordination of Cu^{2+} ions, *Dalton Trans.*, pp. 619–24.

Luczkowski, M., Kozlowski, H., Stawikowski, M., Rolka, K., Gaggelli, E. Valensin, D. and Valensin, G. (2002). Is the monomeric prion octapeptide repeat PHGGGWGQ a specific ligand for Cu^{2+} ions?, *J. Chem. Soc.-Dalton Trans.*, pp. 2269–74.

Marino, T., Russo, N. and Toscano, M. (2007). On the copper(II) ion coordination by prion protein HGGGW pentapeptide model, *J. Phys. Chem. B.*, 111, pp. 635–40.

Mekmouche, Y., Coppel, Y., Hochgräfe, K., Guilloreau, L., Talmard, C., Mazarguil, H. and Faller P. (2005). Characterization of the ZnII binding to the peptide amyloid-beta1–16 linked to Alzheimer's disease. *Chembiochem.*, 6, pp. 1663–71.

Mentler, M., Weiss, A., Grantner, K., Del Pino, P., Deluca, D., Fiori, S., Renner, C., Klaucke, W. M., Moroder, L., Bertsch, U., Kretzschmar, H. A., Tavan, P. and Parak, F. G. (2005). A new method to determine the structure of the metal environment in metalloproteins: investigation of the prion protein octapeptide repeat $Cu^{(2+)}$ complex, *Eur. Biophys. J.*, 34, pp. 97–112.

Millhauser, G. L. (2004). Copper binding in the prion protein, *Acc. Chem. Res.* 37, pp. 79–85.

Millhauser, G. L. (2007). Copper and the prion protein: Methods, structures, function, and disease, *Ann. Rev. Phys. Chem.* 58, pp. 299–320.

Minicozzi, V., Stellato, F., Comai, M., Serra, M. D., Potrich, C., Meyer-Klaucke, W. and Morante, S. (2008). Identifying the minimal copper- and zinc-binding site sequence in amyloid-beta peptides. *J. Biol. Chem.*, 283, pp. 10784–92.

Miura, T., Hori-I, A., Mototani, H. and Takeuchi, H. (1999). Raman spectroscopic study on the copper(II) binding mode of prion octapeptide and its pH dependence, *Biochemistry*, 38, pp. 11560–69.

Osz, K. (2008). A new, model-free calculation method to determine the coordination modes and distribution of copper(II) among the metal binding sites of multihistidine peptides using circular dichroism spectroscopy, *J. Inorg. Biochem.* 102, pp. 2184–95.

Osz, K., Nagy, Z., Pappalardo, G., Di Natale, G., Sanna, D., Micera, G., Rizzarelli, E. and Sovago, I. (2007). Copper(II) interaction with prion peptide fragments encompassing histidine residues within and outside the octarepeat domain: Speciation, stability constants and binding details, *Chem.-Eur. J.* 13, pp. 7129–43.

Paik *et al.* (1999). Copper(II)-induced self-oligomerization of α-synuclein. *Biochem. J.*, 340, pp. 821–28.

Pradines V., Stroia Jurca A. and Faller P. (2008). Amyloid fibrils: modulation of formation and structure by copper(II). *New J. Chem*, 32, pp. 1189–94.

Pushie, M. J. and Rauk, A. (2003). Computational studies of Cu(II)[peptide] binding motifs: Cu[HGGG] and Cu[HG] as models for Cu(II) binding to the prion protein octarepeat region, *J. Biol. Inorg. Chem.*, 8, pp. 53–65.

Pushie, M. J., Rauk, A., Jirik, F. and Vogel, H. J. (2009). Can copper binding to the prion protein generate a misfolded form of the protein? *Biometals,* 22, pp. 159–75.

Pushie, M. J. and Vogel, H. J. (2007). Molecular dynamics simulations of two tandem octarepeats from the mammalian prion protein: fully Cu^{2+}-bound and metal-free forms, *Biophys. J.*, 93, pp. 3762–74.

Pushie, M. J. and Vogel, H. J. (2008). Modeling by assembly and molecular dynamics simulations of the low Cu^{2+} occupancy form of the mammalian prion protein octarepeat region: gaining insight into Cu^{2+}-mediated beta-cleavage, *Biophys. J.*, 95, pp. 5084–91.

Quaglio, E., Chiesa, R. and Harris, D. A. (2001). Copper converts the cellular prion protein into a protease-resistant species that is distinct from the scrapie isoform. *J Biol Chem.*, 276, pp. 11432–38.

Rauk, A. (2009), The chemistry of Alzheimer's disease. *Chem. Soc. Rev.*, 38, pp. 2698–715.

Rasia, R. M., Bertoncini, C. W., Marsh, D., Hoyer, W., Cherny, D., Zweckstetter, M., Griesinger, C., Jovin, T. and Fernandez, C. O. (2005). Structural characterization of copper(II) binding to α-synuclein: Insights into the bioinorganic chemistry of Parkinson's disease. *Proc. Natl. Acad. Sci. U.S.A.*, 102, pp. 4294–99.

Remelli, M., Donatoni, M., Guerrini, R., Janicka, A., Pretegiani, P. and Kozlowski, H.(2005). Copper-ion interaction with the 106–113 domain of the prion protein: a solution-equilibria study on model peptides, *Dalton Trans.*, pp. 2876–85.

Remelli, M., Valensin, D., Bacco, D., Gralka, E., Guerrini, R., Migliorini, C. and Kozlowski, H. (2009). The complex-formation behaviour of His residues in the fifth Cu^{2+} binding site of human prion protein: a close look, *New J. Chem.*, pp. 2300–310.

Riihimäki, E. S., Martínez, J. M. and Kloo, L. (2007). Molecular dynamics simulations of Cu(II) and the PHGGGWGQ octapeptide, *J. Phys. Chem. B.*, 111, pp. 10529–37.

Riihimäki, E. S., Martínez, J. M. and Kloo, L. (2008). Structural effects of Cu(II)-coordination in the octapeptide region of the human prion protein, *Phys. Chem. Chem. Phys.*, 10, pp. 2488–95.

Shearer, J., Soh, P. and Lentz, S. J. (2008). Both Met(109) and Met(112) are utilized for Cu(II) coordination by the amyloidogenic fragment of the human prion protein at physiological pH, *J. Inorg. Biochem.*, 102, pp. 2103–13.

Shearer, J. and Szalai, V. A. (2008). The amyloid-beta peptide of Alzheimer's disease binds Cu(I) in a linear bis-his coordination environment: insight into a possible neuroprotective mechanism for the amyloid-beta peptide. *J. Am. Chem. Soc.*, 130, pp. 17826–35.

Shin, B. K. and Saxena, S. (2008). Direct evidence that all three histidine residues coordinate to Cu(II) in amyloid-beta(1–16), *Biochemistry* 47, pp. 9117–23.

Stanczak, P., Juszczyk, P., Grzonka, Z., Kozłowski, H., (2007). The whole hexapeptide repeats domain from avian PrP displays untypical hallmarks in aspect of the Cu^{2+} complex formation. *FEBS Letters,* 581, pp. 4544–48.

Stanczak, P. and Kozlowski., H. (2007). Can Chicken and human PrPs possess SOD-like activity after β-cleveage? *Biochem. Biophys. Res. Commun.,* 352, pp. 198–202.

Stanczak, P., Valensin, D., Juszczyk, P., Grzonka, Z., Migliorini, C., Molteni, E., Valensin, G., Gaggelli, E. and Kozlowski, H. (2005a). Structure and stability of the CuII complexes with tandem repeats of the chicken prion, *Biochemistry* 44, pp. 12940–54.

Stanczak, P., Valensin, D., Juszczyk, P., Grzonka, Z., Valensin, G., Bernardi, F., Molteni, E., Gaggelli, E. and Kozlowski, H. (2005b). Fine tuning the structure of the Cu^{2+} complex with the prion protein chicken repeat by proline isomerization, *Chem. Commun.*, pp. 3298–300.

Stanczak, P., Valensin, D., Porciatti, E., Jankowska, E., Grzonka, Z., Molteni, E., Gaggelli, E., Valensin, G. and Kozlowski, H. (2006). Tandem repeat-like domain of "similar to prion protein" (StPrP) of Japanese pufferfish binds Cu(II) as effectively as the mammalian protein, *Biochemistry* 45, pp. 12227–39.

Streltsov, V. A., Titmuss, S. J., Epa, V. C., Barnham, K. J., Masters C. L. and Varghese, J. N. (2008). The structure of the amyloid-beta peptide high-affinity copper II binding site in Alzheimer disease. *Biophys J.,* 95, pp. 3447–56.

Streltsov, V. A. and Varghese, J. N. (2008). Substrate mediated reduction of copper-amyloid-beta complex in Alzheimer's disease. *Chem. Commun. (Camb.),* 27, pp. 3169–71.

Syme, C. D., Nadal, R. C., Rigby, S. E. and Viles, J. H. (2004). Copper binding to the amyloid-beta (Abeta) peptide associated with Alzheimer's disease: folding, coordination geometry, pH dependence, stoichiometry, and affinity of Abeta-(1–28): insights from a range of complementary spectroscopic techniques. *J. Biol. Chem.*, 279, pp. 18169–77.

Syme, C. D. and Viles, J. H. (2006). Solution ^1H NMR investigation of Zn^{2+} and Cd^{2+} binding to amyloid-beta peptide (Abeta) of Alzheimer's disease. *Biochim. Biophys. Acta*, 1764, pp. 246–56.

Szyrwiel, L., Jankowska, E., Janicka-Klos, A., Szewczuk, Z., Valensin, D. and Kozlowski, H. (2008). Zn(II) ions bind very efficiently to tandem repeat region of "prion related protein" (PrP-rel-2) of zebra-fish. MS and potentiometric evidence, *Dalton Trans.*, pp. 6117–20.

Talmard, C., Bouzan, A. and Faller, P. (2007). Zinc binding to amyloid-beta: isothermal titration calorimetry and Zn competition experiments with Zn sensors. *Biochemistry*, 46, pp. 13658–66.

Uversky, V. N., Li, J. and Fink, A. L. (2001). Metal-triggered Structural Transformations, Aggregation, and Fibrillation of Human α-Synuclein. *J Biol Chem*, 276, pp. 44284–96.

Valensin, D., Gajda, K., Gralka, E., Valensin, G., Kamysz, W. and Kozlowski, H. (2010). Copper binding to chicken and human prion protein amylodogenic regions: differences and similarities revealed by Ni^{2+} as a diamagnetic probe, *J. Inorg. Biochem.*, 104, pp. 71–78.

Valensin, D., Luczkowski, M., Mancini, F. M., Legowska, A., Gaggelli, E., Valensin, G., Rolka, K. and Kozlowski, H. (2004). The dimeric and tetrameric octarepeat fragments of prion protein behave differently to its monomeric unit, *Dalton Trans.*, 9, pp. 1284–93.

Valensin, D., Szyrwiel, Ł., Camponeschi, F., Rowińska-Zyrek, M., Molteni, E., Jankowska, E., Szymanska, A., Gaggelli, E., Valensin, G. and Kozłowski, H. (2009a). Heteronuclear and homonuclear Cu^{2+} and Zn^{2+} complexes with multihistidine peptides based on zebrafish prion-like protein, *Inorg. Chem.*, 48, pp. 7330–40.

Valensin, G., Molteni, E., Valensin, D., Taraszkiewicz, M. and Kozlowski, H. (2009b). Molecular dynamics study of the Cu^{2+} binding-induced "structuring" of the N-terminal domain of human prion protein, *J. Phys. Chem. B.*, 113, pp. 3277–79.

Viles, J. H., Cohen, F. E., Prusiner, S. B., Goodin, D. B., Wright, P. E. and Dyson, H. J. (1999). Copper binding to the prion protein: structural implications of four identical cooperative binding sites, *Proc. Natl. Acad. Sci. U.S.A.*, 96, pp. 2042–47.

Viles, J. H., Klewpatinond, M. and Nadal, R. C. (2008). Copper and the structural biology of the prion protein. *Biochem Soc. Trans.* 36 pp. 1288–92.

Walter, E. D., Chattopadhyay M. and Millhauser, G. L. (2006). The affinity of copper binding to the prion protein octarepeat domain: Evidence for negative cooperativity, *Biochemistry*, 45, pp. 13083–92.

Walter, E. D., Stevens, D. J., Spevacek, A. R., Visconte, M. P., Rossi, A. D. and Millhauser, G. L. (2009). Copper Binding Extrinsic to the Octarepeat Region in the Prion Protein, *Curr. Prot. Pept. Sci.*, 10, pp. 529–35.

Walter, E. D., Stevens, D. J., Visconte, M. P. and Millhauser, G. L. (2007). The prion protein is a combined zinc and copper binding protein: Zn^{2+} alters the distribution of Cu^{2+} coordination modes, *J. Am. Chem. Soc.*, 129, pp. 15440–41.

Whittal, R. M., Ball, H. L., Cohen, F. E., Burlingame, A. L., Prusiner, S. B. and Baldwin, M. A. (2000). Copper binding to octarepeat peptides of the prion protein monitored by mass spectrometry, *Protein Sci.*, 9, pp. 332– 43.

Zirah, S., Kozin, S. A., Mazur, A. K., Blond, A., Cheminant, M., Ségalas-Milazzo, I., Debey, P., and Rebuffat, S. (2006). Structural changes of region 1–16 of the Alzheimer disease amyloid beta-peptide upon zinc binding and in vitro aging. *J. Biol. Chem. 2006,* 281, pp. 2151–61.

Chapter 4

Mammalian Metallothioneins

Duncan E. K. Sutherland and Martin J. Stillman
Department of Chemistry, The University of Western Ontario, London, Ontario, Canada N6A 5B7

Metallothioneins (MTs), first discovered in 1957 (Margoshes and Vallee, 1957), are a family of small cysteine rich proteins found in all organisms. MTs have been implicated in toxic metal detoxification (Liu *et al.*, 2000), protection against oxidative stress (Kang, 2006) and as metallochaperones (Tapia *et al.*, 2004; Maret, 2008a), supplying both Zn^{2+} and Cu^{+} to their respective apo-enzymes. Zinc and copper are essential for the normal function of the brain and significant amounts are present in both the normal and diseased state. Zinc levels are critical for normal physical and mental development with numerous symptoms being reported in the cases of zinc deficiency. Copper is essential in the redox-based enzyme chemistry, for example the superoxide dismutases. Metallothionein is widely considered to play a significant role in the homeostasis of both these metals in other organs, particularly, in the liver, and *in vitro* studies show well-defined metallation chemistry. Metallothioneins are also implicated in the redox balance of cells as a result of the 20 cysteine thiols present and the variable metallation status possible. Therefore, the metallation of MT is intimately connected with both the concentrations of these metals and the ability to act as a metal-chaperone, aiding in cellular metal buffering. Further, MT is most likely a key component in the mechanistic pathways that describe

Brain Diseases and Metalloproteins
Edited by David R. Brown

ISBN 978-981-4316-01-9 (Hardcover), 978-981-4364-07-2 (eBook)
www.panstanford.com

absorption, function and excretion of zinc and copper. Mammals produce four metallothionein isoforms (MT-1 to MT-4), of which MT-1 and MT-2 are expressed in all organs, MT-3 is found predominantly in the central nervous system, and MT-4 is present in some stratified tissues. However, despite 50 years of intense research, the exact function(s) remain unknown and the mechanistic details of the metallation reactions and subsequent metal transfer reactions are poorly understood.

A number of reports have linked metallothioneins to the homeostasis of zinc in brain tissue. While human MT-1 and MT-2 are found all organs, MT-3 is unique to the brain. A recent study demonstrated metal swapping between zinc bound MT-3 and copper bound amyloid-β peptide ($A\beta_{1\text{-}40}$) led to the formation of a mixed copper-zinc species in MT-3 with a concomitant decrease in reactive oxygen species (ROS) production, which suggests MT-3 may be significant in attenuating cellular damage caused by increased oxidative stress found in Alzheimer's disease (Meloni *et al.*, 2008). Perhaps, more astonishingly, the recent discovery of a weakly bound eighth zinc in human MT-3 may well be a metal exchange intermediate with significant biological consequences (Meloni *et al.*, 2009). From these recent results, it is evident that the study of metallothioneins *in vivo* requires a detailed and complete understanding of its metal binding properties, which can be clearly obtained *in vitro*.

The aim of this chapter is to provide the framework for those *in vivo* studies by describing metallation studies reported for *in vitro* reactions that exploit the different probes of metallation afforded by each metal. The studies included bring together a number of spectroscopic techniques including electronic absorption, circular dichroism, emission and nuclear magnetic resonance, as well as the more recent advances in the mechanistics of metallation by mass spectrometry. Together, these methods allow a portrait of the metallation properties of metallothionein to be constructed with a range of metals. As the studies of human MT-3 cited above show, metallothionein probably acts to bind a variety of metals in mixed-metal configurations. Certainly mass spectral data from human MT-1 shows the presence of mixed metal species, for example mixed Zn^{2+}, Cd^{2+} and Cu^{+} (Merrifield *et al.*, 2002; Sutherland and Stillman, 2008). The mechanisms for metal transfer from metallothionein to a receptor protein, and the reverse, are very poorly understood, yet are critical to the specificity of metallothioneins in general. The following chapter will focus on important experimental techniques used to study the metallation chemistry of MT, followed by a brief description of MT-1 and MT-2, as well as their relation to the neuroinhibitory isoform, MT-3.

4.1 Metallothionein

Metallothioneins (MTs) are characterized by their low mass, high cysteine and metal content, as well as a lack of aromatic amino acids. Owing to the high cysteine content, ~30% of all residues with no disulfide bonds, MT has been implicated in metal ion homeostasis, toxic metal detoxification, and cellular protection against oxidative stress, indeed in much of the cellular chemistry that involves the Group 11 and 12 d^{10} metals. Four MT isoforms are present in humans: MT-1 and MT-2 expressed in virtually all cells; MT-3 expressed in the central nervous system; MT-4 expressed in squamous epithelial cells. MT-1 and MT-2 can be induced by a variety of stimuli, including metal ions, glucocorticoids, cytokines and oxidative stress, however, MT-3 and MT-4 appear relatively nonresponsive to these inducing agents (Uchida *et al.*, 1991; Kramer *et al.*, 1996a; Kramer *et al.*, 1996b; Vasak and Hasler, 2000; Meloni *et al.*, 2006). This lack of responsiveness suggests a more specific role for the latter metallothionein isoforms. With only a handful of studies, little is known about MT-4 and a detailed comparison of this isoform to the other three cannot be performed at this time. Figure 4.1 provides a comparison of the sequences of MT-1a, MT-2 and MT-3. With the exception of two insertions, a single amino acid in the β-domain and a hexapeptide sequence in the α-domain, there is no significant difference between the three sequences. In particular,

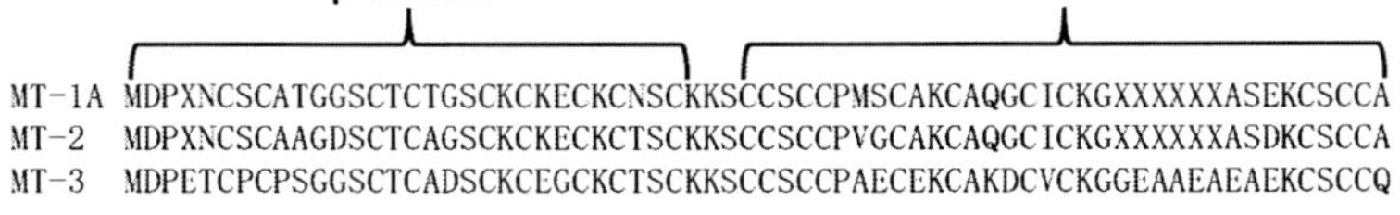

Figure 4.1 Amino acid sequence of human MT-1A, MT-2 and MT-3. Cysteine residues are labeled C and located in invariant positions, while inserts required for alignment are labeled X. There is significant sequence similarity between the three proteins, with an identical linker, KKS, connecting both domains. All cysteine residues are located in invariant positions inside.

the cysteine residues occupy invariant positions, which provide all forms of the protein their metal binding properties.

The binding chemistry of the Group 12 d^{10} metals, Zn^{2+}, Cd^{2+} and Hg^{2+}, to MTs has been a major area of study since the first protein was isolated from horse kidney containing cadmium (Margoshes and Vallee, 1957; Kagi and Vallee, 1961). The remarkably high cysteine content and the sequence organization into Cys-X-Cys and Cys-X-X-Cys motifs make metallothioneins ideal chelators of these soft metals. Both Zn^{2+} and Cd^{2+} bind to four cysteine residues in a tetrahedral manner, with metal-to-protein stoichiometries of 3 and 4 in the β- and α-domain, respectively. Studies on the binding of Cd^{2+} to metallothionein has been intense because, in addition to having a similar coordination number, several properties of Cd^{2+} make it an attractive model to probe these Zn^{2+} binding sites. The properties specific to Cd^{2+} include a red shifted ligand (thiolate)-to-metal charge transfer band in the absorption and CD spectra, a spin ½ nucleus, and finally a binding constant several orders of magnitude higher than that of Zn^{2+} (Kagi and Vallee, 1961; Vasak *et al.*, 1981a; Stillman *et al.*, 1987; Oz *et al.*, 1998; Krezel and Maret, 2007). The binding of Hg^{2+} has also been analyzed with metallation shown to be dependent on a number of variables, including pH and counter-ions present (Lu and Stillman, 1993; Lu *et al.*, 1993; Stillman *et al.*, 2000; Leiva-Presa *et al.*, 2004). Several protein stoichiometries have been reported for Hg-binding, with coordination geometries that change in sequence from the tetrahedral HgS_4 in Hg_7-MT to the trigonal HgS_3 in Hg_{11}-MT and, finally, to a digonal coordination for HgS_2 that has been hypothesized for Hg_{20}-MT, a structure proposed to involve one cysteine residue per Hg^{2+} atom (Lu and Stillman, 1993). Hg_{18}-MT, a structurally more interesting species, has been reported when Hg^{2+} was added to rabbit liver MT-2a at a $pH < 6$ and in the presence of chloride ions (Lu *et al.*, 1993) for which a proposed ladder like structure involving Hg^{2+}, coordinated to bridging thiolates and outlying chlorides, in a pseudotetrahedral geometry. The metallation reactions of Cu^{+} are very complicated and will be discussed separately below.

Zn^{2+} and Cu^{+} binding to MT is a critical part of metal ion homeostasis and may be the most significant property of the metallothioneins in general. The X-ray crystal and NMR solution structures of rat liver MT-2, as well as the NMR solution structures of human MT-2 show identical molecular architecture (Schultze *et al.*,

1988; Messerle *et al.*, 1990; Robbins *et al.*, 1991; Braun *et al.*, 1992; Messerle *et al.*, 1992). These structures demonstrate that both Zn^{2+} and Cd^{2+} bind with a tetrahedral geometry to four cysteinyl-thiolates in two linked but encapsulated binding domains (Fig. 4.2). It should be noted that while mammalian MTs have been shown to bind both Zn^{2+} and Cd^{2+} using exclusively cysteine residues, the distantly related bacterial MT, SmtA, has been shown to coordinate to Zn^{2+} through the use of histidine residues, where this zinc finger motif acts to maintain structure, possibly aiding in protein and/or DNA recognition (Blindauer *et al.*, 2001; Blindauer *et al.*, 2003). While an X-ray crystal structure of mammalian MT bound to copper has not been reported, it has been proposed that Cu^{+} can bind with digonal, trigonal, and tetrahedral coordination (Presta *et al.*, 1995). A recent X-ray crystal structure of Cu^{+} binding to yeast metallothionein showed the coexistence of both digonal and trigonal coordination (Calderone *et al.*, 2005). Spectroscopic studies of rabbit liver MT-2a have demonstrated the presence of both a Cu_{12}- and Cu_{15}-MT species, in which it was proposed that the former involves exclusively trigonal coordination, while the latter requires a mixture of trigonal and diagonal coordination geometries (Presta *et al.*, 1995). A Cu_4-MT intermediate has been isolated from mammalian MT-1 and MT-3, and may be a folding intermediate of unknown coordination (Jensen *et al.*, 1998). To summarize, mammalian metallothioneins (Fig. 4.2), contain two domains, an N-terminal β-domain with 9 cysteine residues traditionally understood to bind 3 Zn^{2+} or Cd^{2+}, or 6 Cu^{+} atoms; and a C-terminal α-domain with 11 cysteine residues traditionally understood to bind 4 Zn^{2+} or Cd^{2+}, or 6 Cu^{+} atoms. The coordination of Zn^{2+} and Cd^{2+} has been observed to be exclusively tetrahedral, while Cu^{+} is likely to have a range of geometries from diagonal to tetrahedral.

At the cellular level the concentration of metal ions, both toxic and nontoxic, is tightly controlled. Estimates of the total cellular zinc pool places the concentration at sub to low millimolar concentrations, which is in stark contrast to free cytosolic Zn^{2+}, which appears to be six orders of magnitude lower, on the order of 1 g atom of Zn^{2+} per cell (Outten and O'Halloran, 2001). The lack of free cytosolic Zn^{2+} would suggest that metallothionein, with a mixture of both strong and weak binding sites, varying by at least 4 orders of magnitude (Krezel and Maret, 2007), acts as a versatile metal storage protein.

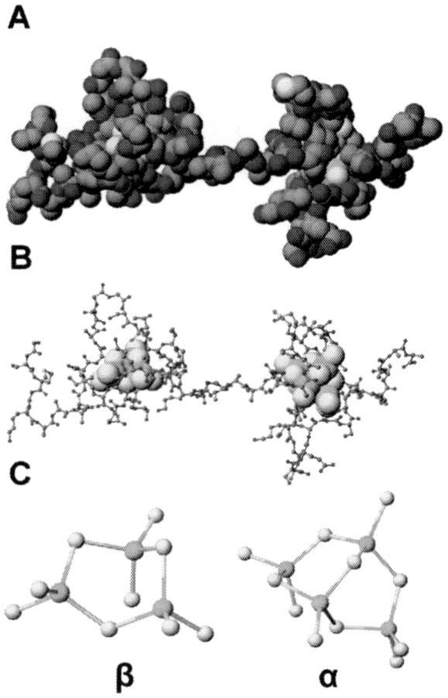

Figure 4.2 Molecular modeling structure of Cd_7-βα-rhMT with the N-terminal β-domain on the left side and the C-terminal α-domain on the right side. (A) A space filling diagram. (B) A ball-and-stick diagram with the domains in space filling form. (C) Ball-and-stick models of Cd_3-β- (left) and Cd_4-α- (right) rhMT cadmium-cysteinyl-thiolate clusters. Cadmium and zinc both bind in an isostructural manner. The fully metallated mammalian metallothionein contains no disulfide bonds. Modelling data adapted from Fowle and Stillman (1997) and Chan *et al.* (2007) and reproduced from Sutherland and Stillman (2008) with permission from Elsevier. See also Colour Insert.

Because of this lack of free cytosolic Zn^{2+}, it has been proposed that metallothionein also functions as a metallochaperone with the transfer of metal ions from Zn-MT to Zn-dependent enzymes, for example, *m*-aconitase (Feng *et al.*, 2005), carbonic anhydrase (Mason *et al.*, 2004), and the prototypical transcription factor, Gal4 (Maret *et al.*, 1997). The removal of zinc from the zinc finger-containing transcription factor Sp1 also been reported (Zeng *et al.*, 1991). In each case, metallothionein appears to be a mediator in zinc controlled chemistry. These studies also show the importance of protein-protein interactions, however, to date, no zinc-exchange intermediate, one involving protein-protein interactions, has been characterized that provides insight into the actual mechanism of metal transfer, with current research aiming to isolate such a species.

The question about the detailed metallation mechanism can be broken down into a two parts: (i) the overall result of metallation and (ii) the single step process of each metal binding to a single binding site.

We start with the overall reaction (i). Of importance in the determination of the overall mechanism of metallation are the putative functions of metallothionein, particularly, that of protecting the cell against oxidative stress. For a long time, the mechanism of metallation of metallothionein was thought to proceed in a cooperative manner, in which the binding of one metal acts to facilitate the binding of subsequent metals (Good *et al.*, 1988; Gehrig *et al.*, 2000). Under this model the only structures of biological significance would be the fully metallated or completely demetallated species, and any partially metallated intermediates would likely be too unstable to have a specific role. However, recent studies have shown the mechanism of metallation to be in fact noncooperative for Cd^{2+} and Zn^{2+} (Palumaa *et al.*, 2002; Rigby-Duncan and Stillman, 2006; Rigby *et al.*, 2006; Rigby-Duncan and Stillman, 2007; Sutherland and Stillman, 2008), as well as for As^{3+} (Ngu and Stillman, 2006; Ngu *et al.*, 2008). It is very likely that noncooperative metallation applies to all metals capable of binding metallothionein although this has yet to be proven. Significantly, a noncooperative mechanism allows for partially metallated and metal exchange intermediates to be stable and, therefore, to be able to take part in cellular chemistry.

If metallothionein metallated in a cooperative manner, one would expect oxidation of any of the thiol groups to facilitate further oxidation. However, in a sequential noncooperative system, oxidation becomes progressively less likely as the oxidation of thiols to disulfides requires the exposure of free thiol groups, which becomes less likely as the association constant of the coordinating metal increases. To elaborate on the latter point, recent research examining the metal release properties of human MT-3 through the oxidation of nitric oxide, has shown that even in conditions of excess oxidant, two α-domain Cd^{2+} atoms persist (Wang *et al.*, 2008). An important consequence of this mechanism is that isolation of partially metallated metallothionein could potentially represent the actual protein in solution, whereas a cooperative mechanism would likely lead to complete oxidation of the protein in question.

Recently, a single nucleotide polymorphism, leading to an Asn27Thr substitution in MT-1a was associated with diabetes and its cardiovascular complications (Giacconi *et al.*, 2008; Maret, 2008b). The protein was less susceptible to intracellular NO-induced release of zinc, while not affecting the intracellular zinc ion availability leading to zinc dyshomeostasis. Because metallation occurs in a noncooperative manner, one can infer that the zinc is sequestered in binding sites with significantly greater association constants, making acquisition of zinc by zinc dependent enzymes more difficult.

Turning now to the stepwise mechanism (ii), it is clear when one considers the ramifications of the conclusions from the As^{3+} metallation studies reported by Ngu *et al.,* (Ngu and Stillman, 2006; Ngu *et al.*, 2008; Ngu *et al.*, 2010b) that metallation is likely a sequential process and that the binding affinities decrease from the first metal bound to the last metal bound such that the last metal bound has a much lower binding affinity. The conclusion is that metals may be dissociated from a fully metallated MT in steps.

In addition to the known metallation chemistries of Ag^{+}, Cu^{+}, Zn^{2+}, Cd^{2+}, Hg^{2+} and As^{3+}, a diverse array of metals are known to bind to metallothionein *in vitro* and including Au^{+}, Co^{2+}, Fe^{2+}, Pb^{2+}, Pt^{2+}, Bi^{3+} and Tc^{5+} (Schmitz *et al.*, 1980; Vasak, 1980; Nielson *et al.*, 1985; Good and Vasak, 1986; Morelock *et al.*, 1988; Palacios *et al.*, 2007; Ngu *et al.*, 2010a). Many of these metals act as spectroscopic probes to analyze the metal binding capabilities of metallothionein, still others may be essential for medicinal purposes. The following sections provide a brief description of techniques currently used to analyze the properties of metal binding to metallothionein.

4.2 Techniques for Studying Metallothioneins

Metallothionein binds a wide array of metals, ranging from physiologically crucial metals, such as Zn^{2+} and Cu^{+}, to the toxic metals, such as Cd^{2+} and Hg^{2+}. We outline several techniques, which have been critical to the understanding of metal binding. Initial studies focused on probing and cataloging both the type and number of metals bound to MT. Following these reports, the next focus has been to understand both the kinetic and mechanistic details of metallation. By determining the above properties, an understanding of how MT behaves *in vivo* becomes possible. Once an understanding of the metallation of MT exists *in vivo*, future work will be aimed at determining both protein binding partners, and the exact function of MT.

4.2.1 *Electronic Absorption Spectroscopy*

Metallothionein exhibits a number of characteristic spectroscopic signatures that depend on the type of metal bound. Because of the lack of either aromatic amino acids or a defined secondary structure, such a β sheet or an α helix, in the apo-form (Rigby-Duncan and Stillman, 2006; Rigby *et al.*, 2006) it is far easier to measure spectroscopic properties that are entirely dependent on both the type and number of metals bound, than is the case with many metalloproteins (Kagi *et al.*, 1984; Kojima, 1991; Kagi, 1993) UV-visible absorption and circular dichroism spectroscopies have been the key techniques in the determination of the relative affinities of metallothioneins for metal ions, the mode of metal coordination and possible stoichiometries of these metals. In particular, the lack of aromatic amino acids allows UV absorption spectroscopy to monitor the ligand-to-metal charge transfer bands above 220 nm without interference. These techniques are widely used to monitor metallation reactions of d^{10} metals, such as Zn^{2+}, Cd^{2+}, Hg^{2+}, Cu^{+} and Ag^{+} (Vasak *et al.*, 1981a; Zelazowski *et al.*, 1984; Presta *et al.*, 1995; Stillman, 1995; Bofill *et al.*, 1999; Leiva-Presa *et al.*, 2004). Typically, a titration with a metal ion leads to a change in the UV- absorption spectrum of the solution demonstrating either binding of the metal to the metal free protein, or, alternatively displacement of an already present metal. The former is used to determine binding stoichiometries for high affinity metals, while the latter can be used to determine binding affinity and possible mixed metal species. If, however, the incoming metal ion has a low binding affinity, then caution should be taken in directly correlating a spectroscopically-determined maximum with a particular metal binding stoichiometry.

Figure 4.3 shows a classical titration of rabbit liver apoMT-2 with Cd^{2+} monitored by UV-absorption spectroscopy between 210 and 290 nm (Stillman *et al.*, 1987). Initially, the absorbance is associated with the protein backbone, the metal-free apo-MT. However, upon

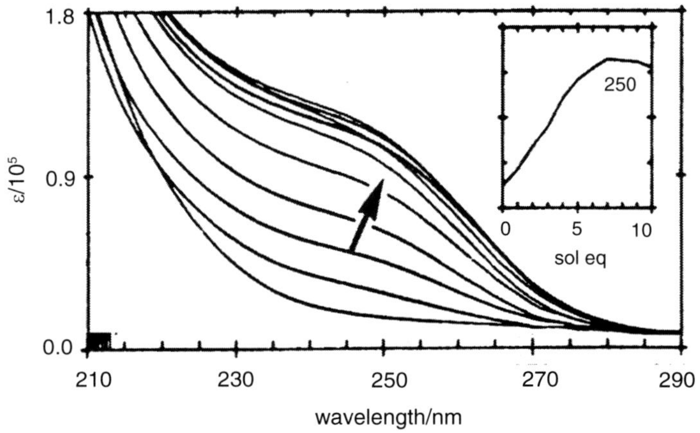

Figure 4.3 UV absorption spectroscopy of rabbit liver metallothionein 2 (βα-rlMT-2) as a function of increasing additions of aliquots of Cd^{2+}. The concentration of protein was 10 μM, while spectra were recorded at pH 7. There are 11 lines: line 1, apo-βα-rlMT-2 at pH 2; line 2, apo-βα-rlMT-2 at pH 7; lines 3-11 for apo-βα-rlMT-2 with different concentrations of Cd^{2+} (1.06, 2.14, 3.5, 5.0, 5.95, 6.05, 6.92, 7.61, and 8.82 molar equivalent). The inset represent changes in the intensity monitored of the ligand-to-metal charge transfer band of cadmium thiolate, monitored at 250 nm, as a function of molar equivalents of Cd^{2+} added. Adapted from Stillman *et al.* (1987).

addition of Cd^{2+} a shoulder forms at 250 nm, which corresponds to the ligand-to-metal charge transfer of cadmium-thiolate bonds. The UV absorbance at 250 nm is at a maximum at 7 Cd^{2+} atoms, corresponding to fully metallated Cd_7-βα-MT-2. Complete binding of the Cd^{2+} ions is the result of the high affinity of cysteinyl sulfurs for Cd^{2+}, $6.0 \times 10^{21} - 10^{25.6}\ M^{-1}$ (Kagi and Vallee, 1961; Zhang *et al.*, 1997). While the stoichiometry of metal binding may be extracted from UV absorption spectra, it is the CD spectra that clearly show changes in the metal-binding sites indicative of a transition from individual CdS_4 sites to the formation of the polynuclear clusters (Willner *et al.*, 1987). The maximum of seven Cd^{2+} per βα-MT has been confirmed by ESI-MS measurements for MTs from many different mammalian sources, for example, human MT (Chan *et al.*, 2007; Sutherland and Stillman, 2008). In the case shown in Fig. 4.3, we see a slight reduction in absorbance at 250 nm, which has been interpreted as due to non-specific Cd^{2+} associating with the protein. That these ions do not change the CD band envelope morphology is strong evidence that the Cd_7S_{20} domain structure is not directly affected. In contrast, formation of supermetallated domains has been reported for human MT-1a, where Cd_8-hMT-1a forms, marked by a slight change in the absorption spectrum and significant changes in the CD spectrum morphology. The use of newer techniques, such as ESI-MS and ^{113}Cd-NMR spectroscopy confirms the presence of the extra Cd^{2+} (Rigby-Duncan *et al.*, 2008; Meloni *et al.*, 2009; Sutherland *et al.*, 2010).

While metallothionein is key in the cellular homeostasis of both Zn^{2+} and Cu^{+}, as d^{10} metals, their spectroscopic signatures are essentially limited to the ligand-to-metal charge transfer bands (thiolate to Zn^{2+} and thiolate to Cu^{+}). On the other hand, metallation of metallothionein using metal ions with an incomplete d shell allows d→d transitions to be monitored. These transitions are sensitive to the coordination geometry of the metal atom, and coupled with information provided by the ligand-to-metal charge transfer bands, can allow assignment of both the coordinating ligands and the metal binding geometry. In the following three examples of Co^{2+}, Ni^{2+} and Fe^{2+}, a maximum metal binding stoichiometry of 7 metal atoms has been determined spectroscopically. The spectral data for the binding of Co^{2+} to metallothionein demonstrates that the protein enforces a distorted tetrahedral arrangement in which the α-domain has more tetrahedral character than the β-domain (Vasak, 1980; Vasak and Kagi, 1981; Vasak *et al.*, 1981b; Good *et al.*, 1991). The spectroscopic data for Ni^{2+} binding to metallothionein does not have distinct d→d transitions, however, the spectrum was similar to Ni^{2+} bound to azurin, and this suggests a tetrahedral coordination. Fe^{2+} has also been shown to bind in a tetrahedral manner, with a d→d transition in the near infrared region (Good and Vasak, 1986). While these metals may not be naturally associated with the protein, their spectroscopic characteristics are useful in determining the flexibility of the protein in accommodating different metals. By tuning the flexibility of both domains, metal selectivity is possible leading to mixed metal species, which have been observed in metal exchange experiments between human MT-3 and amyloid-β protein (Meloni *et al.*, 2008).

4.2.2 *Circular Dichroism Spectroscopy*

CD spectroscopy is an extremely valuable method used to measure the difference in absorption between left- and right-circularly polarized light caused by chirality in the chromophore as a function of the wavelength of light used. This method is highly sensitive to conformations of optically active chiral molecules. Traditionally in biochemistry, the CD technique has been used to determine the extent of secondary structure of a protein, such as α-helices, β-sheets and turns in the protein, or the status of complexation based on the relative differential absorbances of the amino acids in the peptide chain. In this manner the effects of denaturing agents, temperature, potential ligands and mutations can be readily and reliably monitored.

In MT much greater structural sensitivity is found in the CD spectrum, because the metals that bind form structurally significant binding sites that exhibit specific chirality and cause the protein to fold into a predefined structure (Fig. 4.2). These chiral structures are composed of the metal ion and the thiolate donor atoms of the cysteine residues. The ligand to metal charge transfer transitions in MTs are easy to measure above 230 nm because they occur in the region of the electronic spectrum devoid of the usual transitions associated with aromatic amino acids. This window region has allowed detailed spectroscopic analyses of the metallation reactions of MT to be carried out for the d^{10} metals that are otherwise chromophorically-silent. The CD spectrum of MT to the red of 220 nm is completely dependent on the binding of metal ions.

Possibly the first CD spectrum of MT was published for Zn,Cd- βα-MT (Weser *et al.*, 1973). In subsequent studies of the rabbit liver Cd_7-βα-MT-2a, it was reported that the morphology of the CD spectrum resembled a derivative shape with the cross-over point aligned with the 250 nm band of the $S \rightarrow Cd^{2+}$ ligand-to-metal charge transfer (Stillman and Szymanska, 1984; Stillman *et al.*, 1987; Stillman and Zelazowski, 1988). The origin of this morphology was explained by Stillman *et al.* through the interpretation of CD spectral changes that took place following the titration of Cd^{2+} into rabbit liver Zn_7-βα-MT-2a (Stillman *et al.*, 1987). That this CD band envelope represented a specific metallated structure was again demonstrated when the supermetallated Cd_5-α-hMT-1a was formed (Rigby-Duncan *et al.*, 2008). Fig. 4.4 shows the titration of human Cd_4-α-MT-1a with excess Cd^{2+} to form the Cd_5-α- MT-1a species, also the associated ESI-MS and ^{113}Cd NMR support the formation of a new species. It is in this way that CD data may act as a fingerprint for the determination of specific metallated species in MT.

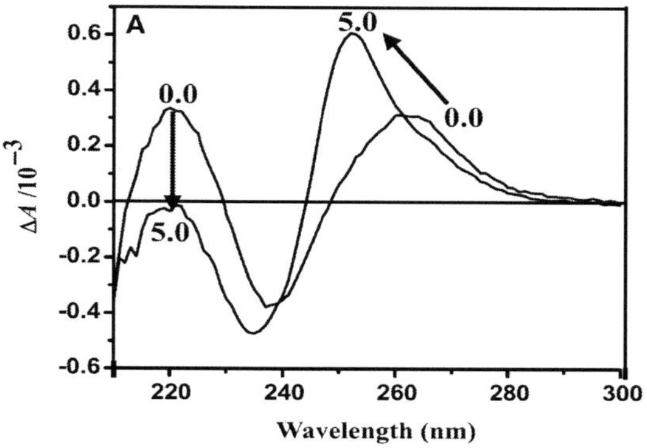

Figure 4.4 Circular dichroism spectral changes observed upon titrating human $Cd_4\alpha$-MT-1a with an additional 5.0 molar equivalents of Cd^{2+} at pH 7.4 and 22°C. Reproduced from Rigby-Duncan *et al.* (2008) with permission from John Wiley and Sons.

Because of the utility of the CD technique, CD spectroscopic studies have been reported for a wide variety of metals (Stillman, 1995), including Ag-MT (Zelazowski *et al.*, 1989; Zelazowski and Stillman, 1992), Cu-MT (Stillman and Szymanska, 1984; Presta *et al.*, 1995; Vaher *et al.*, 2001; Roschitzki and Vasak, 2002; Meloni *et al.*, 2008), Zn-MT (Stillman *et al.*, 1987; Stillman and Zelazowski, 1988; Presta *et al.*, 1995; Meloni *et al.*, 2008; Meloni *et al.*, 2009), Cd-MT (Stillman and Szymanska, 1984; Stillman *et al.*, 1987; Stillman and Zelazowski, 1988; Presta *et al.*, 1995; Vaher *et al.*, 2001), and Hg-MT (Lu and Stillman, 1993; Lu *et al.*, 1993; Leiva-Presa *et al.*, 2004). An example of metal-induced folding monitored by CD spectroscopy is shown in Fig. 4.5 for the binding of Hg^{2+} to rabbit liver Zn_7-$\beta\alpha$-MT. The spectral changes are complicated, which immediately indicates that there are significant structural changes as Hg^{2+} displaces Zn^{2+}, the two metals are not isomorphously connected, unlike Zn^{2+} and Cd^{2+} in MT. The CD spectral envelope changes as a function of the stoichiometry of Hg^{2+} added (the *z*-axis dimension). Three maxima are observed, at 0, 7 and 11 equivalents of Hg^{2+}, and interpreted as corresponding to Zn_7-MT, Hg_7-MT and Hg_{11}-MT. The shift in the location of the maxima is the result of a change from tetrahedral to trigonal coordination. A second titration, using a reduced pH and chloride ions, demonstrates that Hg_{18}-MT can also be formed, which

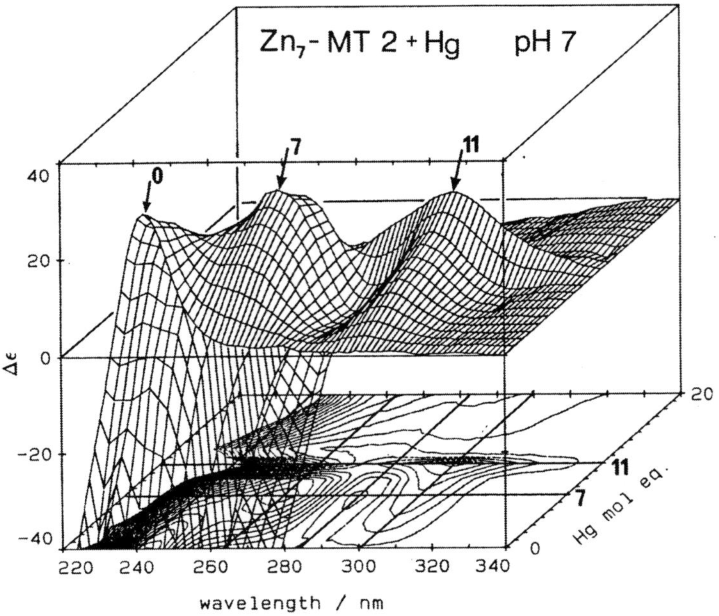

Figure 4.5 Circular dichroism spectra recorded as a 3-D projection plot during a titration of a single sample of rabbit liver Zn_7-MT-2 with Hg^{2+} at pH 7. The band labeled 0 represents the CD spectrum of Zn_7-MT-2, and the 7 and 11 labels mark maxima in the CD spectra. The z-axis is plotted in units of the Hg^{2+} added to the solution of Zn_7-MT-2. The grid lines added to the contour diagram are drawn for Hg^{2+} molar ratios of 7 and 11 and indicate 285 and 300 nm. Reproduced from Lu and Stillman (1993) with permision from the American Chemical Society.

corresponds to a linear coordination (Lu *et al.*, 1993). In this manner, the stoichiometries of well defined structures can be identified by maxima in the ligand-to-metal charge transfer.

4.2.3 *Emission Spectroscopy*

The first emission spectroscopic study reported was of Cu^+ binding to a metallothionein found in the fungus *Neurospora crassa* (Beltramini and Lerch, 1981). Since this study, further characterization of mammalian MTs by equilibrium and kinetic emission studies for both Cu^+ and Ag^+ (Gasyna *et al.*, 1988; Stillman *et al.*, 1988), have been reported. These studies have led to an understanding of the structural dynamics of protein folding because they monitor the change in the intensity as a function of metal stoichiometry. In these studies, excitation at 300 nm leads to a spin forbidden transition of $3d^{10} \rightarrow 3d^9 4s^1$, or alternatively $3d^{10} \rightarrow 3d^9 4p^1$, populating a state that emits at a wavelength significantly red shifted from the initial excitation band (600 nm for Cu^+ and 570 nm for Ag^+). Because the transition from triplet to singlet state is spin-forbidden, the phosphorescent lifetime of this emission is on the order of microseconds and can be quenched by molecular oxygen. In the case of Cu^+, the maximum emission is observed at 12 molar equivalents. After 12 added equivalents, the emission intensity decreases significantly, suggesting the clusters formed have become more solvent exposed leading to a quenching of luminescence through radiationless decay (Fig. 4.6). In addition, Cu-MT has been shown to form a Cu_{15}-MT species requiring a mixture of trigonal and digonal coordination geometries (Presta *et al.*, 1995). Using rabbit liver Zn_7-βα-MT-2, Cu^+ was shown to bind in a distributive manner, followed by a rearrangement that was temperature dependent with higher temperatures increasing the rate of rearrangement. The final product of this rearrangement is the metal center located in the β-domain of the protein (the preferred location of copper binding) (Green *et al.*, 1994), which was later confirmed by domain mixing experiments (Salgado and Stillman, 2004). Unlike Cu^+, Ag^+ has three distinct emission maxima, corresponding to Ag_6-, Ag_{12}- and Ag_{18}-MT species (Zelazowski *et al.*, 1989), while Ag_{12}- and Cu_{12}-MT are likely analogous structures, in order to accommodate additional Ag^+ ions, a change in coordination from trigonal to digonal is probable. When comparing maximum metal binding of the individual domains, the preference for digonal coordination can also be seen with each domain capable of coordinating 7 Ag^+, as compared to 6 Cu^+ (Salgado *et al.*, 2007).

Kinetic determinations of metal binding to the metallothioneins have been very hard to obtain due in part to the lack of resolved spectral signatures for metals as they bind one-by-one. However, when Cu^+ binds, significant time-dependent spectral changes are measured; these changes are associated with the thermodynamic pressures of adopting the lowest energy binding site rather than the initial binding reaction. Green and Stillman, and later Salgado and Stillman (Green *et al.*, 1994; Salgado and Stillman, 2004), have reported the details of the time-dependent rearrangement of Cu^+ after binding to MTs. The important and significant feature

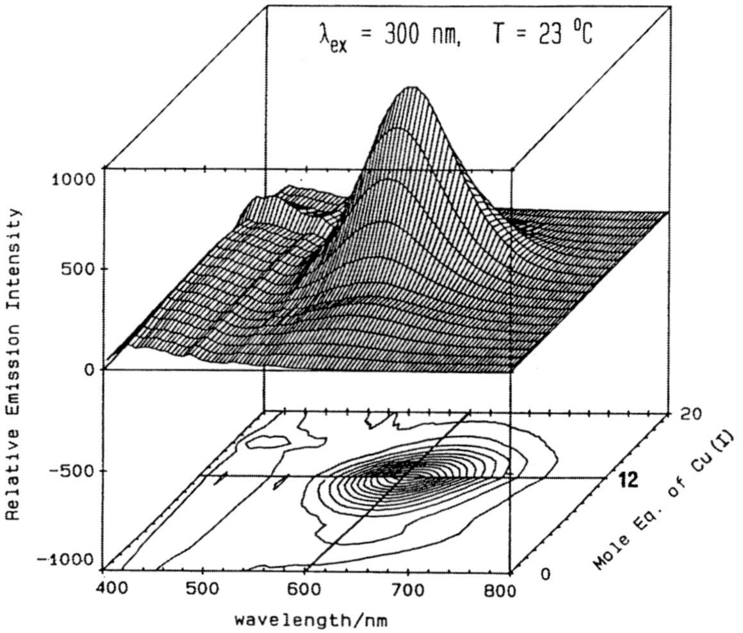

Figure 4.6 Three-dimensional plot showing changes in the emission spectrum as a function of the Cu^+:MT molar ratio for a single sample of rabbit liver Zn-MT-2 at 23°C and pH 6.6 following excitation at 300 nm. The grid line drawn across the contour diagram represents a Cu^+ molar ratio of 12, and the grid lines parallel to the *z*-axis represents the 600 nm point in the spectra. Note that the contour diagram shows that the change in emission intensity for Cu_0Zn_7-to Cu_6Zn_4-MT is much less than that between Cu_6Zn_4-MT and Cu_{12}-MT. Reproduced from Green *et al.* (1994) with permission from the American Chemical Society.

of this work is that it establishes that the binding site structures are dominated by both kinetic and thermodynamic properties. Following binding, each Cu^+ added rearranges to a different site. The emission intensity is dependent on the exact site occupied: as the Cu^+ occupation changes, the emission intensity changes. In addition to the intensity changing, it was shown that the wavelength of the emission was also time-dependent representing each site occupied. While complete details of these studies are beyond this review, the data indicate how the metals in MT are likely to move within the binding domains.

The selectivity of emission spectroscopy makes it a potential tool for monitoring the *in vivo* location of metal atoms. Specifically, the metal distribution of rat liver, exposed to excess copper chloride, and human liver, from a patient with Wilson's disease, were monitored (Stillman *et al.*, 1989a). The presence of Cu-MT was directly observable for both liver types, and increased emission was associated with increased copper exposure. In addition to Cu^+ and Ag^+, emission spectroscopy may be measured for Au^+ and Pt^{2+} containing MTs (Stillman *et al.*, 1989b). Since Au^+ and Pt^{2+} are used in anti-arthritic and anti-cancer drugs, respectively, monitoring the emission properties could lead to a better understanding of the interaction between drugs that use these metal ions and metallothionein. In neurobiology, emission spectroscopy may be used in conjunction with confocal microscopy to localize copper-containing metallothionein. To summarize this section, emission spectroscopy is an extremely sensitive technique, allowing the monitoring of the metal coordination environment, the metal-binding stoichiometry and it also enables one to correlate the environment of the metal with the *in vivo* localization.

4.2.4 *Nuclear Magnetic Resonance Spectroscopy (NMR Spectroscopy)*

NMR spectroscopy is one of the most powerful techniques used to analyze the structure and dynamics of metalloproteins. Specifically, structural studies of metallothionein have focused largely on $^{111/113}Cd$ binding. There are several advantages to using Cd^{2+} to study the metallation of metallothionein including the natural presence of Cd^{2+} in metallothionein (Margoshes and Vallee, 1957), Cd^{2+} having a similar coordination geometry to Zn^{2+}, and enhanced resistance to oxidation due to Cd^{2+} having a higher affinity for thiolates.

Both isotopes ($^{111/113}Cd$) have a spin of ½: ^{111}Cd has a natural abundance of 12.75% but a relative sensitivity to ^1H ($^1H = 1$) of only 9.54×10^{-3}, while ^{113}Cd has a natural abundance of 12.26% and a relative sensitivity of 1.09×10^{-2}. The increased relative sensitivity of ^{113}Cd compared with ^{111}Cd makes the former isotope more widely used. The chemical shift range of ^{113}Cd has been recorded for a number of structurally characterized ^{113}Cd-substituted metalloproteins (Oz *et al.*, 1998). With a chemical shift range of roughly 900 ppm (relative to 0.1 M $Cd(ClO_4)_2$), each ^{113}Cd is extremely sensitive to the coordinating ligands: heptacoordinate oxygen ligands are the most shielded, while tetrathiolate ligands are the most deshielding.

Isotopic enrichment with ^{113}Cd leads to an eight-fold enhancement of sample sensitivity compared to a sample constituted with natural abundance cadmium. This enhancement allows for data acquisition on low concentration (mM) samples in a reasonable amount of time. Samples can be prepared by demetallation at low pH, followed by removal of metal ions using gel filtration, and finally, remetallation using isotopically enriched $^{113}CdCl_2$. It should be stressed at this point that the high concentrations of a typical NMR sample have been reported to cause dimerization. Further complicating this issue are reports suggesting that demetallation through the use of strong acid leads to structural changes in metallothionein itself. As, such NMR spectral data should be correlated with other spectroscopic techniques (above), as well as spectrometric techniques, such as ESI-MS (below), which do not necessarily rely on either acid-induced demetallation or the large concentrations necessary for this technique (Gan *et al.*, 1995; Ejnik *et al.*, 2003; Namdarghanbari *et al.*, 2010).

Initial studies of MT using proton NMR spectroscopy suggested the metal free form existed as an open random coil like structure, and that metallation increased the rigidity of the structure (Rupp *et al.*, 1974; Galdes *et al.*, 1978; Vasak *et al.*, 1980). Key to the rigidity of the metallated protein, the structures of the polynuclear clusters were determined through analysis of both the cadmium chemical shifts and splitting patterns. These experiments identified the coordinating ligands as cysteinyl-thiolates and formulated the polynuclear centers as Cd_3S_9 and Cd_4S_{11} (Sadler *et al.*, 1978; Otvos and Armitage, 1979; 1980; Boulanger and Armitage, 1982). Several complete solution structures of MT have been elucidated: including mammalian (human, rat, rabbit and mouse) (Arseniev *et al.*, 1988; Schultze *et al.*, 1988; Messerle *et al.*, 1990; 1992; Zangger *et al.*, 1999), plant (Peroza *et al.*, 2009) and cyanobacteria (Blindauer *et al.*, 2001). In the case of mammalian metallothionein, the overall structure of the cadmium-thiolate clusters was known, but the combined solution structures provided the absolute connectivity of each cadmium atom and its relation to all cysteine residues. The final critical step in determining the structure of mammalian MT, was when the X-ray crystallographic structure (Robbins *et al.*, 1991) showed identical molecular architecture to the previously determined NMR structure (Braun *et al.*, 1992).

The reactivity of each domain has also been investigated to give insight into the potential functions of MT. It is generally assumed that the domains of MT do not interact due to the similarity in the NMR spectra of both the isolated α-domain and the full protein (Boulanger *et al.*, 1982). With little domain-domain interaction, the selective distribution of different metal ions could be critical to the role of metallothionein in the body. Isolation of a mixed metal calf liver MT (3 Cu^{+} and 4 Zn^{2+}) could have only the Zn^{2+} present selectively displaced by $^{113}Cd^{2+}$. Subsequent NMR analysis using $^{113}Cd^{2+}$ led to the conclusion that the Cu^{+} ions were selectively located in the β-domain, while Zn^{2+} and Cd^{2+} were selectively located in the α-domain (Briggs and Armitage, 1982). These results strongly suggest that metallothionein, key to metal ion homeostasis, is capable of distributing Zn and Cu ions into separate domains, where they can be made available to apoenzymes in a selective manner. One hypothesis is that the β-domain, being more fluctional, would make metal ions coordinated to this cluster more readily available for use in cellular chemistry. And indeed it has been shown that the β-domain is more reactive towards metal ion chelators, such as EDTA (Gan *et al.*, 1995), as well oxidizing agents, such as nitric oxide (Wang *et al.*, 2008).

The experimentally-determined features of the metal binding properties of MTs bring us to the most important of the known metal-binding domains specificities. Namely, that the β-domain of MT-3 has been shown to be critical to the function of MT-3 as a neuronal growth inhibitor (Sewell *et al.*, 1995). It has also been shown that MT-1 can be altered to inhibit neuronal growth by the addition of Thr5, and mutations of S6P and S8P, which is thought to increase cluster fluctionality (Romero-Isart *et al.*, 2002). Based on these results, it becomes evident that β-domain-centered chemistry, which has not been well studied in an isolated form, may be critical to many functions of metallothionein.

In addition to the well known domain specificity of the MTs, it has been recently shown through ESI-MS, CD and NMR spectral studies of MT-1 and MT-3, that both isoforms are capable of binding an additional Cd^{2+} ion (Rigby-Duncan *et al.*, 2008; Meloni *et al.*, 2009; Sutherland *et al.*, 2010). The spectroscopic data suggests that, interestingly, this new metal binding site is not present in MT-2. However, it is not clear why this should be the case because MT-1 and MT-2 are more closely related to each other than to MT-3, the former having diverged before distinction of the mammalian orders while the latter, more structurally distinct, evolved much earlier. The α- and β- domains of MT-1 have each been shown to bind exactly one extra Cd^{2+} ion (Figs. 4.7 and 4.8), while the full protein MT-3 has been shown to bind only a single additional Cd^{2+} ion.

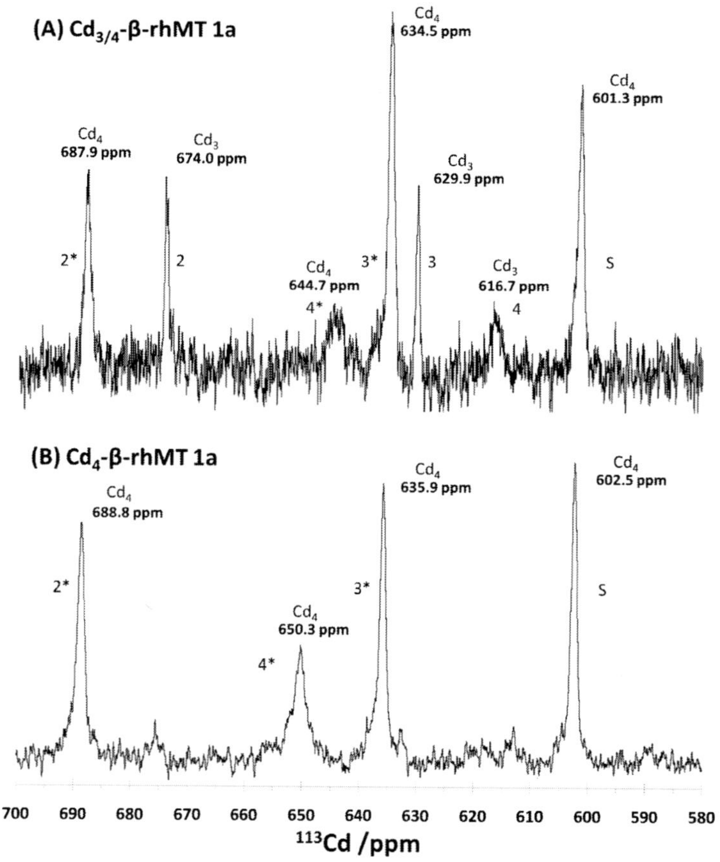

Figure 4.7 Direct 1D $^{113}Cd[^1H]$ NMR spectrum (133 MHz) of (A) mixture of Cd_3-β-rhMT-1a and Cd_4-β-rhMT-1a and (B) Cd_4-β-rhMT-1a formed by addition of excess $^{113}CdCl_2$ to Cd_3-β-rhMT-1a. The $Cd_{3/4}$-β-rhMT-1a was prepared in 10 mM ammonium formate at pH 7.4 and buffer exchanged into 90% D_2O. The spectra of $Cd_{3/4}$-β-rhMT-1a and Cd_4-β-rhMT-1a were acquired at 10°C and 25°C, respectively. Reproduced from Sutherland *et al.* (2010) with permission from the American Chemical Society.

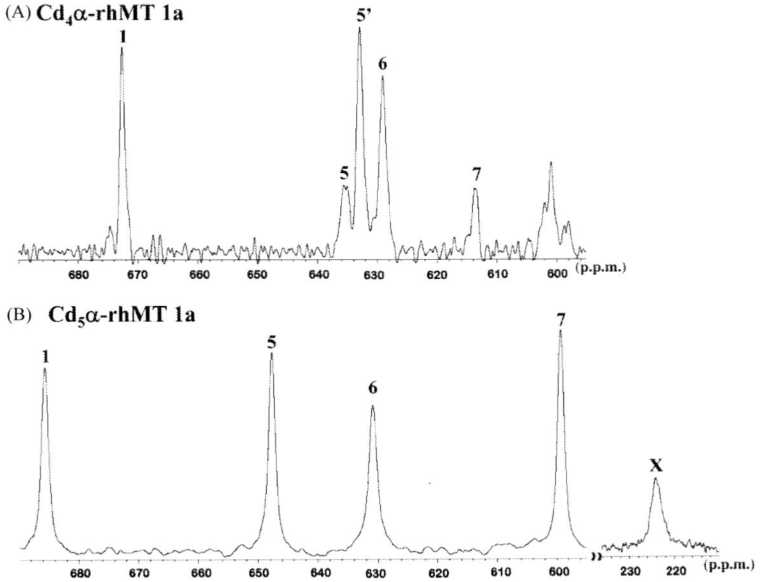

Figure 4.8 Direct 1D $^{113}Cd[^{1}H]$ NMR spectrum (133 MHz) of (A) α-rh MT-1a following isolation and purification of the recombinant protein from *E. coli* with the natural isotopic abundance of ^{113}Cd, showing primarily the $^{113}Cd_4$α-rhMT-1a species. (B) $^{113}Cd_4$-rhMT-1a titrated with an additional 10.0 molar equivalents of $^{113}Cd^{2+}$ to form $^{113}Cd_5$-α-rhMT-1a. The spectrum of $^{113}Cd_5$-α-rhMT-1a (B) is a combination of two separate spectra acquired in the regions of 585-705 ppm and 220-245 ppm. Samples were prepared in 10 mM Tris/HCl pH 7.4, and buffer-exchanged into 10% D_2O for the Cd_4-α-rhMT1a sample and >70% for the $^{113}Cd_5$-α-rhMT-1a sample. The spectra were acquired at 25°C. Reproduced from Rigby-Duncan *et al.* (2008) with permission from John Wiley and Sons.

Given the recently determined importance of the β-domain regarding neuronal growth inhibition and reactivity toward metal chelators, the binding of a fourth Cd^{2+} ion to the cluster could have significant consequences for the function of MT-1 and MT-3. The supermetallated form of the β-domain of human MT-1 was found to have four signals between 600 and 700 ppm, with the supermetallated peak located at 602.5 ppm. Interestingly, the location of the peak suggests tetrahedral coordination to four thiolates, or three thiolates and a single water molecules/chloride ion. In the case of the α-domain of MT-1, the signal associated with the supermetallated form appears at 224 ppm, which corresponds to octahedrally coordinated $Cd(SR)_2(OH_2)_4$. It was suggested that this additional metal binding site is located at the cluster crevice, where a number of cysteinyl-sulfurs are solvent exposed. In the case of MT-3, supermetallation leads to a decrease in the Stoke's radius, a result of the protein assuming a more compact form. It is quite likely that both domains are involved in the coordination of the eighth metal ion. In this case, supermetallation of the individual domains is the result of the retention of some metal binding capacity. However, based on the chemical shifts of the peaks associated with both the α- and β-domain, one can infer that coordination to the β-domain requires more thiolates, which would act to stabilize the bound metal. Consequently, while both domains are capable of binding an additional metal ion, the bulk of the chemistry related to this new metal atom is likely associated with the β-domain, while the α-domain aids in the stabilization of the structure.

The NMR spectra of several species, including human Cd_7-βα-MT isoforms 1, 2 and 3, the two isolated fragments Cd_3-β-rhMT-1a and Cd_4-α-rhMT-1a, as well as their respective supermetallated counter parts have been reported. The data in Fig. 4.9 demonstrate that Cd_4-β-rhMT-1a has a typical ^{113}Cd NMR spectral signature that falls within the range of other Cd-MT spectral data; while Cd_5-α-rhMT-1a has a single peak located significantly upfield from other observed species. There is significant overlap between the spectra of Cd_7-βα-rhMT-1a and Cd_4-β-rhMT-1a possibly due to the former containing a slight amount of supermetallated Cd_8-βα-rhMT-1a. Electrospray ionization mass spectrometry, a technique able to determine the mass of a particular species, has been of great value in determining the existence of supermetallation. This technique, in conjunction with NMR spectroscopy, allowed speciation to be determined and was used to confirm NMR evidence of supermetallated forms of the protein.

This additional metal could be an exchange intermediate essentially frozen by the higher binding affinity of Cd^{2+} compared to that of Zn^{2+}. The hypothesis proposed to account for the presence of the intermediate suggests that it represents the mechanism to allow both the metallation/demetallation of apoenzymes, as well aiding in the sequestration of toxic metals with concomitant release of Zn^{2+} to upregulate the production of more metallothionein. This proposal would imply that the extra metal is located on the outside of the

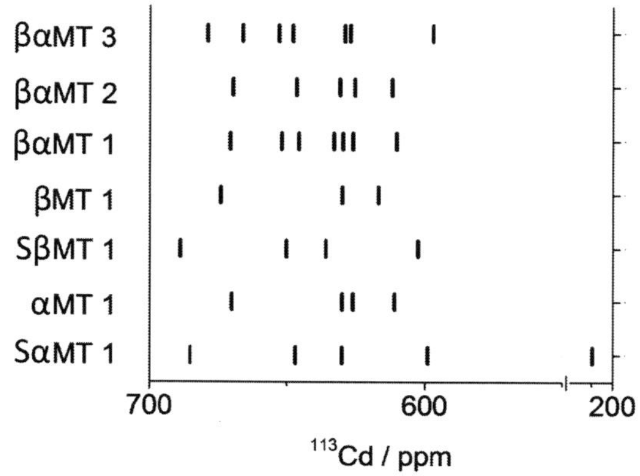

Figure 4.9 Comparison of the ^{113}Cd NMR resonances for human MTs: Cd_7-βα-hMT-3, Cd_7-βα-hMT-2, Cd_7-βα-hMT-1a, Cd_3-β-rhMT-1a, supermetallated Cd_4-β-rhMT-1a, Cd_4-α-rhMT-1a, and supermetallated Cd_5-α-rhMT-1a. We note that in this diagram, for the spectrum of Cd_7-βα-hMT-2, the resolution precludes separation of overlapping resonance 1 and 2 (near 670 ppm) and 3 and 4 (near 645 ppm), so that in total there are seven resonances. Reproduced from Sutherland *et al.* (2010) with permission from the American Chemical Society.

cluster; however, there exists ambiguity in the exact location of the additional metal binding site. Future work includes determination of the solution structure, competition experiments to determine the relative affinity of the isolated domains compared to the whole protein, as well as mutational studies that will aid in determining essential amino acids and their effect on protein chemistry.

4.2.5 *Electrospray Ionization Mass Spectrometry (ESI-MS)*

The electrospray ionization (ESI) mass spectrometry (MS) technique measures the mass-to-charge ratio (m/z) of analytes, and because the ionization method allows the infusion of the analyte directly from solution, this method has become critical to many metallothionein binding experiments. Unique to this technique is its ability to determine not only the number of metals bound but also the types of metals bound, which allows for a direct correlation between the spectroscopy and speciation of a sample. Recently, ESI-MS data have been used in determining the kinetics of metal binding and to discover new low affinity binding sites. A brief description of the technique, followed by how data is interpreted, will allow for a better understanding of its usefulness in studying the complex metal binding properties of metallothioneins.

In the ESI-MS experiment, a sample in solution is introduced, under atmospheric pressure, into the tip of the electrospray capillary. This tip is electrically charged to several thousand volts and causes the solution to accumulate positive charge. Enrichment of the tip with positive ions leads to the formation of a Taylor cone, which ejects small positively charged droplets. Solvent evaporation decreases the radius of the droplet, while the charge remains constant. When Coulombic forces, due to the nearness of positively charged ions, overcome the surface tension, fission of the droplet occurs. Evaporation-fission events repeat until very small charged droplets are produced. Eventual formation of the multiply charged protein is thought to occur via the charged residue model (CRM) (Felitsyn *et al.*, 2002; Kebarle and Verkerk, 2009). The CRM mechanism of gas phase ion formation states that the small highly charged droplet will evaporate to dryness leaving any unpaired ions as adducts to the protein. Since each droplet will have varying numbers of charges, a distribution of mass-to-charge ratios will be observed. The charge state distribution is directly related to both the number of basic sites on the protein, as well as the size of the protein, as such any alterations to this size, or site exposure, through folding or through denaturation, will lead to changes in the charge state distribution.

Charged proteins are separated by differences in the m/z ratio; common analyzers include the quadrupole and time-of-flight mass analyzers, allowing detection of species even with very similar m/z values. The ESI-MS technique allows for the introduction of samples in aqueous solution, at physiological pH, and allows one to quantify the sample composition. Folding experiments can be conducted, in which the charge state distribution can be directly related to changes in sample conditions, such as pH or metal added, and changes in metallation status can be monitored by mass changes. By knowing the concentration of protein, from spectroscopic techniques, accurate quantification is possible through a comparison of the deconvoluted peaks. This, however, assumes that all species in solution ionize at the same rate, and may not always be true.

In the case of MT, metal induced folding experiments may be performed by the addition of small aliquots of metal ions to apo-MT (Rigby-Duncan and Stillman, 2007; Sutherland and Stillman,

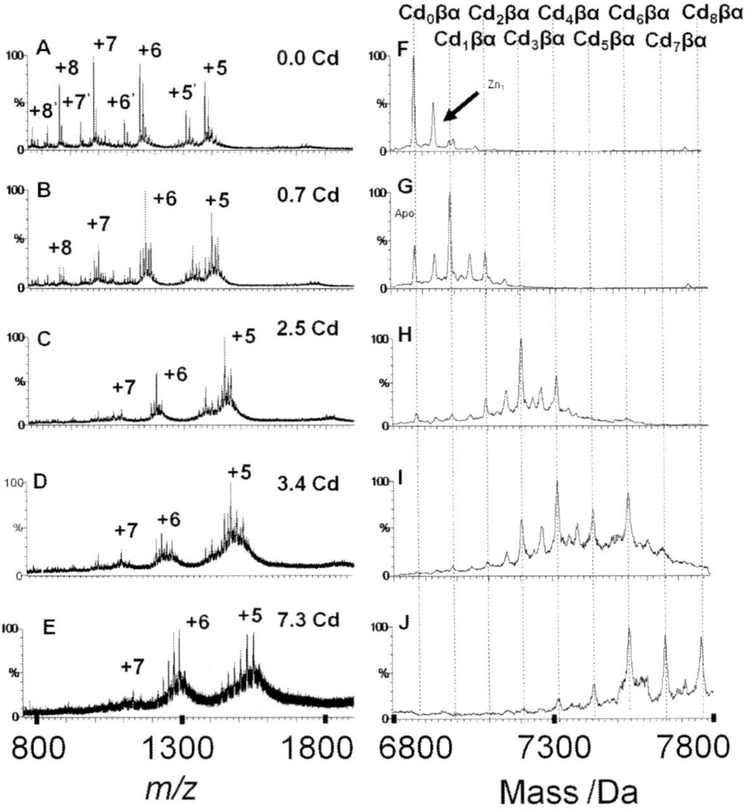

Figure 4.10 ESI mass spectra recorded during the titration of a solution of 18.4 μM human apo-βα-MT-1a, at pH 8.4, with $CdSO_4$. Spectral changes were recorded as aliquots of Cd^{2+} (7.1 mM) were titrated into each solution at 22°C. Spectra of human βα-MT-1a (A-E) and the respective deconvoluted spectral (F-J) were recorded at Cd^{2+} molar equivalents of 0.0, 0.7, 2.5, 3.4, 7.3. Charge states +5' to +8' correspond to a less prevalent truncated species. Reproduced from Sutherland and Stillman (2008) with permission from Elsevier.

2008). In this way both the number of metals bound and changes in the folding of the protein may be observed. In detail, the data in Fig. 4.10 show the stepwise titration of Cd^{2+} with human apo-MT-1a. Initially charge states range from +8 to +5 indicating an open structure. As increasing amounts of Cd^{2+} are added to the sample, the +8 charge state disappears and a significant decrease in the intensity of the +7 charge state is observed. At roughly 1 equivalents of Cd^{2+}

the disappearance of the +8 charge state shows that the protein has adopted a more folded conformation. Further titration to roughly 3 equivalents of Cd^{2+} leads to a significant decrease in the intensity of the +7 charge state. Interestingly, at 7.3 equivalents an equilibrium exists between Cd_6-, Cd_7- and Cd_8-MT-1a. An increase in the +6 charge state relative to the +5 charge state suggests the supermetallated form (Cd_8-MT-1a) is a more open conformation, leading to the exposure of more basic residues, and is likely the result of the protein accommodating the eighth Cd^{2+} ion.

The speciation of MT in solution may also be observed from the deconvoluted spectra, Fig. 4.10 (right). These spectra are the result of the analysis of the *m/z* data, and allow slight shifts in the charge state of the protein to be interpreted as the relative populations of differently metallated species. Because a series of species are observed ranging from Cd_1- to Cd_8-MT-1a, the mechanism of metal binding is shown to be noncooperative. If the mechanism of metallation were cooperative, one would expect only the existence of metal free and the fully metallated MT. However, the existence of intermediate metallation states, which track with the added equivalents of Cd^{2+}, indicates that partially metallated species are stable. These initial titrations were the first evidence of supermetallation and lead to the NMR investigation described above (Rigby-Duncan and Stillman, 2007; Rigby-Duncan *et al.*, 2008; Sutherland and Stillman, 2008; Sutherland *et al.*, 2010).

If a sufficiently slow binding metal, such as As^{3+}, is monitored, then kinetic studies can be performed to determine the rate of metallation (Ngu and Stillman, 2006; Ngu *et al.*, 2008; Ngu *et al.*, 2010b). In the case of the human MT-1a, these kinetic studies have shown that the rate of metallation is directly dependent on the number of binding sites, with the isolated domains metallating slowest and the genetically engineered triple–domains metallating fastest. By analyzing both temperature and time-dependent ESI-MS experiments, the mechanism of metallation has been determined to proceed in a sequential and noncooperative manner, in which each As^{3+} is capable of binding 3 cysteine residues. Fig. 4.11A shows the time resolved relative abundance of each species with the smooth lines corresponding to the reaction modeled as a series of nonreversible sequential reactions. While Fig. 4.11B shows the rate constants of both the isolated domains, with and without purification tag, as well

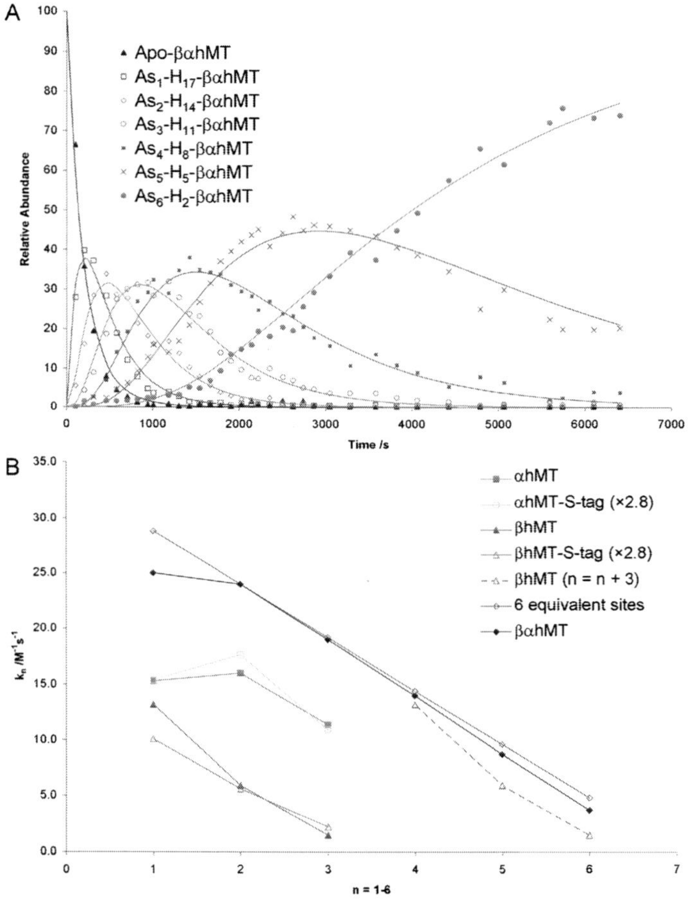

Figure 4.11 (A) Time-resolved ESI-MS relative abundances of human apo-βα-MT-1a following reaction with As^{3+} at 25°C and pH 3.5 to form As_n-MT(n = 1–6). The reaction was carried out with an As^{3+}: MT stoichiometric ratio of 11:1. The relative abundances of each of the species are shown as the data points on the graph. The lines were calculated by fitting all the data to a series of sequential bimolecular reactions. (B) Comparison of the rate constants calculated from the time-resolved ESI-MS measurements for As^{3+}-metallation of α-MT-1a-s-tag, β-MT-1a-stag, α-MT-1a, β-MT-1a and the trend in rate constant values for six equivalents sites where k_1 = 28.8 $M^{-1}s^{-1}$. The dashed line represents rate constant data for the β-MT-1a redrawn with the value of *n* shifted by three illustrate the similarity to the rate constant trend for the final three As^{3+} binding to the βα-MT-1a. Reproduced from (Ngu *et al.*, 2008) with permission from the American Chemical Society. See also Colour Insert.

as the full protein. Complete metallation of the 9 cysteine β-domain and 11 cysteine α-domain each requires 3 As^{3+} and metallation of the full protein requires 6 As^{3+}. The activation energies of all six specific rate constants for the full protein have been calculated and it was determined that the rate constant is dependent on the number of available sites (Fig. 4.11).

Studies using ESI-mass spectral data allow the determination of both equilibrium and kinetic information. This has furthered our knowledge by providing conclusive evidence that the metallation of metallothionein proceeds in a noncooperative manner. A great strength of this technique is allowing the correlation of spectroscopic data with speciation in solution. By analysis of the deconvoluted masses, one can determine the identity of the metal binding to metallothionein, the stoichiometry of metal binding, and the relative abundance of differently metallated species in solution.

4.3 MT-1 and MT-2: Inducible Metallothioneins

Both MT-1 and MT-2 are induced by a number of stresses, such as cytokines, glucocorticoids, reactive oxygen species and metal ions (Miles *et al.*, 2000). The most potent inducers of mammalian metallothionein are the metals Cd^{2+} and Zn^{2+}(Balamurugan and Schaffner, 2009). Zn^{2+} is the dominant metal found in human MT, Cu^{+} is also found, particularly in fetal MT, and finally also Cd^{2+}, due to environmental exposure, but in no case is MT found exclusively bound to either Cu^{+} or Cd^{2+} (Li and Maret, 2008). In mammals, MT-1 and MT-2 are found in all organs, and appear to be related to requirements for metal ion homeostasis, toxic metal detoxification, and protection against oxidative stress. Induction of MT-1 and MT-2 requires both the metal response element binding transcription factor (MTF-1) and the metal response elements (MREs) to interact. MREs are *cis*-acting DNA sequences, while MTF-1 responds to the concentration of cytosolic Zn^{2+}. Having six zinc-fingers, Cys_2His_2, MTF-1 is exquisitely sensitive to the concentration of zinc in a cell (Lindert *et al.*, 2009). Through the use of cell free transcription experiments, it has been shown that a number of metallothionein inducers, such as Cd^{2+}, Cu^{+} and H_2O_2, function by displacing naturally bound Zn^{2+} from MT, which then binds to MTF-1 (Zhang *et al.*, 2003). When cells are in an unstressed state, MTF-1 is located in the cytoplasm. Upon exposure to a stressor, such as a reactive oxygen species, oxidation of the thiols of metallothionein led to a release of Zn^{2+}. Freed Zn^{2+} subsequently binds MTF-1 causing translocation to the nucleus where its interaction with MRE leads to upregulation of metallothionein. In this way, the organism is capable of handling the stressor.

The facility to bind biologically necessary Zn^{2+} and Cu^{+} with very high binding affinities, suggests that MT-1 and MT-2 are essential for metal ion homeostasis. In support of this function, MT has een shown to be capable of transferring Zn^{2+} from Zn_7-βα-MT to m-aconitase (Feng *et al.*, 2005), carbonic anhydrase (Mason *et al.*, 2004), and the prototypical transcription factor, Gal4 (Maret *et al.*, 1997). While the proposed metallochaperone function of MT (transfer of either Cu^{+} or Zn^{2+} to metalloproteins) is not essential for an organism's viability under normal conditions (Palmiter, 1998), it does appear important under the extremes of zinc exposure. MT-1 and MT-2 knockout mice exposed to these extremes showed delayed kidney development as pups when fed severely deficient diets, while adult knockout mice challenged with increased Zn^{2+} had a greater incidence of pancreatic acinar cell degeneration(Kelly *et al.*, 1996). In humans a single substitution of MT-1a, Asn27Thr, is correlated with diabetes type 2 (Giacconi *et al.*, 2008; Maret, 2008b). These results are precedent setting as they show for the first time that of the ten functional MTs identified(West *et al.*, 1990; Li and Maret, 2008), a single mutation in isoform 1a is associated with a diseased state. As individuals age, MT appears to become more critical for continued health(Cipriano *et al.*, 2006; Malavolta *et al.*, 2008). In terms of zinc ion homeostasis, it is likely that under physiologically normal conditions MT is not required for cell viability, however, when an organism is stressed, either through an excess or a deficit of Zn^{2+}, MT is required.

Copper homeostasis is also critical to an organism's health, because of its redox properties homeostatic mechanisms involving copper are controlled by a number of metallochaperones. Two well known chaperones include the copper chaperone for superoxide dismutase (CCS), and the cytochrome *c* oxidase copper chaperone (COX17). The former donates copper to superoxide dismutase, critical in protecting the organism from damage produced by superoxides (O_2^-) and the latter donates copper to other metallochaperones necessary for the synthesis of cytochrome *c* oxidase, which is the enzyme responsible for the final step in the electron transport chain. Because many crucial enzymes rely on a constant supply of copper, homeostatic imbalance results in disease (Prohaska, 2008). A significant amount of regulation of copper through the human body is the result of two homologous copper transport proteins: ATP7A and ATP7B.

ATP7A, also known as Menkes protein (MNK) (Crisponi *et al.*, 2010), is composed of three domains: an ATP binding region, six or eight chains that form a channel for the metal ion and finally six copper-binding sites. Menkes protein is expressed in all tissues except the liver, specifically in the brain it is expressed in astrocytes, neurons, cerebrovascular endothelial cells that comprise the blood-brain barrier, and the choroid plexus (Kodama and Fujisawa, 2009). Under normal conditions ATP7A is located in the trans-Golgi membrane and acts to transport copper from the cytosol into the golgi apparatus. In the case of intestinal copper absorption this leads to a transfer of copper from the gut epithelial cells to the portal circulation (Mercer, 1998). Mutations to this protein can result in Menkes disease, where copper accumulates in the kidney and intestinal wall, while the brain, serum and liver do not receive adequate copper. The lack of copper uptake causes a reduced activity of Cu-enzymes, such as superoxide dismutase and cytochrome c oxidase (Kodama and Fujisawa, 2009). Treatment using subcutaneous injections of copper-histidine can ameliorate the low activity of the enzymes, except in the case of neurons where copper accumulates in the blood-brain barrier and does not transfer. Most patients die by the age of three. A milder form of Menkes disease, occipital horn syndrome, has slight residual activity of ATP7A and treatment is similar to that of Menkes disease with patients surviving into adulthood. A murine model of Menkes disease involving MT-1 and -2 knockout mice has shown enhanced sensitivity to copper toxicity caused by a loss in the copper efflux protein (Kelly and Palmiter, 1996). Because there is no known copper storage protein, other than metallothionein, it has been suggested that one of the roles of MT is to control transient fluctuations in copper levels (Prohaska, 2008). In the case of Menkes disease, the loss of a copper efflux protein can be somewhat alleviated by the storage capacity provided by MT. Macular mice, another model for Menkes disease, accumulated Cu-MT in the kidneys as demonstrated by fluorescent emission spectroscopy (Suzuki-Kurasaki *et al.*, 1997).

Interestingly, Cu-MT appeared localized in the proximal convoluted tubule, the primary site of nephrotoxicity caused by administration of Cd-MT (Klaassen and Liu, 1997), with a large concentration found in the proximal tubule. It is quite probable that like Cd-MT, which is considered toxicologically inert when intracellularly stored, Cu-MT is stored in much the same way. From these models it becomes clear that the functions of MT likely include both the storage and the detoxification of copper.

The second well studied disease that implicates copper and metallothionein is Wilson's disease. Wilson's disease, a result of a mutation to the copper transport protein ATP7B, is a genetic disorder characterized by copper accumulation in the liver leading to cellular damage through the production of reactive oxygen species and release of free copper into the blood serum. It is expressed primarily in the liver, kidneys and placenta. In the liver ATP7B, localized in the trans-Golgi network, is involved with the removal of copper through biliary excretion and packaging of copper into holoceruloplasmin for transport to other parts of the body (Kodama and Fujisawa, 2009). The liver of patients with Wilson's disease has been shown to have a marked increase in the amount of Cu-MT(Nartey *et al.*, 1987; Stillman *et al.*, 1989a). More specifically, this MT is saturated, Cu_{12}-MT. MT can act to temporarily control a significant influx of copper, by sequestration using its 20 thiols (Green *et al.*, 1994; Kang, 2006). *In vitro* experiments have demonstrated the ability for zinc bound MT to be readily displaced by copper, which would lead to MT upregulation. An added advantage of overexpressing MT in Wilson's disease is MT's ability to act as an antioxidant, which will mitigate some of the cellular damage caused by reactive oxygen species (Kang, 2006; Chiaverini and DeLey, 2010). In fact cultured embryonic mice cells deficient in both MT-1 and -2 are hypersensitive to oxidative stress (Lazo *et al.*, 1995). Treatment of WD involves the use of chelating agents, such as penicillamine, trientine and zinc. This last treatment, zinc, is a strong inducer of MT and further underscores its importance in the treatment of WD.

Briefly metallothionein is used by the cell as protection against oxidative stress. The upregulation of MT by ROS occurs through the antioxidant response element (ARE), a promoter region on the MT gene, ARE-binding transcription factors, as well as MTF-1 (Chiaverini and DeLey, 2010). The twenty thiols found on mammalian MT

(Fig. 4.1) make it an ideal molecule to interact with, and inhibit, reactive oxygen species (ROS). This interaction is best exemplified by *in vitro* experiments using MT, since a common difficulty is the maintenance of stringently anaerobic conditions meant to keep the protein in a reduced state. In addition to oxygen, nitric oxide has also been show to effect the intracellular release of zinc through the oxidation of Zn-MT (St Croix *et al.*, 2002). All three isoforms of MT have been analysed, including MT-3 resulting in the formation of a long lived Cd_2-α-domain species (Wang *et al.*, 2008), MT-1 having exclusively β-domain metals released and MT-2 with a more distributive release of metals (Zangger *et al.*, 2001; Khatai *et al.*, 2004). It is probable that these differences in metal release are the result of each specific MT isoform having a distinct function under nitrosative stress. Cell culture studies of Zn-MT with H_2O_2 have also shown that MT is capable of acting as an antioxidant, oxiziding its thiols to disulfides with concomitant release of Zn^{2+} (Quesada *et al.*, 1996; Elgohary *et al.*, 1998). The release of Zn^{2+} is likely to also aid in combating oxidants by upregulation MT through MTF-1. Metals, such as copper or iron, may also be the cause of ROS through the production of hydroxyl radicals. To inhibit this species, Zn-MT is capable of sequestering reactive oxygen species and has been shown to inhibit the production of copper catalyzed hydroxyl radical production *in vitro* (Cai *et al.*, 1995). One of the causes of this inhibition is likely the high affinity of MT for copper, which would act to sequester the reactive copper center. Animal studies using cardiac specific metallothionein overexpressing transgenic mice have also supported the role of MT in protecting the organism against oxidative stress, specifically in the inhibition of ischemia/reperfusion-induced myocardial injury (Kang *et al.*, 1999; Kang *et al.*, 2003). A metallothionein redox cycle, connecting oxidants with the redox silent Zn^{2+}, has been suggested to account for the protective role of MT in the cell. Initially a ROS reacts with MT leading to Zn^{2+} release. This Zn^{2+} then acts to up regulates Zn-dependent proteins, through MTF-1, and subsequently reduction of MT using glutathione, or replacement with *de novo* synthesized MT and reestablishment of zinc homeostasis (Kang, 2006; Maret, 2008a).

MT has also been implicated in metal detoxification, since a lack of MT-1 and MT-2 expression has been shown to enhance the nephrotoxicity of $CdCl_2$ and $NaAsO_2$ in mice models (Liu *et al.*, 2000).

Interestingly, tissue retention of As^{3+} was found to be much less than that of Cd^{2+} suggesting that As^{3+} may have a lower binding affinity for MT. However, kinetic studies on the binding of MT to As^{3+} have been recently performed and contrary to this hypothesis it was shown that As^{3+} in fact has a higher affinity for MT than Cd^{2+}. However, the Cd^{2+} binds very much faster to MT than does the As^{3+} (<4 ms for Cd^{2+} vs. several minutes for As^{3+}) (Ejnik *et al.*, 2002; Ngu and Stillman, 2006; Ngu *et al.*, 2008). Since these studies were performed on metal free MT, it is likely that the presence of fully metallated Zn_7-MT-(1 or 2) will further impede the rate of metallation leading to a significant decrease in As^{3+} retention by the kidneys. The production of Zn-MT, ZnT-1 (Zn^{2+} efflux protein) and glutathione are all controlled by MTF-1 (Andrews, 2001). Knockout studies on MTF-1 have shown that embryonic mice become more susceptible to both cadmium and H_2O_2 with embryonic death occurring during the 14th day of gestation (Gunes *et al.*, 1998). Under normal conditions, MT would preferentially bind cadmium resulting in its sequestration from the cellular environment, while the antioxidant properties of MT would react with H_2O_2. In both cases, the stressor would upregulate metallothionein through Zn^{2+} release with subsequent Zn^{2+} binding to MTF-1.

The lethality of metals may be modulated by a pretreatment of the organism with an inducer of MT. This inducer acts to upregulate MT beyond the levels normally observed yielding a higher antioxidant/metal buffering capacity. To illustrate: the liver of mice with MT-1 and MT-2 inactivated, experience enhanced hepatic Cd^{2+} poisoning, however, under normal laboratory conditions the mice were viable and reproduced normally (Masters *et al.*, 1994a). In the case of a pretreatement of Cd^{2+} followed by an acute dose, the LD_{50} of wild-type mice experienced a seven-fold increase, whereas the viability of MT knockout mice was unaffected. This clearly shows the protective role of MT, when an otherwise lethal dose may be dealt with by an increase in chelation ability of the organism (Klaassen *et al.*, 2009).

In the case of chronic cadmium poisoning, the kidneys are significantly affected. It has been traditionally believed that a transfer of Cd-MT from the liver to the kidneys and subsequent degradation leads to a high local concentration of Cd^{2+} causing kidney damage. One would predict from this theory that if MT were not produced by an organism, then preferential accumulation of Cd^{2+} in the kidneys would not be possible. Indeed liver transplant studies aimed at monitoring Cd-MT levels showed a time dependent decrease in the amount of Cd-MT present in the liver with a concomitant increase in kidney Cd-MT (Chan *et al.*, 1993). However, recent studies have demonstrated that MT knockout mice, which accumulated 7% Cd^{2+} compared to wild-type mice, are hypersensitive to Cd^{2+}. These results would suggest that nephrotoxicity is greater for cadmium salts, such as $CdCl_2$, compared to Cd-MT. In fact nephrotoxicity may be caused by both Cd-MT and the cadmium salt itself, but in the case of Cd-MT the MT acts as a heavy metal sink, leading to the production of biologically inert metal-thiol complexes, which act to maintain the health of the organism (Liu *et al.*, 1998).

The expression of all three MT isoforms in a mouse brain has been determined: in general MT-1 has the highest expression, MT-3 has intermediate expression and MT-2 has the lowest expression (Choudhuri *et al.*, 1995). The local concentration of MT-1 and MT-2 are usually found to be highest in glial and ependymal cells, while MT-3 is preferentially expressed in neurons, however, each isoform is expressed in all parts of the brain. Highest expression of all isoforms has been reported to be in the olfactory bulb, with significant expression of MT-1 and MT-2 also occurring in the cerebellum. Variations in the expression of MT suggest that there exists a specific function for each MT isoform. The induction profile of each MT isoform has been determined for mouse neurons *in vitro*, using Pb^{2+}, $MeHg^{+}$, dexamethasone, Cd^{2+}, Zn^{2+}, and Hg^{2+} (Kramer *et al.*, 1996b). Neither Pb^{2+} nor $MeHg^{+}$ had any effect on the expression profiles of any MT isoform, while dexamethasone, Cd^{2+}, Zn^{2+}, and Hg^{2+} enhanced MT-1 and MT-2 expression with a concomitant decrease in the expression of MT-3. Similar to neurons, the induction profile of astrocytes *in vitro* using dexamethasone, Cd^{2+}, Zn^{2+}, and Hg^{2+} showed enhanced MT-1 and MT-2 levels, while MT-3 was relatively unresponsive to these toxicological insults (Kramer *et al.*, 1996a). While the blood-brain barrier plays a crucial role in determining MT levels, it cannot be the sole determining factor in MT levels and MT enhancement must be controlled by other mechanisms. Specifically, the increased enhancement of astrocyte MT relative to neurons may be related to the positioning of astrocytes between the capillary endothelium (blood brain barrier) and the neurons. This would allow astrocytes to effectively protect the neurons against toxic metals.

The neurotoxicity of methylmercury is also related to the expression of MT (West *et al.*, 2008). Research suggests that astrocytes play a fundamental role in $MeHg^+$ mediated death. Astrocytes, which express three times as much MT as neurons, could potentially function to protect the neurons from toxic insult through an enhanced expression of MT (Kramer *et al.*, 1996a; Kramer *et al.*, 1996b). To support this hypothesis, exposure of primates to $MeHg^+$ results in preferential accumulation in astrocytes and reactive glia (Charleston *et al.*, 1994; Charleston *et al.*, 1996). An analysis of binding affinities suggests that $MeHg^+$, binding thiols with an association constant ranging from 10^{15} to 10^{23} M^{-1} (West *et al.*, 2008), would be able to effectively outcompete zinc, which has an association constant ranging from 10^7 to 10^{12} M^{-1} (Krezel and Maret, 2007). However, the induction of MT *in vivo* and *in vitro* with both astrocytes and neurons appears unresponsive to $MeHg^+$ (Yasutake *et al.*, 1998). While $MeHg^+$ may not be able to directly displace Zn-MT, upregulation of MT in both neurons and astrocytes has been observed for zinc, cadmium and mercury (Kramer *et al.*, 1996a; Kramer *et al.*, 1996b; Aschner *et al.*, 2006). Further a pretreatment of either zinc or cadmium is capable of aiding in metal-induced resistance to neurotoxicity (Rising *et al.*, 1995; Aschner *et al.*, 1998). Several recent papers on the mechanistic details of metal binding may be able to account for this discrepancy since it has been shown that the metallation of metallothionein occurs in a noncooperative fashion (Palumaa *et al.*, 2002; Rigby-Duncan and Stillman, 2007; Sutherland and Stillman, 2008). This mechanism is critical since it affords stability to partially metallated species in a cellular environment. These partially metallated species, having several non-coordinating thiols, would likely be more susceptible to $MeHg^+$ binding and account for $MeHg^+$ resistance. Administration of $MeHg^+$ to mice has been found to enhance oxidative stress, due at least in part to the inhibition of superoxide dismutase, observed in brain tissue (Yee and Choi, 1994). With respect to $MeHg^+$, it is likely that MT expression is buffering $MeHg^+$, and providing the secondary benefit of deactivating various ROS formed.

To conclude inducible MT-1 and MT-2 are capable of binding a variety of essential and toxic metals and in this fashion act as a buffer to maintain cellular homeostasis. The reactive thiols also allow MT to deactivate ROS species protecting the cell from damage, and through the release of Zn^{2+} modulate cellular response. Many of these systems are intertwined and only through a thorough understanding of the mechanistic details of the metallation and subsequent demetallation reactions, as well as an understanding of the effects of oxidative stress, can the functions of MT be understood at a cellular level.

4.4 MT-3: A Central Nervous System Metallothionein

MT-3 was first isolated as a growth inhibitory factor natively containing 3 Zn^{2+} and 4 Cu^{+} ions per protein that was down-regulated in the brain's of individuals with Alzheimer's disease (Uchida *et al.*, 1991). Expressed mainly in the central nervous system, MT-3 consists of 68 amino acids compared to 61–62 amino acids of MT-1 and MT-2 and has approximately 70% sequence homology. In the case of Cd^{2+} and Zn^{2+}, metal stoichiometries suggest the existence of two distinct clusters (a β-domain Me_3S_9 and an α-domain Me_4S_{11}) with tetrahedral coordination (Faller and Vasak, 1997; Hasler *et al.*, 1998). In the case of copper binding, an intermediate Cu_4 species is produced in both domains with subsequent saturation occurring at Cu_6S_9 and Cu_6S_{11}. It is likely that copper is bound to MT-3 in the β-domain as Cu_4-β-rhMT-3, because of similarities in air stability and coordination environment (Bogumil *et al.*, 1998), while Zn^{+} is likely to be located in the α-domain (Vasak *et al.*, 2000). Unlike MT-1 and MT-2, MT-3 is unresponsive to induction by typical agents, such as Zn^{2+}, Cd^{2+} and Hg^{2+} (Kramer *et al.*, 1996a; Kramer *et al.*, 1996b). Further differences between MT-3 and inducible MT-1 can be seen in ectopic overexpressing transgenic mice. In this case, the overexpression of MT-3 lead to progressive degeneration of pancreatic acinar cells and ultimately death (Quaife *et al.*, 1998). While the exact reason for this degeneration is unknown, overexpressing MT-1 transgenic mice were viable, and it is likely that this response is the result of the specific function of MT-3. Structural differences between MT-3 and MT-1/MT-2 include an additional threonine (Thr5) in a ^{5}TCPCP^{9} motif (Romero-Isart *et al.*, 2002), and a glutamate rich hexapeptide (EAAEAE) (Zheng *et al.*, 2003) near the C-terminus critical to the function and labilization of Zn^{2+} atoms, respectively.

All three MT isoforms are constitutively expressed throughout the central nervous system, and are abundant in the olfactory bulb (Aschner, 1996). MT-3 is predominantly expressed in neurons, specifically those that sequester zinc in their synaptic vesicles, and the choroid plexus epithelium as well as regions with a high concentration of vascular zinc, particularly the hippocampus, piriform cortex, and the amygdale (Masters *et al.*, 1994b; Aschner, 1996). Both MT-1 (MT-2 is assumed to express identically) and MT-3 are expressed in the olfactory bulb, cortex, hippocampus, brainstem, and spinal cord. MT-3 is largely absent from white matter enriched with glial cells, with both MT-1 and MT-2 being common in these cells types. The localization of MT-3 would suggest that MT-3 specifically functions in conjunction with Zn^{2+}-homeostasis.

The neuroinhibitory activity of MT-3 has been mapped to the β-domain (Sewell *et al.*, 1995; Romero-Isart *et al.*, 2002). Analysis of the metal binding properties of MT-3, and its individual domains, has shown that there is little difference in the reactivity or metal binding stoichiometry of the metal ions compared to MT-1 and -2, but that mutation of the CPCP motif to either CSCA, found in MT-1 and -2, or CTCT leads to a loss in the neuroinhibitory activity of the protein. Further mutations of unreactive MT-1 (S6P, S8P and the insertion of Thr5) imbue it with neuronal inhibitory properties (Romero-Isart *et al.*, 2002). An analysis of the overall fold of the protein, and proline mutants (P7S, P9A and P7S/P9A), shows distinct alteration of the overall conformations adopted by MT-3 (Hasler *et al.*, 2000). Thr5 mutation studies, specifically T5S, T5A and ΔT5, have measured the reactivity of both EDTA, a metal chelating agent, and DTNB, a thiol reactant (Cai *et al.*, 2006). The presence of Thr5 inhibits the ability of both EDTA and DTNB to react when compared to a T5A mutant, while both ΔT5 and T5S are not significantly affected. This reactivity suggests that the hydroxyl group does not significantly mobilize the metals bound to MT-3. Molecular dynamics simulations of the β-domain of MT-3 suggest that both proline residues will be parallel to each other, facing outward and the trans-/trans-isomer of ^{6}CPCP^{9} is energetically favored leading to the constraint of the first 13 residues of the β-domain (Ni *et al.,* 2007). While Thr5 loosens the overall structure by forming a hydrogen bond with Asp2, otherwise disrupting the hydrogen bonding between Asp2 and Lys25 observed in MT-2, leading to an increase in distance between the first four amino acids and residues 23 to 26(Ni *et al.*, 2007). It therefore follows that crucial differences in structure, not metal reactivity, between MT-3 and the other isoforms are likely to provide the protein with a receptor recognition site necessary for its biological functions.

While poorly understood, the glutamate rich hexapeptide (EAAEAE) has also been shown to impact the neuroinhibitory activity of MT-3. NMR structural studies have shown this insert to have rapid internal motion unrestricted by the metal-thiolate cluster(Oz *et al.*, 2001; Wang *et al.*, 2006). Deletion of the hexapeptide results in significant loss of neuronal growth inhibition, as determined by the ability to inhibit neuronal neurite extension, while also inhibiting the reactivity towards SNOC (an NO donor) (Cai *et al.*, 2009). Association constants have also been determined through a pH titration and demonstrated a decrease in the affinity of the α-domain for Cd^{2+} when the hexapeptide was present (α-domain 8.4×10^{19} vs. 2.6×10^{20}, β-domain 7.9×10^{19} vs. 2.8×10^{19} with and without hexapeptide, respectively)(Zheng *et al.*, 2003). Molecular dynamics simulations of MT-3 have shown that a single hydrogen bond between Lys31 and Glu41 brings the two domains closer together, while simultaneously leading to significant structural changes in the β-domain. Deletion of the EAAEAE inset resulted in the destruction of the Lys31-Glu41 and Ser33-Cys38 hydrogen bonds, which caused the domains to separate. Taken together these results suggest that the overall metal-thiolate structure of the α-domain of MT-3 is more solvent exposed, has a lower affinity for metals and leads to significant structural rearrangement that may be critical to its cellular chemistry. Two possible functions for the insert include: 1) formation of an acidic surface critical in protein-protein interactions, 2) through domain-domain interactions, affect the metal release properties of the β-domain. It is likely that both functions contribute somewhat to the overall activity of MT-3 in the cell.

Perhaps most famous is the report that MT-3 is downregulated in Alzheimer's disease (AD)(Uchida *et al.*, 1991). The amyloid-β peptide, through interaction with metal ions, produces ROS and aggregates. Specifically, ROS are thought to be generated through the redox cycling of copper coordinated to Aβ aggregates using readily available reducing agents, such as ascorbate, glutathione, dopamine and cholesterol (Opazo *et al.*, 2002; Dikalov *et al.*, 2004). *In vitro* experiments have shown that Zn_7-βα-MT-3 is capable of exchanging metal ions with Cu-Aβ forming the oxidatively stable Zn_4Cu_4-βα-MT-3 (Meloni *et al.*, 2008). This exchange can then eradicate ROS production by deactivation of the redox active Cu-Aβ and formation of redox inert Zn-Aβ (Cuajungco *et al.*, 2000). In addition to AD, Parkinson's disease (PD) characterized by the aggregation of α-synuclein, the main component of neuronal and glial cytoplasmic inclusions (Lewy bodies), is accelerated by Cu^{2+} (Rasia *et al.*, 2005). The 20 thiol groups, and high affinity of MT for Cu^{+} would make it an ideal group to reduce and subsequently sequester any excess Cu that may be contributing to PD. Interestingly, MT-3 is downregulated in PD, as well as other neurological diseases, and it may be that changes in the regulation of MT-3 is critical in their development (Sogawa *et al.*, 2001). The expression of MT-3 can also be affected by injury to the central nervous system. Studies using rat cortical-ablation demonstrated that one day after cortical ablation a decrease in expression in the cortex ipsilateral to the injury, increasing markedly four days after injury and again increasing in the surrounding tissue 2 to 3 weeks after injury. These results may be interpreted as an initial inhibition of MT-3 to allow neurite regeneration, while subsequent upregulation of MT-3 is to inhibit over growth of the neurons (Yuguchi *et al.*, 1995).

MT-3 has many of the same features of MT-1 and -2 including the ability to bind a variety of essential and toxic metals. Interestingly, MT-3 does not respond to induction with metals and as such is not likely the primary agent in metal ion buffering. However, the reactive thiols also allow MT to deactivate ROS species, which would lead to cellular damage, and Zn^{2+} exchange may modulate AD suggesting a significant role as an antioxidant. MT-3 is highly fluctional, and the presence of two conserved proline residues, as well as an acidic hexapeptide, suggests that it functions as a binding partner for an as yet unknown protein.

Acknowledgements

We gratefully acknowledge the financial support from NSERC of Canada to MJS (operating and equipment funds) and DEKS (Alexander Graham Bell Canada Graduate Scholarship) and the Academic Development Fund at the University of Western Ontario (equipment).

References

Andrews, G. K. (2001). Cellular zinc sensors: MTF-1 regulation of gene expresssion. *Biometals*, 14, pp. 223–237.

Arseniev, A., Schultze, P., Worgotter, E., Braun, W., Wagner, G., Vasak, M., Kagi, J. H. R. and Wuthrich, K. (1988). Three-dimensional structure of rabbit liver [Cd_7]metallothionein-2a in aqueous solution determined by nuclear magnetic resonance. *J. Mol. Biol.*, 201, pp. 637–657.

Aschner, M. (1996). The functional significance of brain metallothioneins. *FASEB J.*, 10, pp. 1129–1136.

Aschner, M., Conklin, D. R., Yao, C. P., Allen, J. W. and Tan, K. H. (1998). Induction of astrocyte metallothioneins (MTs) by zinc confers resistance against the acute cytotoxic effects of methylmercury on cell swelling, Na^+ uptake, and K^+ release. *Brain Res.*, 813, pp. 254–261.

Aschner, M., Syversen, T., Souza, D. O. and Rocha, J. B. T. (2006). Metallothioneins: Mercury species-specific induction and their potential role in attenuating neurotoxicity. *Exp. Biol. Med.*, 231, pp. 1468–1473.

Balamurugan, K. and Schaffner, W. (2009). Chapter 2: Regulation of metallothionein gene expression. Cambridge, Royal Society of Chemistry.

Beltramini, M. and Lerch, K. (1981). Luminescence properties of *Neurospora* copper metallothionein. *FEBS Lett.*, 127, pp. 201–203.

Blindauer, C. A., Harrison, M. D., Parkinson, J. A., Robinson, A. K., Cavet, J. S., Robinson, N. J. and Sadler, P. J. (2001). A metallothionein containing a zinc finger within a four-metal cluster protects a bacterium from zinc toxicity. *Proc. Natl. Acad. Sci. USA*, 98, pp. 9593–9598.

Blindauer, C. A., Polfer, N. C., Keiper, S. E., Harrison, M. D., Robinson, N. J., Langridge-Smith, P. R. R. and Sadler, P. J. (2003). Inert site in a protein zinc cluster: Isotope exchange by high resolution mass spectrometry. *J. Am. Chem. Soc.*, 125, pp. 3226–3227.

Bofill, R., Palacios, O., Capdevila, M., Cols, N., Gonzalez-Duarte, R., Atrian, S. and Gonzalez-Duarte, P. (1999). A new insight into the Ag^+ and Cu^+ binding sites in the metallothionein β domain. *J. Inorg. Biochem.* , 73, pp. 57–64.

Bogumil, R., Faller, P., Binz, P.-A., Vasak, M., Charnock, J. M. and Garner, C. D. (1998). Structural characterization of Cu(I) and Zn(II) sites in neuronal-growth-inhibitory factor by extended x-ray absorption fine structure (EXAFS). *Eur. J. Biochem.*, 255, pp. 172–177.

Boulanger, Y. and Armitage, I. M. (1982). ^{113}Cd nmr study of the metal cluster structure of human liver metallothionein. *J. Inorg. Biochem.*, 17, pp. 147–153.

Boulanger, Y., Armitage, I. M., Miklossy, K.-A. and Winge, D. R. (1982). ^{113}Cd NMR study of a metallothionein fragment: Evidence for a two domain structure. *J. Biol. Chem.*, 257, pp. 13717–13719.

Braun, W., Vasak, M., Robbins, A. H., Stout, C. D., Wagner, G., Kagi, J. H. R. and Wuthrich, K. (1992). Comparison of the NMR solution structure and the x-ray crystal structure of rat metallothionein-2. *Proc. Natl. Acad. Sci. USA*, 89, pp. 10124–10128.

Briggs, R. W. and Armitage, I. M. (1982). Evidence for site-selective metal binding in calf liver metallothionein. *J. Biol. Chem.*, 257, pp. 1259–1262.

Cai, B., Ding, Z.-C., Zhang, Q., Ni, F.-Y., Wang, H., Zheng, Q., Wang, Y., Zhou, G.-M., Wang, K.-Q., Sun, H.-Z., Wu, H.-M. and Huang, Z.-X. (2009). The structural and biological signficance of the EAAEAE insert in the α-domain of human neuronal growth inhibitory factor. *FEBS J.*, 276, pp. 3547–3558.

Cai, B., Zheng, Q., Teng, X.-C., Chen, D., Wang, Y., Wang, K.-Q., Zhou, G.-M., Xie, Y., Zhang, M.-J., Sun, H.-Z. and Huang, Z.-X. (2006). The role of Thr5 in human neuron growth inhibitory factor. *J. Biol. Inorg. Chem.*, 11, pp. 476–482.

Cai, L., Koropatnick, J. and Cherian, M. G. (1995). Metallothionein protects DNA from copper-induced but not iron-induced cleavage in vitro. *Chem-Biol. Interact.* , 96, pp. 143–155.

Calderone, V., Dolderer, B., Hartmann, H.-J., Echner, H., Luchinat, C., DelBianco, C., Mangani, S. and Weser, U. (2005). The crystal structure of yeast copper thionein: The solution of a long-lasting enigma. *Proc. Natl. Acad. Sci. USA*, 102, pp. 51–56.

Chan, H. M., Zhu, L.-F., Zhong, R., Grant, D., Goyer, R. A. and Cherian, M. G. (1993). Nephrotoxicity in rats following liver transplantation from cadmium-exposed rats. *Toxicol. Appl. Pharmacol.*, 123, pp. 89–96.

Chan, J., Huang, Z., Watt, I., Kille, P. and Stillman, M. J. (2007). Characterization of the conformational changes in recombinant human metallothioneins using ESI-MS and molecular modeling. *Can. J. Chem.*, 85, pp. 898–912.

Charleston, J. S., Body, R. L., Bolender, R. P., Mottet, N. K., Vahter, M. E. and Burbacher, T. M. (1996). Changes in the number of astrocytes and microglia in the thalamus of the monkey *Macaca fascicularis* following long-term subclinical methylmercury exposure. *Neurotoxicology*, 17, pp. 127–138.

Charleston, J. S., Bolender, R. P., Mottet, N. K., Body, R. L., Vahter, M. E. and Burbacher, T. M. (1994). Increases in the number of reactive glia in the visual cortex of *Macaca fascicularis* following subclinical long-term methyl mercury exposure. *Toxicol. Appl. Pharmacol.*, 129, pp. 196–206.

Chiaverini, N. and DeLey, M. (2010). Protective effect of metallothionein on oxidative stress-induced DNA damage. *Free Radic. Res.*, 44, pp. 605–613.

Choudhuri, S., Kramer, K. K., Berman, N. E. J., Dalton, T. P., Andrews, G. K. and Klaassen, C. D. (1995). Constitutive expression of metallothionein genes in mouse brain. *Toxicol. Appl. Pharmacol.*, 131, pp. 144–154.

Cipriano, C., Malavolta, M., Costarelli, L., Giacconi, R., Muti, E., Gasparini, N., Cardelli, M., Monti, D., Mariani, E. and Mocchegiani, E. (2006). Polymorphisms in MT1a gene coding region are associated with longevity in italian central female population. *Biogerontology*, 7, pp. 357–365.

Crisponi, G., Nurchi, V. M., Fanni, D., Gerosa, C., Nemolato, S. and Faa, G. (2010). Copper-related diseases: From chemistry to molecular pathology. *Coord. Chem. Rev.* , 254, pp. 876–889.

Cuajungco, M. P., Goldstein, L. E., Nunomura, A., Smith, M. A., Lim, J. T., Atwood, C. S., Huang, X., Farrag, Y. W., Perry, G. and Bush., A. I. (2000). Evidence that the β-amyloid plaques of alzheimer's disease represent the redox-silencing and entombment of Aβ by zinc. *J. Biol. Chem.*, 275, pp. 19439–19442.

Dikalov, S. I., Vitek, M. P. and Mason, R. P. (2004). Cupric-amyloid β peptide complex stimulates oxidation of ascorbate and generation of hydroxyl radical. *Free Radical Bio. Med.*, 36, pp. 340–347.

Ejnik, J., Robinson, J., Zhu, J., Forsterling, H., Shaw-III, C. F. and Petering, D. H. (2002). Folding pathway of apo-metallothionein induced by Zn^{2+}, Cd^{2+} and Co^{2+}. *J. Inorg. Biochem.*, 88, pp. 144–152.

Ejnik, J. W., Munoz, A., DeRose, E., Shaw-III, C. F. and Petering, D. H. (2003). Structural consequences of metallothionein dimerization: solution structure of the isolated Cd_4-α-domain and comparison with the holoprotein dimer. *Biochemistry*, 42, pp. 8403–8410.

Elgohary, W. G., Sidhu, S., Krezoski, S. O., Petering, D. H. and Byrnes, R. W. (1998). Protection of DNA in HL-60 cells from damage generated by hydroxyl radicals produced by reaction of H_2O_2 with cell iron by zinc-metallothionein. *Chem. Biol. Interact.*, 115, pp. 85–107.

Faller, P. and Vasak, M. (1997). Distinct metal-thiolate clusters in the N-terminal domain of neuronal growth inhibitory factor. *Biochemistry*, 36, pp. 13341–13348.

Felitsyn, N., Peschke, M. and Kebarle, P. (2002). Origin and number of charges observed on multiply-protonated native proteins produced by ESI. *Int. J. Mass Spectrom.*, 219, pp. 39–62.

Feng, W., Cai, J., Pierce, W. M., Franklin, R. B., Maret, W., Benz, F. W. and Kang, Y. J. (2005). Metallothionein transfers zinc to mitochondrial aconitase through a direct interaction in mouse hearts. *Biochem. Biophys. Res. Commun.*, 332, pp. 853–858.

Fowle, D. A. and Stillman, M. J. (1997). Comparison of the structures of the metal-thiolate binding site in Zn(II)-, Cd(II)-, and Hg(II)-metallothionein using molecular modelling techniques. *J. Biomol. Struct. Dyn.*, 14, pp. 393–406.

Galdes, A., Vasak, M., Hill, H. A. O. and Kagi, J. H. R. (1978). ^1H NMR spectra of metallothioneins. *FEBS Lett.*, 92, pp. 17–21.

Gan, T., Munoz, A., Shaw-III, C. F. and Petering, D. H. (1995). Reaction of $^{111}Cd_7$-metallothionein with EDTA. A reappraisal. *J. Biol. Chem.*, 270, pp. 5339–5345.

Gasyna, Z., Zelazowski, A., Green, A. R., Ough, E. and Stillman, M. J. (1988). Luminescence decay from copper(I) complexes of metallothionien. *Inorg. Chim. Acta*, 153, pp. 115–118.

Gehrig, P. M., You, C., Dallinger, R., Gruber, C., Brouwer, M., Kagi, J. H. R. and Hunziker, P. E. (2000). Electrospray ionization mass spectrometry of zinc, cadmium, and copper metallothioneins: Evidence for metal-binding cooperativity. *Protein Sci.*, 9, pp. 395–402.

Giacconi, R., Bonfigli, A. R., Testa, R., Sirolla, C., Cipriano, C., Marra, M., Muti, E., Malavolta, M., Costarelli, L., Piacenza, F., Tesei, S. and Mocchegiani,

E. (2008). +647 A/C and +1245 MT1A polymorphisms in the susceptibility of diabetes mellitus and cardiovascular complications. *Mol. Genet. Metab.*, 94, pp. 98–104.

Good, M., Hollenstein, R., Sadler, P. J. and Vasak, M. (1988). ^{113}Cd NMR studies on metal-thiolate cluster formation in rabbit Cd(II)-metallothionein: Evidence for a pH dependence. *Biochemistry*, 27, pp. 7163–7166.

Good, M., Hollenstein, R. and Vasak, M. (1991). Metal selectivity of clusters in rabbit liver metallothionein. *Eur. J. Biochem.*, 197, pp. 655–659.

Good, M. and Vasak, M. (1986). Iron(II)-substituted metallothionein: Evidence for the existence of iron-thiolate clusters. *Biochemistry*, 25, pp. 8353–8356.

Green, A. R., Presta, A., Gasyna, Z. and Stillman, M. J. (1994). Luminescent probe of copper-thiolate cluster formation within mammalian metallothionein. *Inorg. Chem.*, 33, pp. 4159–4168.

Gunes, C., Heuchel, R., Georgiev, O., Muller, K.-H., Lichtlen, P., Bluthmann, H., Marino, S., Aguzzi, A. and Schaffner, W. (1998). Embryonic lethality and liver degeneration in mice lacking the metal-responsive transcriptional activator MTF-1. *EMBO J.* , 17, pp. 2846–2854.

Hasler, D. W., Faller, P. and Vasak, M. (1998). Metal-thiolate clusters in the C-terminal domain of human neuronal growth inhibitory factor (GIF). *Biochemistry*, 37, pp. 14966–14973.

Hasler, D. W., Jensen, L. T., Zerbe, O., Winge, D. R. and Vasak, M. (2000). Effect of the two conserved prolines of human growth inhibitory factor (metallothionein-3) on its biological activity and structure fluctuation: Comparison with a mutant protein. *Biochemistry*, 39, pp. 14567–14575.

Jensen, L. T., Peltier, J. M. and Winge, D. R. (1998). Identification of a four copper folding intermediate in mammalian copper metallothionein by electrospray ionization mass spectrometry. *J. Biol. Inorg. Chem.*, 3, pp. 627–631.

Kagi, J. H. R. (1993). Evolution, structure and chemical activity of class I metallothioneins: An overview. Berlin, Birkhauser-Verlag.

Kagi, J. H. R. and Vallee, B. L. (1961). Metallothionein: A cadmium and zinc-containing protein from equine renal cortex. *J. Biol. Chem.*, 236, pp. 2435–2442.

Kagi, J. H. R., Vasak, M., Lerch, K., Gilg, D. E. O., Hunziker, P., Bernhard, W. R. and Good, M. (1984). Structure of mammalian metallothionein. *Environ. Health Perspect.*, 54, pp. 93–103.

Kang, Y. J. (2006). Metallothionein redox cycle and function. *Exp. Biol. Med.*, 231, pp. 1459–1467.

Kang, Y. J., Li, G. and Saari, J. T. (1999). Metallothionein inhibits ischemia–reperfusion injury in mouse heart. *Am. J. Physiol.*, 276, pp. H993–H997.

Kang, Y. J., Li, Y., Sun, X. and Sun, X. (2003). Antiapoptotic effect and inhibition of ischemia/reperfusion-induced myocardial injury in metallothionein-overexpressing transgenic mice. *Am. J. Pathol.*, 163, pp. 1579–1586.

Kebarle, P. and Verkerk, U. H. (2009). Electrospray: From ions in solution to ions in the gas phase, what we know now. *Mass Spectrom. Rev.*, 28, pp. 898–917.

Kelly, E. J. and Palmiter, R. D. (1996). A murine model of Menkes disease reveals a physiological function of metallothionein. *Nat. Genet.*, 13, pp. 219–222.

Kelly, E. J., Quaife, C. J., Froelick, G. J. and Palmiter, R. D. (1996). Metallothionein I and II protect against zinc deficiency and zinc toxicity in mice. *J. Nutr.*, 126, pp. 1782–1790.

Khatai, L., Goessler, W., Lorencova, H. and Zangger, K. (2004). Modulation of nitric oxide-mediated metal release from metallothionein by the redox state of glutathione *in vitro*. *Eur. J. Biochem.*, 271, pp. 2408–2416.

Klaassen, C. D. and Liu, J. (1997). Role of metallothionein in cadmium-induced hepatotoxicity and nephrotoxicity. *Drug Metab. Rev.*, 29, pp. 79–102.

Klaassen, C. D., Liu, J. and Diwan, B. A. (2009). Metallothionein protection of cadmium toxicity. *Toxicol. Appl. Pharmacol.*, 238, pp. 215–220.

Kodama, H. and Fujisawa, C. (2009). Copper metabolism and inherited copper transport disorders: Molecular mechanisms, screening, and treatment. *Metallomics*, 1, pp. 42–52.

Kojima, Y. (1991). Introduction [2] Definitions and nomenclature of metallothioneins. San Diego, Academic Press, Inc.

Kramer, K. K., Liu, J., Choudhuri, S. and Klaassen, C. D. (1996a). Induction of metallothionein mRNA and protein in murine astrocyte cultures. *Toxicol. Appl. Pharmacol.*, 136, pp. 94–100.

Kramer, K. K., Zoelle, J. T. and Klaassen, C. D. (1996b). Induction of metallothionein mRNA and protein in primary murine neuron cultures. *Toxicol. Appl. Pharmacol.*, 141, pp. 1–7.

Krezel, A. and Maret, W. (2007). Dual nanomolar and picomolar Zn(II) binding properties of metallothionein. *J. Am. Chem. Soc.*, 129, pp. 10911–10921.

Lazo, J. S., Kondo, Y., Dellapiazza, D., Michalska, A. E., Choo, K. H. and Pitt, B. R. (1995). Enhanced sensitivity to oxidative stress in cultured embryonic cells from transgenic mice deficient in metallothionein I and II genes. *J. Biol. Chem.*, 270, pp. 5506–5510.

Leiva-Presa, A., Capdevila, M. and Gonzalez-Duarte, P. (2004). Mercury(II) binding to metallothioneins: Variables governing the formation and structural features of the mammalian Hg-MT species. *Eur. J. Biochem.*, 271, pp. 4872–4880.

Li, Y. and Maret, W. (2008). Human metallothionein metallomics. *J. Anal. At. Spectrom.*, 23, pp. 1055–1062.

Lindert, U., Cramer, M., Meuli, M., Georgiev, O. and Schaffner, W. (2009). Metal-responsive transcription factor 1 (MTF-1) activity is regulated by a nonconventional nuclear localization signal and a metal-responsive transactivation domain. *Mol. Cell. Biol.*, 29, pp. 6283–6293.

Liu, J., Liu, Y., Habeebu, S. M., Waalkes, M. P. and Klaassen, C. D. (2000). Chronic combined exposure to cadmium and arsenic exacerbates nephrotoxicity, particularly in metallothionein-I/II null mice. *Toxicology*, 147, pp. 157–166.

Liu, J., Liu, Y., Habeebu, S. S. and Klaassen, C. D. (1998). Susceptibility of MT-null mice to chronic $CdCl_2$-induced nephrotoxicity indicates that renal injury is not mediated by the CdMT complex. *Toxicol. Sci.*, 46, pp. 197–203.

Lu, W. and Stillman, M. J. (1993). Mercury-thiolate clusters in metallothionein. Analysis of circular dichroism spectra of complexes formed between α-metallothionein, apometallothionein, zinc metallothionein, and cadmium metallothionein and Hg^{2+}. *J. Am. Chem. Soc.*, 115, pp. 3291–3299.

Lu, W., Zelazowski, A. J. and Stillman, M. J. (1993). Mercury binding to metallothioneins: Formation of the Hg_{18}-MT species. *Inorg. Chem.*, 32, pp. 919–926.

Malavolta, M., Cipriano, C., Costarelli, L., Giacconi, R., Tesei, S., Muti, E., Piacenza, F., Pierpaoli, S., Larbi, A., Pawelec, G., Dedoussis, G., Herbein, G., Monti, D., Jajte, J., Rink, L. and Mocchegiani, E. (2008). Metallothionein downregulation in very old age: A phenomenon associated with cellular senescence? *Rejuvenation Res.*, 11, pp. 455–459.

Maret, W. (2008a). Metallothionein redox biology in the cytoprotective and cytotoxic functions of zinc. *Exp. Geront.*, 43, pp. 363–369.

Maret, W. (2008b). A role for metallothionein in the pathogenesis of diabetes and its cardiovascular complications. *Mol. Genet. Metab.*, 94, pp. 1–3.

Maret, W., Larsen, K. S. and Vallee, B. L. (1997). Coordination dynamics of biological zinc "clusters" in metallothioneins and in the DNA-binding domain of the transcription factor Gal4. *Proc. Natl. Acad. Sci. USA*, 94, pp. 2233–2237.

Margoshes, M. and Vallee, B. L. (1957). A cadmium protein from equine kidney cortex. *J. Am. Chem. Soc.*, 79, pp. 4813–4814.

Mason, A. Z., Perico, N., Moeller, R., Thrippleton, K., Potter, T. and Lloyd, D. (2004). Metal donation and apo-metalloenzyme activation by stable isotopically labeled metallothionein. *Mar. Environ. Res.*, 58, pp. 371–375.

Masters, B. A., Kelly, E. J., Quaife, C. J., Brinster, R. L. and Palmiter, R. D. (1994a). Targeted disruption of metallothionein I and II genes increases sensitivity to cadmium. *Proc. Natl. Acad. Sci. USA*, 91, pp. 584–588.

Masters, B. A., Quaife, C. J., Erickson, J. C., Kelly, E. J., Froelick, G. J., Zambrowicz, B. P., Brinster, R. L. and Palmiter, R. D. (1994b). Metallothionein III is expressed in neurons that sequester zinc in synaptic vesicles. *J. Neurosci.*, 14, pp. 5844–5857.

Meloni, G., Polanski, T., Braun, O. and Vasak, M. (2009). Effects of Zn^{2+}, Ca^{2+}, and Mg^{2+} on the structure of Zn_7metallothionein-3: Evidence for an additional zinc binding site. *Biochemistry*, 48, pp. 5700–5707.

Meloni, G., Sonois, V., Delaine, T., Guilloreau, L., Gillet, A., Teissie, J., Faller, P. and Vasak, M. (2008). Metal swap between Zn_7-metallothionein-3 and amyloid-β-Cu protects against amyloid-β toxicity. *Nat. Chem. Biol.*, 4, pp. 366–372.

Meloni, G., Zovo, K., Kazantseva, J., Palumaa, P. and Vasak, M. (2006). Organization and assembly of metal-thiolate clusters in epithelium-specific metallothionein-4. *J. Biol. Chem.*, 281, pp. 14588–14595.

Mercer, J. F. B. (1998). Menkes syndrome and animal models. *Am. J. Clin. Nutr.*, 67, pp. 1022S–1028S.

Merrifield, M. E., Huang, Z., Kille, P. and Stillman, M. J. (2002). Copper speciation in the α and β domains of recombinant human metallothionein by electrospray ionization mass spectrometry. *J. Inorg. Biochem.*, 88, pp. 153–172.

Messerle, B. A., Schaffer, A., Vasak, M., Kagi, J. H. R. and Wuthrich, K. (1990). Three-dimensional structure of human [$^{113}Cd_7$]metallothionein-2 in solution determined by nuclear magnetic resonance spectroscopy. *J. Mol. Biol.*, 214, pp. 765–779.

Messerle, B. A., Schaffer, A., Vasak, M., Kagi, J. H. R. and Wuthrich, K. (1992). Comparison of the solution conformations of human [Zn_7]-metallothionein-2 and [Cd_7]-metallothionein-2 using nuclear magnetic resonance spectroscopy. *J. Mol. Biol.*, 225, pp. 433–443.

Miles, A. T., Hawksworth, G. M., Beattie, J. H. and Rodilla, V. (2000). Induction, regulation, degradation, and biological significance of mammalian metallothioneins. *Crit. Rev. Biochem. Mol. Biol.*, 35, pp. 35–70.

Morelock, M. M., Cormier, T. A. and Tolman, G. L. (1988). Technetium metallothioneins. *Inorg. Chem.*, 27, pp. 3137–3140.

Namdarghanbari, M. A., Meeusen, J., Bachowski, G., Giebel, N., Johnson, J. and Petering, D. H. (2010). Reaction of the zinc sensor FluoZin-3 with Zn_7-metallothionein: Inquiry into the existence of a proposed weak binding site. *J. Inorg. Biochem.*, 104, pp. 224–231.

Nartey, N. O., Frei, J. V. and Cherian, M. G. (1987). Hepatic copper and metallothionein distribution in wilson's disease (hepatolenticular degeneration). *Lab Invest.*, 57, pp. 397–401.

Ngu, T. T., Easton, A. and Stillman, M. J. (2008). Kinetic analysis of arsenic-metalation of human metallothionein: Significance of the two-domain structure. *J. Am. Chem. Soc.*, 130, pp. 17016–17028.

Ngu, T. T., Krecisz, S. and Stillman, M. J. (2010a). Bismuth binding studies to the human metallothionein using electrospray mass spectrometry. *Biochem. Biophys. Res. Commun.*, 396, pp. 206–212.

Ngu, T. T., Lee, J. A., Pinter, T. B. J. and Stillman, M. J. (2010b). Arsenic-metalation of triple-domain human metallothioneins: Support for the evolutionary advantage and interdomain metalation of multiple-metal-binding domains. *J. Inorg. Biochem.*, 104, pp. 232–244.

Ngu, T. T. and Stillman, M. J. (2006). Arsenic binding to human metallothionein. *J. Am. Chem. Soc.*, 128, pp. 12473–12483.

Ni, F.-Y., Cai, B., Ding, Z.-C., Zheng, F., Zhang, M.-J., Wu, H.-M., Sun, H.-Z. and Huang, Z.-X. (2007). Structural prediction of the β-domain of metallothionein-3 by molecular dynamics simulation. *Proteins: Struct. Funct. Bioinf.*, 68, pp. 255–266.

Nielson, K. B., Atkin, C. L. and Winge, D. R. (1985). Distinct metal-binding configurations in metallothionein. *J. Biol. Chem.*, 260, pp. 5342–5350.

Opazo, C., Huang, X., Cherny, R. A., Moir, R. D., Roher, A. E., White, A. R., Cappai, R., Masters, C. L., Tanzi, R. E., Inestrosa, N. C. and Bush, A. I. (2002). Metalloenzyme-like activity of Alzheimer's disease β-amyloid: Cu-dependent catalytic conversion of dopamine, cholesterol, and biological reducing agents to neurotoxic H_2O_2. *J. Biol. Chem.*, 277, pp. 40302–40308.

Otvos, J. D. and Armitage, I. M. (1979). ^{113}Cd NMR of metallothionein: direct evidence for the existence of polynuclear metal binding sites. *J. Am. Chem. Soc.*, 101, pp. 7734–7736.

Otvos, J. D. and Armitage, I. M. (1980). Structure of the metal clusters in rabbit liver metallothionein. *Proc. Natl. Acad. Sci. USA*, 77, pp. 7094–7098.

Outten, C. E. and O'Halloran, T. V. (2001). Femtomolar sensitivity of metalloregulatory proteins controlling zinc homeostasis. *Science*, 292, pp. 2488–2492.

Oz, G., Pountney, D. L. and Armitage, I. M. (1998). NMR spectroscopic studies of I = 1/2 metal ions in biological systems. *Biochem. Cell Biol.*, 76, pp. 223–234.

Oz, G., Zangger, K. and Armitage, I. M. (2001). Three-dimensional structure and dynamics of a brain specific growth inhibitory factor: Metallothionein-3. *Biochemistry*, 40, pp. 11433–11441.

Palacios, O., Leiva-Presa, A., Atrian, S. and Lobinski, R. (2007). A study of the Pb(II) binding to recombinant mouse Zn_7-metallothionein 1 and its domains by ESI TOF MS. *Talanta*, 72, pp. 480–488.

Palmiter, R. D. (1998). The elusive function of metallothioneins. *Proc. Natl. Acad. Sci. USA*, 95, pp. 8428–8430.

Palumaa, P., Eriste, E., Njunkova, O., Pokras, L., Jornvall, H. and Sillard, R. (2002). Brain-specific metallothionein-3 has higher metal-binding capacity than ubiquitous metallothioneins and binds metals noncooperatively. *Biochemistry*, 41, pp. 6158–6163.

Peroza, E. A., Schmucki, R., Guntert, P., Freisinger, E. and Zerbe, O. (2009). The β_E-domain of wheat E_c-1 metallothionein: A metal-binding domain with a distinctive structure. *J. Mol. Biol.*, 387, pp. 207–218.

Presta, A., Green, A. R., Zelazowski, A. and Stillman, M. J. (1995). Copper binding to rabbit liver metallothionein: Formation of a continuum of copper(I)-thiolate stoichiometric species. *Eur. J. Biochem.*, 227, pp. 226–240.

Prohaska, J. R. (2008). Role of copper transporters in copper homeostasis. *Am. J. Clin. Nutr.*, 88, pp. 826S–829S.

Quaife, C. J., Kelly, E. J., Masters, B. A., Brinster, R. L. and Palmiter, R. D. (1998). Ectopic expression of metallothionein-III causes pancreatic acinar cell necrosis in transgenic mice. *Toxicol. Appl. Pharmacol.*, 148, pp. 148–157.

Quesada, A. R., Byrnes, R. W., Krezoski, S. O. and Petering, D. H. (1996). Direct reaction of H_2O_2 with sulfhydryl groups in HL-60 cells: Zinc-metallothionein and other sites. *Arch. Biochem. Biophys.*, 334, pp. 241–250.

Rasia, R. M., Bertoncini, C. W., Marsh, D., Hoyer, W., Cherny, D., Zweckstetter, M., Griesinger, C., Jovin, T. M. and Fernandez, C. O. (2005). Structural characterization of copper(II) binding to α-synuclein: Insights into the bioinorganic chemistry of parkinson's disease. *Proc. Natl. Acad. Sci. USA*, 102, pp. 4294–4299.

Rigby-Duncan, K. E., Kirby, C. W. and Stillman, M. J. (2008). Metal exchange in metallothioneins - a novel structurally significant Cd_5 species in the alpha domain of human metallothionein 1a. *FEBS J.*, 275, pp. 2227–2239.

Rigby-Duncan, K. E. and Stillman, M. J. (2006). Metal-dependent protein folding: Metallation of metallothionein. *J. Inorg. Biochem.*, 100, pp. 2101–2107.

Rigby-Duncan, K. E. and Stillman, M. J. (2007). Evidence for noncooperative metal binding to the α domain of human metallothionein. *FEBS J.*, 274, pp. 2253–2261.

Rigby, K. E., Chan, J., Mackie, J. and Stillman, M. J. (2006). Molecular dynamics study on the folding and metallation of the individual domains of metallothionein. *Proteins: Struct. Funct. Bioinf.*, 62, pp. 159–172.

Rising, L., Vitarella, D., Kimelberg, H. K. and Aschner, M. (1995). Metallothionein induction in neonatal rat primary astrocyte cultures protects against methylmercury cytotoxicity. *J. Neurochem.*, 65, pp. 1562–1568.

Robbins, A. H., McRee, D. E., Williamson, M., Collett, S. A., Xuong, N. H., Furey, W. F., Wang, B. C. and Stout, C. D. (1991). Refined crystal structure of Cd, Zn metallothionein at 2.0 Å resolution. *J. Mol. Biol.*, 221, pp. 1269–1293.

Romero-Isart, N., Jensen, L. T., Zerbe, O., Winge, D. R. and Vasak, M. (2002). Engineering of metallothionein-3 neuroinhibitory activity into the inactive isoform metallothionein-1. *J. Biol. Chem.*, 277, pp. 37023–37028.

Roschitzki, B. and Vasak, M. (2002). A distinct Cu_4-thiolate cluster of human metallothionein-3 is located in the N-terminal domain. *J. Biol. Inorg. Chem.*, 7, pp. 611–616.

Rupp, H., Voelter, W. and Weser, U. (1974). 270 MHz proton magnetic resonance spectra of metallothionein. *FEBS Lett.*, 40, pp. 176–179.

Sadler, P. J., Bakka, A. and Beynon, P. J. (1978). ^{113}Cd nuclear magnetic resonance of metallothionien: Non-equivalent CdS_4 sites. *FEBS Lett.*, 94, pp. 315–318.

Salgado, M. T., Bacher, K. L. and Stillman, M. J. (2007). Probing structural changes in the α and β domains of copper- and silver-substituted metallothionein by emission spectroscopy and electrospray ionization mass spectrometry. *J. Biol. Inorg. Chem.*, 12, pp. 294–312.

Salgado, M. T. and Stillman, M. J. (2004). Cu^+ distribution in metallothionein fragments. *Biochem. Biophys. Res. Commun.*, 318, pp. 73–80.

Schmitz, G., Minkel, D. T., Gingrich, D. and Shaw-III, C. F. (1980). The binding of gold(I) to metallothionein. *J. Inorg. Biochem.*, 12, pp. 293–306.

Schultze, P., Worgotter, E., Braun, W., Wagner, G., Vasak, M., Kagi, J. H. R. and Wuthrich, K. (1988). Conformation of $[Cd_7]$-metallothionein-2 from rat liver in aqueous solution determined by nuclear magnetic resonance spectroscopy. *J. Mol. Biol.*, 203, pp. 251–268.

Sewell, A. K., Jensen, L. T., Erickson, J. C., Palmiter, R. D. and Winge, D. R. (1995). Bioactivity of metallothionein-3 correlates with its novel β domain sequence rather than metal binding properties. *Biochemistry*, 34, pp. 4740–4747.

Sogawa, C. A., Asanuma, M., Sogawa, N., Miyazaki, I., Nakanishi, T., Furuta, H. and Ogawa, N. (2001). Localization, regulation, and function of metallothionein-III/growth inhibitory factor in the brain. *Acta. Med. Okayama*, 55, pp. 1–9.

St. Croix, C. M., Wasserloos, K. J., Dineley, K. E., Reynolds, I. J., Levitan, E. S. and Pitt, B. R. (2002). Nitric oxide-induced changes in intracellular zinc homeostasis are mediated by metallothionein/thionein. *Am. J. Physiol. Lung Cell Mol. Physiol.*, 282, pp. L185–L192.

Stillman, M. J. (1995). Metallothioneins. *Coord. Chem. Rev.*, 144, pp. 461–511.

Stillman, M. J., Cai, W. and Zelazowski, A. J. (1987). Cadmium binding to metallothioneins: Domain specificity in reactions of α and β fragments, apometallothionein, and zinc metallothionein with Cd^{2+}. *J. Biol. Chem.*, 262, pp. 4538–4548.

Stillman, M. J., Gasyna, Z. and Zelazowski, A. J. (1989a). A luminescence probe for metallothionein in liver tissue: emission intensity measured directly from copper metallothionein induced in rat liver. *FEBS Lett.*, 257, pp. 283–286.

Stillman, M. J. and Szymanska, J. A. (1984). Absorption, circular dichroism, magnetic circular dichroism and emission study of rat kidney Cd,Cu-metallothionein. *Biophys. Chem.* , 19, pp. 163–169.

Stillman, M. J., Thomas, D., Trevithick, C., Guo, X. and Siu, M. (2000). Circular dicroism, kinetic and mass spectrometric studies of copper(I) and mercury(II) binding to metallothionein. *J. Inorg. Biochem.*, 79, pp. 11–19.

Stillman, M. J. and Zelazowski, A. J. (1988). Domain specificity in metal binding to metallothionein. A circular dichroism and magnetic circular dichroism study of cadmium and zinc binding at temperature extremes. *J. Biol. Chem.*, 263, pp. 6128–6133.

Stillman, M. J., Zelazowski, A. J. and Gasyna, Z. (1988). Luminescent Ag_{12}-metallothionein: dependence of emission intensity on silver-thiolate cluster formation. *FEBS Lett.*, 240, pp. 159–162.

Stillman, M. J., Zelazowski, A. J., Szymanska, J. and Gasyna, Z. (1989b). Luminescent metallothioneins: Emission properties of copper, silver, gold and platinum complexes of MT. *Inorg. Chim. Acta*, 161, pp. 275–279.

Sutherland, D. E. K. and Stillman, M. J. (2008). Noncooperative cadmium(II) binding to human metallothionein 1a. *Biochem. Biophys. Res. Commun.*, 372, pp. 840–844.

Sutherland, D. E. K., Willans, M. J. and Stillman, M. J. (2010). Supermetalation of the β domain of human metallothionein 1a. *Biochemistry*, 49, pp. 3593–3601.

Suzuki-Kurasaki, M., Okabe, M. and Kurasaki, M. (1997). Copper-metallothionein in the kidney of macular mice: A model for Menkes disease. *J. Histochem. Cytochem.*, 45, pp. 1493–1501.

Tapia, L., Gonzalez-Aguero, M., Cisternas, M. F., Suazo, M., Cambiazo, V., Uauy, R. and Gonzalez, M. (2004). Metallothionein is crucial for safe intracellular copper storage and cell survival at normal and supra-physiological exposure levels. *Biochem. J.*, 378, pp. 617–624.

Uchida, Y., Takio, K., Titani, K., Ihara, Y. and Tomonaga, M. (1991). The growth inhibitory factor that is dificient in the alzheimer's disease brain is a 68 amino acid metallothionein-like protein. *Neuron*, 7, pp. 337–347.

Vaher, M., Romero-Isart, N., Vasak, M. and Palumaa, P. (2001). Reactivity of Cd_7-metallothionein with Cu(II) ions: Evidence for a cooperative formation of Cd_3, $Cu(I)_5$-metallothionein. *J. Inorg. Biochem.*, 83, pp. 1–6.

Vasak, M. (1980). Spectroscopic studies on cobalt(II) metallothionein: Evidence for pseudotetrahedral metal coordination. *J. Am. Chem. Soc.*, 102, pp. 3953–3955.

Vasak, M., Galdes, A., Hill, H. A. O., Kagi, J. H. R., Bremner, I. and Young, B. W. (1980). Investigation of the structure of metallothioneins by proton nuclear magnetic resonance spectroscopy. *Biochemistry*, 19, pp. 416–425.

Vasak, M. and Hasler, D. W. (2000). Metallothioneins: New functional and structural insights. *Curr. Opin. Chem. Biol.* , 4, pp. 177–183.

Vasak, M., Hasler, D. W. and Faller, P. (2000). Metal-thiolate clusters in neuronal growth inhibitory factor (GIF). *J. Inorg. Biochem.*, 79, pp. 7–10.

Vasak, M. and Kagi, J. H. R. (1981). Metal thiolate clusters in cobalt(II)-metallothionein. *Proc. Natl. Acad. Sci. USA*, 78, pp. 6709–6713.

Vasak, M., Kagi, J. H. R. and Hill, H. A. O. (1981a). Zinc(II), cadmium(II), and mercury(II) thiolate transitions in metallothionein. *Biochemistry*, 20, pp. 2852–2856.

Vasak, M., Kagi, J. H. R., Holmquist, B. and Vallee, B. L. (1981b). Spectral studies of cobalt(II)- and nickel(II)-metallothionein. *Biochemistry*, 20, pp. 6659–6664.

Wang, H., Li, H., Cai, B., Huang, Z.-X. and Sun, H. (2008). The effect of nitric oxide on metal release from metallothionein-3: Gradual unfolding of the protein. *J. Biol. Inorg. Chem.*, 13, pp. 411–419.

Wang, H., Zhang, Q., Cai, B., Li, H., Sze, K.-H., Huang, Z.-X., Wu, H.-M. and Sun, H. (2006). Solution structure and dynamics of human metallothionein-3 (MT-3). *FEBS Lett.*, 580, pp. 795–800.

Weser, U., Rupp, H., Donay, F., Linnemann, F., Voelter, W., Voetsch, W. and Jung, G. (1973). Characterisation of Cd, Zn-thionein (metallothionein) isolated from rat and chicken liver. *Eur. J. Biochem.*, 39, pp. 127–140.

West, A. K., Hidalgo, J., Eddins, D., Levin, E. D. and Aschner, M. (2008). Metallothionein in the central nervous system: Roles in protection, regeneration and cognition. *Neurotoxicology*, 29, pp. 489–503.

West, A. K., Stallings, R., Hildebrand, C. E., Chiu, R., Karin, M. and Richard, R. I. (1990). Human metallothionein genes: Structure of the functional locus at 16q13. *Genomics*, 8, pp. 513–518.

Willner, H., Vasak, M. and Kagi, J. H. R. (1987). Cadmium-thiolate clusters in metallothionein: Spectrophotometric and spectropolarimetric features. *Biochemistry*, 26, pp. 6287–6292.

Yasutake, A., Nakano, A. and Hirayama, K. (1998). Induction by mercury compounds of brain metallothionein in rats: Hg^0 exposure induces long-lived brain metallothionein. *Arch. Toxicol.*, 72, pp. 187–191.

Yee, S. and Choi, B. H. (1994). Methylmercury poisoning induces oxidative stress in the mouse brain. *Exp. Mol. Pathol.*, 60, pp. 188–196.

Yuguchi, T., Kohmura, E., Yamada, K., Sakaki, T., Yamashita, T., Otsuki, H., Kataoka, K., Tsuji, S. and Hayakawa, T. (1995). Expression of growth inhibitory factor mRNA following cortical injury in rat. *J. Neurotrauma*, 12, pp. 299–306.

Zangger, K., Oz, G., Haslinger, E., Kunert, O. and Armitage, I. M. (2001). Nitric oxide selectively releases metals from the amino-terminal domain of metallothioneins: Potential role at inflammatory sites. *FASEB J.*, 15, pp. 1303–1305.

Zangger, K., Oz, G., Otvos, J. D. and Armitage, I. M. (1999). Three-dimensional solution structure of mouse [Cd_7]-metallothionein-1 by homonuclear and heteronuclear NMR spectroscopy. *Protein Sci.*, 8, pp. 2630–2638.

Zelazowski, A. J., Gasyna, Z. and Stillman, M. J. (1989). Silver binding to rabbit liver metallothionein. Circular dichroism and emission study of silver-thiolate cluster formation with apometallothionein and the α and β fragments. *J. Biol. Chem.*, 264, pp. 17091–17099.

Zelazowski, A. J. and Stillman, M. J. (1992). Silver binding to rabbit liver zinc metallothionein and zinc α and β fragments. Formation of silver metallothionein with Ag(I):protein ratios of 6,12 and 18 observed using circular dichroism spectroscopy. *Inorg. Chem.*, 31, pp. 3363–3370.

Zelazowski, A. J., Szymanska, J. A., Law, A. Y. C. and Stillman, M. J. (1984). Spectroscopic properties of the α fragment of metallothionein. *J. Biol. Chem.*, 259, pp. 12960–12963.

Zeng, J., Heuchel, R., Schaffner, W. and Kagi, J. H. R. (1991). Thionein(apometallothionein) can modulate DNA binding and transcription activation by zinc finger containing factor Sp1. *FEBS Lett.* , 279, pp. 310–312.

Zhang, B., Georgiev, O., Hagmann, M., Gunes, C., Cramer, M., Faller, P., Vasak, M. and Schaffner, W. (2003). Activity of metal-responsive transcription factor 1 by toxic heavy metals and H_2O_2 in vitro is modulated by metallothionein. *Mol. Cell. Biol.*, 23, pp. 8471–8485.

Zhang, B. L., Sun, W. Y. and Tang, W. X. (1997). Determination of the association constant of platinum(II) to metallothionein. *J. Inorg. Biochem.*, 65, pp. 295–298.

Zheng, Q., Yang, W.-M., Yu, W.-H., Cai, B., Teng, X.-C., Xie, Y., Sun, H.-Z., Zhang, M.-J. and Huang, Z.-X. (2003). The effect of the EAAEAE insert on the property of human metallothionein-3. *Protein Eng.*, 16, pp. 865–870.

Chapter 5

Copper Transporting P-Type ATPases in the Brain

Sharon La Fontaine,[a] James Camakaris,[b] and Julian Mercer[a]

[a]*Strategic Research Centre for Molecular and Medical Research and School of Life and Environmental Sciences, Deakin University, Burwood, VIC. 3125, Australia*

[b]*Department of Genetics, The University of Melbourne, Parkville, VIC. 3010, Australia*

Mammals have two transmembrane copper transporting ATPases, which show a high degree of homology and play crucial roles in copper homeostasis in various tissues including the brain (La Fontaine and Mercer, 2007; Lutsenko *et al.*, 2007). They catalyse transport of copper across membranes and can function in both the delivery of copper to copper-dependent enzymes and proteins and in the efflux of copper from cells. Copper is essential and potentially toxic and these transporters are pivotal in maintaining copper homeostasis in individual tissues and in the whole organism. ATP7A encodes the Menkes copper transporter whilst ATP7B encodes the Wilson copper transporter. Severe neurological symptoms arise from a deficiency of either transporter. Mutations in *ATP7A* lead to Menkes disease (OMIM 309400), which has serious neurological symptoms resulting from severe copper deficiency, whilst mutations in *ATP7B* lead to Wilson disease (OMIM 277900) where hepatic and neurological symptoms are due to copper toxicosis (La Fontaine and Mercer, 2007; Lutsenko

Brain Diseases and Metalloproteins
Edited by David R. Brown

ISBN 978-981-4316-01-9 (Hardcover), 978-981-4364-07-2 (eBook)
www.panstanford.com

et al., 2007). The function of these transporters is largely regulated by their sub-cellular localisation. The distribution, developmental changes, and function of these copper transporters in the brain are discussed in this chapter.

5.1 ATP7A in the Brain

The Menkes copper-translocating P_{1B}-type ATPase (ATP7A) is a transmembrane protein and belongs to the P-type ATPase family with characteristic phosphatase, acyl phosphorylation and ATP binding domains (Lutsenko *et al.,* 2007) (Fig. 5.1). The N-terminal domain of ATP7A in humans has six metal binding sites consisting of repeating CxxC motifs (metal binding sites) whilst the C-terminal domain has important localisation and trafficking motifs (Fig. 5.1). The widespread expression of ATP7A and the systemic defects caused by its absence in Menkes disease have pointed to a house-keeping role for ATP7A. ATP7A has a critical role in supplying copper to cuproenzymes in the secretory pathway, in copper efflux, and in translocation of copper across epithelial and endothelial cell barriers.

5.1.1 *Disease Causing Mutations in ATP7A*

"Severe" mutations in ATP7A, associated with loss of function, lead to Menkes disease in humans, which is an X-linked recessive copper

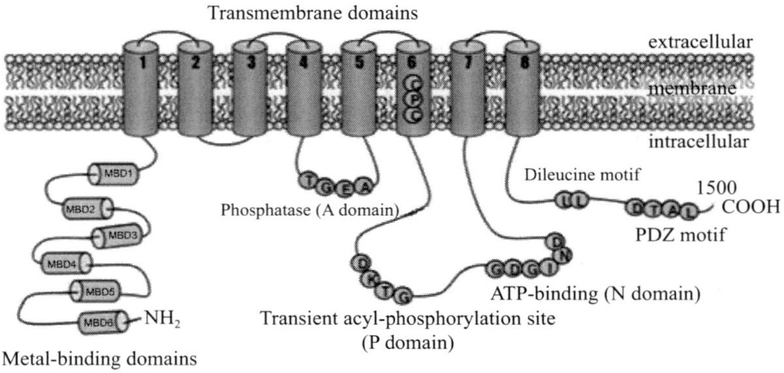

Figure 5.1 Schematic diagram of ATP7A. Key conserved domains and motifs are shown. ATP7B has a similar structure, but lacks the PDZ motif.

deficiency disorder characterised by major connective tissue and neurological abnormalities. "Milder" mutations in ATP7A where there is significant residual function in ATP7A lead to a primarily connective tissue disease with sparing of neurological disease (Mercer, 2001). A third clinical phenotype associated with mutations in ATP7A has recently been described (Kennerson *et al.*, 2010). This variant is associated with progressive distal motor neuropathy with minimal or no sensory symptoms. The mutations described in this variant were missense mutations in highly conserved amino acids which do not disrupt any of the known functional domains (Kennerson *et al.*, 2010). The clinical phenotype suggests that motor neurons are particularly sensitive to mild copper deficiency. ATP7A is responsible for copper delivery to several key enzymes in the brain—cytochrome *c* oxidase, dopamine beta hydroxylase, peptidylglycine alpha-amidating monooxygenase and tyrosinase, and deficiencies in these enzymes as well as other cupro-enzymes such as SOD1 leads to the brain pathology seen in Menkes disease (Prohaska and Gybina, 2004). In classical Menkes disease there is cerebral atrophy and dysmyelination often accompanied by seizures (Tang *et al.*, 2008). NMDA receptor activation mediates copper release from hippocampal neurons via ATP7A (as discussed below). NMDA–mediated excitotoxic hippocampal neuronal cell death increases under copper deficiency conditions, which may provide a target for treating seizures in Menkes disease patients (Schlief *et al.*, 2006).

Menkes disease infants have been treated with daily copper injections, which should commence within days of birth (Kaler *et al.*, 2008). Copper therapy has been successful with some mouse ATP7A mutant alleles and this appears to be most successful in the early post-natal period at times of maximal ATP7A expression (Mann *et al*,. 1979; Fujii *et al.*, 1990; Niciu *et al.*, 2006). However, pre- and post- natal stages of brain development are different in humans and rodents (Clancy *et al.*, 2007) and this could result in post-natal therapy of affected mice being more successful than that in humans. Kaler *et al.*, (2008) reported that commencement of copper treatment early in the neonatal period resulted in improved survival, and response to early copper treatment appears to be dependent on mutant ATP7A alleles that encode a protein with some residual copper transport function (Kaler *et al.*, 2008). The role of ATP7A in transporting copper across the blood brain barrier provides challenges in delivering sufficient copper to the developing brain of Menkes disease infants (Kaler, 1998). Madsen and Gitlin (2008) reported exciting findings that a zebrafish ATP7A splice site mutant could be rescued using antisense oligonucleotides (Madsen *et al.*, 2008). This opens up the possibility of *in utero* treatment of Menkes disease in humans.

5.1.2 *ATP7A and Brain Tumours*

ATP7A has recently been shown to be a novel target of retinoic acid receptor beta 2 in neuroblastoma cells (Bohlken *et al.*, 2009). This finding has important implications for the mechanism of anti-cancer action of retinoids and their use in the treatment of neuroblastoma and other cancers. ATP7A transcription was induced by retinoids in neuroblastoma cells and this was associated with increases in ATP7A protein. The authors proposed that important anti-tumour effects of retinoic acid are mediated by ATP7A. In this model the malignant neuroblastoma cells require increased copper levels for cell viability and retinoid induced differentiation of these cells results in lowering of intracellular copper via the action of ATP7A in promoting copper efflux.

5.1.3 *Expression of ATP7A in the Brain and Changes During Development*

Kodama *et al.* (1991) reported that copper accumulates in cultured astrocytes from the brain of the macular mouse, a mouse model of Menkes disease, and hence concluded that ATP7A in this cell type plays an important role in the transport of copper across the blood-brain barrier (Kodama *et al.*, 1991). Moreover, mouse cerebrovascular endothelial cells that comprise the blood-brain barrier were shown to express ATP7A (Qian *et al.*, 1998), and strong ATP7A expression in the choroid plexus was observed in mouse brain by *in situ* hybridisation (Nishihara *et al.*, 1998). Hence, lack of expression of ATP7A in these cells would be expected to be an important contributor to copper deficiency in the brain of Menkes disease patients. Choi and Zheng (2009) observed high expression of ATP7A in the blood-cerebrospinal fluid barrier whilst ATP7B was expressed highly in the blood-brain-barrier (Choi and Zheng, 2009). The functional significance of differential expression of the two copper ATPases in the brain barriers remains to be elucidated and this will be important in understanding the regulation of copper levels in the brain under normal conditions and in diseases such as Alzheimer's disease where there is copper dyshomeostasis (Bush and Curtain, 2008).

In situ hybridisation studies demonstrated widespread expression of ATP7A mRNA in neurons and ependymal cells in the developing brain (Kuo *et al.,* 1997; Murata *et al.,* 1997). This widespread expression during embryonic and postnatal development suggests a house-keeping role for ATP7A in the brain and central nervous system (CNS). Using a highly specific ATP7A antibody (rabbit polyclonal serum) a detailed study of ATP7A protein expression in the mouse brain was carried out (Niciu *et al.,* 2006). ATP7A protein expression was most abundant in the early postnatal period. It peaked in the neocortex and cerebellum at P4. In the developing and adult brain, ATP7A levels were highest in the choroid plexus/ependymal cells of the lateral and third ventricles. ATP7A expression levels declined from P0 to adult in most brain areas, and this decline was observed to be larger in the hippocampus and cerebellum than in the hypothalamus. Niciu *et al.* (2006) suggest that responsiveness to postnatal copper injections, in particular in the mottled mouse mutants (murine models of human Menkes disease), when administered in the early postnatal period, may be due to ATP7A expression being highest in the brain at this time ie. a consequence of increased levels of mutant ATP7A with significantly reduced catalytic activity. Interestingly these authors observed that ATP7A protein expression in CA2 hippocampal pyramidal cells and cerebellar Purkinje neurons increases during postnatal development during a period when there is a general decline in ATP7A levels in other brain areas. As the CA2 region is relatively resistant to epileptogenesis they suggest that the high levels of ATP7A may contribute to the seizure resistance. Studies on ATP7A mutant mice revealed a novel function of ATP7A in neurodevelopment at the time of axon extension and synaptogenesis (El-Meskini *et al.,* 2005; El Meskini *et al.,* 2007). It was suggested that failure of these functions may be an important contributor to the neuropathology observed in Menkes disease. Barnes *et al.* (2005) provided evidence for distinct roles of ATP7A and ATP7B in adult and developing cerebellum. They observed cell-specific expression of these transporters in adult cerebellum. ATP7B was continuously expressed in Purkinje neurons where it delivers copper to ceruloplasmin, whilst during development, ATP7A expression switched from Purkinje neurons to Bergmann glia. The latter cells have an important role in adult cerebellum in supporting Purkinje neurons including supply of trophic factors. It was proposed that ATP7A in Bergmann glia export copper to supply Purkinje neurons, consistent with the functional role of ATP7A in the gut in supplying copper for systemic utilisation (Barnes *et al.,* 2005). It was also suggested that the more rapid catalytic kinetics of ATPA compared to ATP7B facilitates the homeostatic role of ATP7A in the cerebellum.

Peptidylglycine alpha-amidating monooxygenase (PAM) is responsible for amidating over 50% of all neuropeptides (Steveson *et al.,* 2003). The importance of ATP7A in supplying copper to PAM for its function was elegantly demonstrated by observing normal levels of PAM protein yet reduced levels of amidated peptides in pituitary and brain extracts of ATP7A mutant ($Mo^{Br/y}$) mice. Reduced levels of these amidated neuropeptides that are important during crucial periods of neuronal growth and development, potentially account for the developmental defects associated with Menkes disease. PAM and ATP7A were co-localised in the *trans*-Golgi network (TGN) suggesting that ATP7A provides copper to PAM in this compartment (Steveson *et al.,* 2003).

Data from our laboratory using transgenic mice overexpressing ATP7A shed some light on the role of this protein in brain copper homeostasis (Ke *et al.,* 2006). The transgenic mice expressed between 9 and 40 fold higher levels of ATP7A than non-transgenic animals. The protein was primarily expressed in the CA2 region of the hippocampus, the Purkinje cells of the cerebellum, and in the choroid plexus. The overexpression of ATP7A resulted in a reduction of brain copper concentrations, which was more obvious in younger adult animals than in old mice. It is probable that the reduction in overall brain copper was related to the higher levels of ATP7A in the choroid plexus, as this area was shown to be highly active in copper uptake, and expresses higher levels of endogenous ATP7A than other brain regions (Choi and Zheng, 2009). The reduction in brain copper in the transgenic mice suggests that ATP7A in the choroid plexus may function to efflux copper back into the circulation, which is consistent with the suggested role of the choroid plexus in tightly regulating copper entry into the CSF (Choi and Zheng, 2009).

The mottled mice are a series of mouse mutants with mutations in the murine orthologue of *ATP7A* (Levinson *et al.*, 1994; Mercer *et al.*, 1994). These mutants exhibit a diverse range of phenotypes, replicating the variable clinical severity of the human disease. The mottled brindled mouse ($Mo^{Br/y}$) most closely resembles classical Menkes disease, with postnatal lethality and neurological defects (Mercer, 1998). In this mutant, an in-frame deletion of six nucleotides results in the loss of two highly conserved amino acids (Ala^{799}-Leu^{800}) (Grimes *et al.*, 1997), and an ATP7A protein that although is expressed at close to normal levels, has severely reduced copper transport activity (La Fontaine *et al.*, 1999; Steveson *et al.*, 2003). Residual activity of the mutant protein is thought to account for the ability of early copper treatment (<10 days postnatal) to rescue the mice (Mann *et al.*, 1979; Royce *et al.*, 1982; Phillips *et al.*, 1986), and for the postnatal lethality of the brindled mouse compared with prenatal lethality of the mottled dappled ($Mo^{dp/y}$) and mottled 9H ($Mo^{9H/y}$) mutants that have complete loss of ATP7A (Mercer, 1998).

A detailed study of ATP7A expression in the CNS of the brindled mouse revealed that although total ATP7A levels were unaltered, there were marked differences in the cell types that expressed ATP7A (Niciu *et al.*, 2007). ATP7A expression was decreased in cerebellar Purkinje neurons and hippocampal pyramidal neurons. The decreased ATP7A expression in Purkinje cells was associated with impaired synaptogenesis and cytoskeletal disruption, likely contributing to Purkinje cell degeneration in Menkes disease and mouse models. In contrast, increased ATP7A expression in endothelial cells, and increased association of astrocytes and microglia with the blood-brain barrier, potentially represent compensatory mechanisms to facilitate copper transport across the blood-brain barrier, and may contribute to defining the time period during which early copper treatment is therapeutically effective. A positive correlation between ATP7A levels and intracellular copper levels in diverse neuronal cell types suggested that copper may regulate ATP7A gene expression (Yoshimura, 1994; Murata *et al.*, 1998; Niciu *et al.*, 2007). There was no compensatory increase in ATP7B expression in the $Mo^{Br/y}$ brain (Niciu *et al.,* 2007).

To determine if the genetic defect in the brindled mice could be corrected, we crossed male transgenic mice expressing human ATP7A described above (Ke *et al.,* 2006), with females heterozygous for the $Mo^{Br/y}$ mutation (Llanos *et al.,* 2006) (Fig. 5.2). Expression of the transgene was increased in several tissues, including the brain, relative to expression of the endogenous gene. In the brain, the transgene was expressed in the Purkinje cell layer of the cerebellum, the CA1 and CA2 regions of the hippocampus, the mitral layer of the olfactory bulb, and some expression could be detected in the vascular endothelium and the choroid plexus, but ATP7A could not be detected in astrocytes. By twelve days of age, the severe copper deficiency in the brain of the uncorrected mutant was completely rescued by expression of the transgene in the corrected mice. This correction persisted to at least sixty days, by which age the uncorrected mutant male mice had died. Although ATP7A expression in the vascular endothelium was variable and undetectable in astrocytes, we concluded that some level of ATP7A expression in these regions must have contributed to copper transport across the blood-brain barrier, and that restoration of brain copper was likely to be critical to the survival of the corrected mice (Fig. 5.2).

We have used a Drosophila model system to elucidate the functions of copper-transporting ATPases during development and in various tissues (Norgate et al, 2007). Drosophila has a sole orthologue, DmATP7 (Norgate *et al.,* 2007). DmATP7 appears to play an important role in the developing Drosophila brain. It is strongly expressed in the larval brain at different developmental stages (Burke *et al.,* 2008). Strong expression was observed in the ventral ganglion but was absent from most of the optic lobes (Burke *et al.,* 2008).

5.1.4 *Sub-Cellular Localisation and Intracellular Trafficking of ATP7A*

The function of ATP7A is largely regulated via its sub-cellular localisation. Normally, ATP7A is localised in the TGN where it is involved in delivering copper to copper-dependent enzymes in the

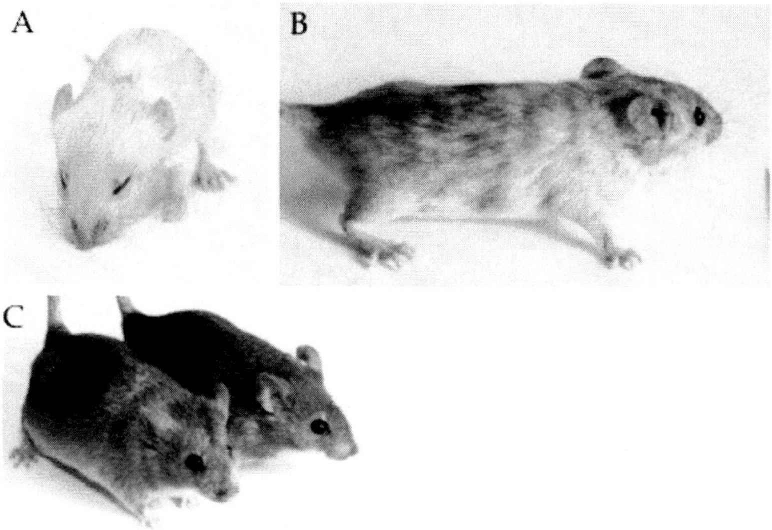

Figure 5.2 Mottled mutant mice and corresponding gene-corrected mutants. A) Uncorrected $Mo^{Br/y}$ mutant mouse (12 days of age). B) Heterogygote female. C) Corrected $Mo^{Br/y}$ male mouse at 135 days of age (left) compared with normal male littermate (right). See also Colour Insert.

secretory pathway. When copper levels are elevated ATP7A traffics to vesicular compartments and the plasma membrane. It also exhibits constitutive recycling between the TGN and the plasma membrane (Petris *et al.*, 1996). The copper-responsive re-localisation facilitates restoration of intracellular copper homeostasis *via* efflux of copper at the plasma membrane. ATP7B also exhibits copper-responsive trafficking towards the plasma membrane (Roelofsen *et al.*, 2000; Guo *et al.*, 2005; Cater *et al.*, 2006). In polarised cells ATP7A traffics to specific membrane domains (Pase *et al.*, 2004; Monty *et al.*, 2005; Nyasae *et al.*, 2007), which facilitates physiological processes such as systemic absorption in gut epithelial cells and transport of copper across the blood brain barrier. Specific trafficking motifs regulate ATP7A copper-responsive trafficking and sub-cellular localisation. For example in gut epithelial cells a dileucine motif proximal to the C-terminus is responsible for targeting of ATP7A to the basolateral membrane and for endocytic retrieval when the transporter recycles, whilst a C-terminal PDZ motif is required for basolateral localisation (Petris *et al.* 1998; Petris and Mercer 1999; Greenough *et al.* 2004; Stephenson *et al.* 2005). Recently we identified twenty serine residues clustered in the C- and N- terminal domains of ATP7A suggesting that kinase signalling is an important regulator of ATP7A function (Veldhuis *et al.*, 2009b). Phosphorylation of eight sites was copper-responsive suggesting that these sites regulate copper-responsive trafficking and/or catalytic activity. One of these sites, S1469, when mutated led to mislocalisation of ATP7A when copper was elevated providing strong evidence that this site is involved in regulating copper-responsive trafficking of ATP7A (Veldhuis *et al.*, 2009b). Candidate phosphorylation regions and sites have also been identified for ATP7B although specific roles for these sites in ATP7B remains to be elucidated (Bartee *et al.*, 2009; Pilankatta *et al.*, 2009).

It is intriguing that both ATP7A and ATP7B are co-expressed in several cells and in these circumstances one may have a physiological role whilst the other a homeostatic role with this being largely determined by whether localisation is at the apical or basolateral membrane (Veldhuis *et al.*, 2009a). Further intrigue has been added by findings that ATP7A traffics in response to physiological stimuli in addition to copper, such as hormones. In placenta and mammary tissue ATP7A traffics in response to the hormones insulin and oestrogen and this is likely to regulate its role in these tissues (La Fontaine and Mercer, 2007; Veldhuis *et al.*, 2009a).

ATP7A also traffics in hippocampal cells in response to NMDA receptor activation (Schleif *et al.*, 2005). This trafficking response is not copper-dependent and it leads to release of copper. Release of copper at the synaptic cleft is important in physiological and pathological processes. It had previously been reported that the concentration of copper in the synaptic cleft is 100 μM and that this copper is involved in controlling neuronal excitability (Kardos *et al.*, 1989). Importantly, it was found that chelating copper exacerbated NMDA-mediated cytotoxic cell death whilst copper addition was protective (Schleif *et al.*, 2005; Schlief *et al.*, 2006). Further substantiation was obtained by the failure to observe ATP7A trafficking from the TGN in neurons from brindled mouse mutants, which have a loss of function ATP7A mutation, and these neurons showed increased sensitivity to excitotoxic injury (Schleif *et al.*, 2005; Schlief *et al.*, 2006). Lack of copper regulation of NMDA receptor mediated excitotoxicity may explain the seizures in Menkes patients and suggests possible therapeutic approaches by blockade of excitatory neuronal pathways (Schlief *et al.*, 2006). These findings also link ATP7A and copper with modulation of memory and learning (Schlief *et al.*, 2006). However, increased amounts of copper and beta amyloid (the amyloidogenic cleavage product of Amyloid Precursor Protein, APP) in the synaptic cleft may promote the formation of toxic beta amyloid-copper oligomers and hence Alzheimer's disease (Hung *et al.*, 2010).

The role of copper in the pathophysiology of Alzheimer's disease is a matter of robust debate with whether "too much or too little" leads to pathology being a central issue (Quinn *et al.*, 2009). The "metals hypothesis" of Alzheimer's disease (Bush and Tanzi 2008) proposes copper and zinc dyshomeostasis leads to Alzheimer's disease pathology. In the case of copper excess, copper in the synapse would promote the aggregation of beta amyloid whilst the sequestering of large amounts of copper would lead to a functional intra-neuronal copper deficiency with both these outcomes promoting Alzheimer's disease pathology. Metal complexing by "ionophore" like compounds would restore copper homeostasis by releasing copper bound to beta amyloid, making it available for neuronal function. Such compounds have met with some success in Alzheimer's disease clinical trials (Faux *et al.*, 2010). This contrasts with approaches involving strong metal chelators, which would be predicted to lead to an overall copper deficiency. ATP7A is likely to play a central role in these processes via its functions in regulating copper availability through the blood brain barrier, its role in delivering copper to cupro-enzymes in the secretory pathway to enzymes involved in neurotransmission (e.g., DBH, PAM), and via its role in the NMDA responses discussed above.

5.2 ATP7B in the Brain

The Wilson P_{1B}-type ATPase, ATP7B, shares approximately 60% amino acid sequence identity with ATP7A. Although structurally and functionally similar, ATP7B has a more restricted expression pattern compared toATP7A, with high expression levels in the liver, and lower levels in the kidney, placenta, brain, heart and lungs (Bull *et al.,* 1993; Tanzi *et al.,* 1993; Vulpe *et al.,* 1993). This restricted expression suggests more specialized functions for ATP7B in regulating copper physiology, such as biliary copper excretion (Terada *et al.,* 1999). ATP7B also has a biosynthetic role, supplying copper to cuproenzymes such as ceruloplasmin (Terada *et al.,* 1998), and where co-expressed with ATP7A, it often has a specific role (La Fontaine *et al.,* 2010), for example in copper secretion to milk during lactation (Michalczyk *et al.,* 2008), or in fine-tuning intracellular copper balance in the kidney (Linz *et al.,* 2008; Barnes *et al.,* 2009).

5.2.1 *ATP7B Expression in the Brain*

While ATP7B expression the brain has been reported (Bull *et al.,* 1993; Tanzi *et al.,* 1993), its expression patterns in the brain and contribution to brain copper homeostasis have been less well characterized than that of ATP7A. A study of ATP7B in the developing mouse embryo failed to detect expression of the protein in the brain, suggesting either that expression was too low to be detected prenatally or that significant expression begins postnatally (Kuo *et al.,* 1997). Consistent with the latter possibility, brain copper levels in an ATP7B null mouse continued to increase slightly throughout adult life (Buiakova *et al.,* 1999). Saito *et al.* (1999) reported the distribution of ATP7B protein and mRNA in the rat brain (Saito *et al.,* 1999). In this study, ATP7B could be detected in the hippocampus, in the granular cells of the dentate gyrus and pyramidal cells of the CA1 to CA4 layers, in the glomerular cell layer of the olfactory bulbs, in Purkinje cells of the cerebellum, in pyramidal neurons of the cerebral cortex, and in cores of several nuclei (e.g., pontine nuclei and lateral reticular nuclei) in the brainstem (Saito *et al.,* 1999). In these brain regions both ATP7B mRNA and copper distribution, the latter determined by staining with the copper chelator bathocuproine disulphonic acid (BCS), correlated with that of the protein. Similar distribution patterns of cuproenzymes such as dopamine beta hydroxylase (DBH) and Cu-Zn SOD, as well as abnormal catecholamine synthesis in the LEC rat model of Wilson disease (Saito *et al.,* 1996; Okabe *et al.,* 1998), led these authors to speculate that ATP7B-mediated control of copper homeostasis in these brain regions is important in regulating DBH activity.

5.2.2 *Role of ATP7B in the Brain and Central Nervous System: Insights from Wilson Disease*

Wilson disease is an autosomal recessively inherited copper toxicity disorder caused by mutation of the *ATP7B* gene (Bull *et al.,* 1993; Petrukhin *et al.,* 1993; Tanzi *et al.,* 1993). The disease manifests primarily in the liver and brain. The liver is the major organ that regulates the copper status of the body. Defective ATP7B leads to impaired biliary copper excretion (Terada *et al.,* 1999) that results in hepatic copper overload, acceleration of apoptotic cell death, liver damage, and spillage of copper into the plasma and CSF (Weisner *et al.,* 1987; Kodama *et al.,* 1988; Strand *et al.,* 1998; Gitlin, 2003). Consequently, copper accumulates in extrahepatic tissues, notably the brain, kidneys and cornea (Danks, 1995; Culotta and Gitlin, 2001). In a significant proportion of Wilson disease patients, neurodegeneration and neurological presentation clearly reveals an important role for ATP7B in maintaining neuronal copper homeostasis.

Symptoms of neurological Wilson disease present in approximately 40 to 50% of cases and typically have a later onset than the liver disease, presenting in the second or third decade (Das and Ray, 2006; Gouider-Khouja, 2009). Although ATP7B is expressed in several brain regions, brain copper accumulation in Wilson disease appears to be secondary to the liver disease, because it can be reversed by transplantation (Emre *et al.,* 2001; Schumacher *et al.,* 2001). The psychiatric symptoms also are reversible with chelation therapy (Madsen and Gitlin, 2007). Despite these observations, there is broad correlation between patterns of neurodegeneration in affected patients, and what is known about ATP7B expression patterns in the brain (as described above). For example, the main areas of the brain affected in Wilson disease are the lenticular nuclei within the basal ganglia where copper deposition causes it to appear brown, the brainstem, thalamus, cerebellum and cerebral cortex (Das and Ray, 2006) (Fig. 5.3). While the neuropathology of neurological Wilson disease can be complex and varied, there is some correspondence with functional disturbance in these brain regions. For example, bulbar symptoms are common initially and include speech and swallowing difficulties and drooling; Parkinsonian symptoms such as rigidity and tremor are consistent with neurodegeneration within the basal ganglia; cerebellar features

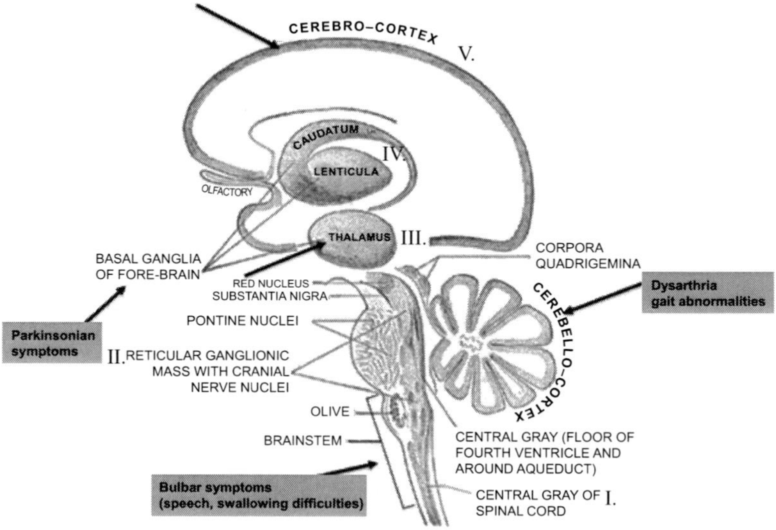

Figure 5.3 Brain regions affected in Wilson disease. Arrows indicate regions of the brain affected in Wilson disease. Associated neurological symptoms are also indicated. Adapted from Gray (1918).

may include dysarthria; and gait abnormalities may develop with both cerebellar and extrapyramidal patterns (Das and Ray, 2006; Pfeiffer, 2007). Intriguingly neurological effects were not evident in the *toxic milk* mouse model of Wilson disease despite elevated brain copper levels following copper loading for three months (Allen *et al.*, 2006).

Neuroimaging is important for diagnosis and monitoring of therapy in Wilson disease. Magnetic resonance imaging (MRI) is commonly used and particularly sensitive for revealing brain abnormalities in Wilson disease (Das and Ray, 2006) Fig. 5.3. Abnormalities are commonly described as changes in the signal intensity of the grey and white matter. T2-weighted scans reveal high signal intensities in the basal ganglia, white matter, thalamus or brainstem. These lesions can be due to neuronal loss, gliosis, demyelinization and edema associated with increased water content in the brain (Das and Ray, 2006). In one study, some of these lesions were reversed following copper chelation therapy (Fig. 5.4) (Kim *et al.*, 2006).

The precise mechanisms mediating neuronal injury in Wilson disease are not clear, but are likely to arise from a combination of increased extracellular copper and disturbed copper homeostasis

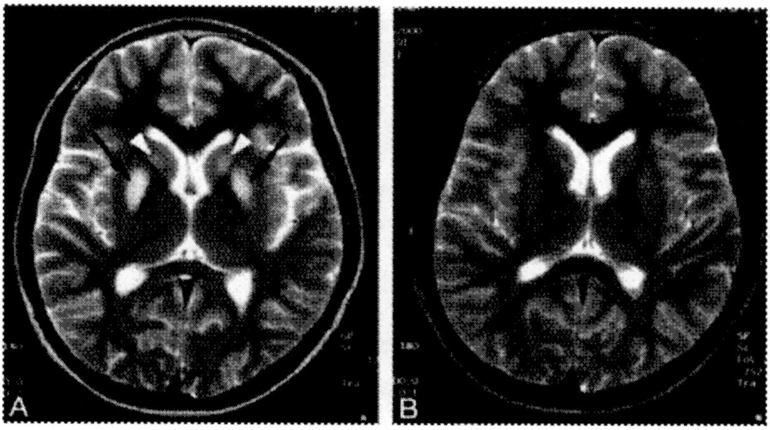

Figure 5.4 Wilson disease in a 14-year-old girl with dysarthria. A) Initial T2-weighted axial MR image shows increased signal intensity in both caudate nuclei (arrowheads) and putamen (arrows). B) Follow-up T2-weighted axial MR image obtained after 3 years shows complete resolution of the lesions. Patient's neurologic symptoms were also improved (from Kim *et al.*, 2006)).

in those regions of the brain suffering loss of ATP7B function. For example, impaired DBH synthesis may explain the predominant abnormalities of the basal ganglia.

5.2.3 *Role of ATP7B in Neuronal Copper Homeostasis*

Earlier studies identified a novel splice variant of ATP7B (PINA) that appeared to function as a weak copper transporter in the *Saccharomyces cerevisiae ccc2Δ* complementation assay (Borjigin *et al.*, 1999). This protein was expressed at night in the pineal gland and retina, under the control of the retina-specific protein, cone rod homeobox (CRX), suggesting a role for this protein and light-regulated copper metabolism in pineal and retinal circadian function (Li *et al.*, 1998; Borjigin *et al.*, 1999). As mentioned earlier, Barnes *et al.*, (2005) found continuous expression of ATP7B in Purkinje neurons (PN), but a postnatal switch in expression of ATP7A from PN to Bergmann glia (BG). Based on studies in mice lacking ATP7B, these authors proposed a homeostatic role for ATP7A in maintaining intracellular copper to a certain level, and a biosynthetic role for ATP7B mediating the synthesis of copper-dependent enzymes such as ceruloplasmin.

There has been limited information in relation to the subcellular localization and trafficking of the copper-ATPases within cells of the CNS. Both ATP7A and ATP7B are expressed within retinal pigment epithelium (RPE) while only ATP7A was also present within the neurosensory retina. The proteins were localized to a perinuclear region that overlapped with TGN and Golgi markers and increased copper levels led to redistribution of ATP7B to a diffuse cytoplasmic compartment in an immortalized human RPE cell line (Krajacic *et al.*, 2006). The authors suggested that the RPE, which forms a blood-brain barrier, may control copper transport to the outer retina via ATP7A and/or ATP7B. Their location within the Golgi area also suggests they may deliver copper to tyrosinase for melanogenesis within the RPE, and to ceruloplasmin and hephaestin, both of which are produced within the RPE, to maintain iron homeostasis. Retinal degeneration is observed in both Menkes and Wilson diseases (Ferreira *et al.*, 1998) as a likely consequence of both abnormal systemic copper levels and loss of either of the copper-ATPases.

In Alzheimer's disease, Aβ-metal interactions play a central role in promoting the neuropathogenic effects of the Aβ peptide according to the "Metal Hypothesis of Alzheimer's disease" (Bush and Tanzi, 2008). In Alzheimer's disease, copper distribution in the brain is abnormal, with amyloid plaques containing excessive amounts of copper and neighbouring cells suffering copper deficiency (Bush and Tanzi, 2008). Transgenic APP mice (CRND8) that were homozygous for the tx^J mutation, and therefore lacking a functional ATP7B protein, exhibited elevated brain copper levels, but a markedly reduced number of amyloid plaques and decreased plasma Aβ levels (Phinney *et al.*, 2003). The mechanism of this beneficial effect of the tx^J mutation is not clear, but the authors proposed that it was due to increased clearance of peripheral Aβ pools. An alternative explanation is that the CRND8 mice are copper deficient and the elevated extracellular copper resulting from the tx^J mutation made copper available for uptake by neuronal cells, thus correcting the copper deficiency. Correcting the copper deficiency then would lead to increased APP production and Cu, Zn-SOD activity, and decreased Aβ production and oxidative stress (Cater *et al.*, 2008).

Until recently the *ApoE ε4* allele was the only established genetic risk factor for Alzheimer's disease (van Es and van den Berg, 2009). It is also associated with poor outcome following traumatic neuronal injury (Alberts *et al.*, 1995; Teasdale *et al.*, 1997). In contrast, the *ApoE ε3* allele has been associated with neuroprotection. In a study of 121 Wilson disease patients with the most common H1069Q mutation, this allele also was associated with later onset of neurological symptoms compared with other *ApoE* genotypes (Schiefermeier *et al.*, 2000). ApoE binds to and promotes Aβ clearance from the brain by facilitating its proteolytic degradation within microglia and extracellularly. The ApoE4 isoform is less effective in carrying out this function than the other isoforms, ApoE2 and ApoE3 (Jiang *et al.*, 2008).

Together with ApoE, clusterin (ApoJ) is the other main escorting protein for β-amyloid in the brain, promoting its clearance (DeMattos *et al.*, 2004). We recently identified a link between clusterin and the copper-ATPases, which may provide insight into the functional significance of the associations between ApoE/clusterin and neurological disease (Materia *et al.*, 2011). Clusterin is a heterodimeric, secreted glycoprotein that interacts with a wide variety of proteins. It is ubiquitously expressed with highest expression in the brain (Aronow *et al.*, 1993). Clusterin is similar to the small heat shock proteins with chaperone activity, binding to stressed and misfolded proteins to maintain them in a state competent for re-folding by other chaperones and to prevent their stress-induced aggregation (Humphreys *et al.*, 1999; Poon *et al.*, 2000; Trougakos and Gonos, 2006). Clusterin has been well characterized as an extracellular chaperone (although intracellular forms also exist) (Trougakos and Gonos, 2006; Wilson *et al.*, 2008), and has been found associated with all amyloid deposits tested (Wilson *et al.*, 2008). The emerging consensus is that clusterin interacts with β-amyloid promoting its clearance by mediating its uptake (via the cell surface megalin receptor) and subsequent degradation (Nuutinen *et al.*, 2009). Whether its association with β-amyloid is protective or neurotoxic depends on the clusterin:β-amyloid ratio (Yerbury *et al.*, 2007; Nuutinen *et al.*, 2009). Studies in Alzheimer's disease mouse models revealed that both clusterin and ApoE cooperatively regulate the deposition and clearance of β-amyloid (DeMattos *et al.*, 2004). Recently, two independent genome-wide association studies of Alzheimer's disease both identified single nucleotide polymorphisms (SNPs) at the *CLU* gene locus as having significant association with Alzheimer's disease (Harold *et al.*, 2009;

Lambert *et al.*, 2009). Therefore, both clusterin and ApoE may act as modifying genes that influence the onset and/or clinical expression of Alzheimer's disease.

Clusterin has been widely implicated in pathological conditions in which oxidative stress plays a central role such as aging, neurodegenerative diseases, cancer progression, inflammation, diabetes among many others. Based on its large repertoire of unrelated binding partners, it has also been implicated in diverse physiological processes, such as lipid transport, cell differentiation, regulation of apoptosis, clearance of cellular debris and stabilization of misfolded proteins (Trougakos and Gonos, 2006). Our recent data demonstrated that clusterin also interacts with ATP7A and ATP7B, and the amount of interacting clusterin increases when cells are stressed by altered copper levels, or by treating cells with oxidizing agents such as H_2O_2 or diamide. Clusterin also showed a stronger interaction with unstable or misfolded ATP7B variants, and appears to facilitate the degradation of these molecules predominantly via the lysosome (Materia *et al.*, 2011). Both Menkes and Wilson diseases exhibit a high degree of clinical variability (de Bie *et al.*, 2007; Tumer and Moller, 2009), with reports of identical mutations, even among siblings, conferring variable clinical expression (Duc *et al.*, 1998; Borm *et al.*, 2004). Hence, these observations implicate other factors in determining the clinical phenotype. Variations in clusterin, conferred by genetic variations such as SNPs, may represent one such factor that could affect the clinical presentation of these (and other) diseases.

Collectively these data are revealing functional similarities between clusterin and ApoE that shed light on their modes of action in neuroprotection, and provide a functional basis for the association of their variant alleles with neurological disease. Moreover, ApoE binds copper with high affinity which was suggested to be associated with its antioxidant activity (Miyata and Smith, 1996). Intriguingly, clusterin also features two CxxC motifs and so potentially also may bind copper. Together these observations point to a potential role for these molecules in modifying the expression of neurological disease such as Alzheimer's, Menkes and Wilson diseases, but whether they function cooperatively or redundantly, remains to be established. Conceivably, variations in the *clusterin* and *ApoE* alleles, together with environmental factors, could contribute to the variability in the clinical expression of Menkes and Wilson diseases, and may influence the onset and severity of Alzheimer's disease.

These data raise a number of questions to be addressed in elucidating the combined roles of the ApoE isoforms and clusterin in copper homeostasis, copper transporter stability, Aβ clearance and neuroprotection. For example, is the copper-binding ability of ApoE allele-specific? Is the neuroprotection afforded by the ApoE ε3 allele in delaying the onset of neurological symptoms in Wilson disease associated with its copper sequestering ability? Does clusterin bind and sequester copper ions? Is neurological Wilson disease also associated with variations at the clusterin gene locus? Does clusterin play a role in reducing the neuropathogenic effects of Alzheimer's disease in a tx^J background? Does ApoE also bind and regulate the stability of the copper transporters? ApoE and clusterin may facilitate the degradation of mutated copper transporters intracellularly as part of the cell's quality control mechanisms, and/or mediate the clearance and degradation of extracellular Aβ, while also sequestering the metal ions accumulating as a consequence of defective transporters, or released from amyloid plaques, respectively. These mechanisms would represent a highly sophisticated and integrated means of conferring protection against copper-mediated oxidative stress in neurological disease.

5.3 Conclusion

In the last seventeen years since their discovery, significant progress has been made towards understanding the role of the copper-ATPases, ATP7A and ATP7B in normal copper homeostasis. Copper plays a central role in a complex network of signalling pathways that regulates a myriad of physiological processes such as development and neurological processes, as well as pathophysiological processes including tumour growth, cancer resistance, and oxidative stress that contributes to inflammation and neurodegenerative diseases. Accordingly, emerging data reveals reveals a variety of physiological and pathophysiological processes that require ATP7A and ATP7B (Veldhuis *et al.*, 2009a; La Fontaine *et al.* 2010). However, given the complexity of the CNS, our knowledge and understanding of the role of ATP7A and ATP7B in neurological development and neurological processes are in their infancy and much remains to be learned. Understanding the factors that affect the regulation of their expression, post-translational modification and activity is progressing, and will provide insight into their involvement and adaptive capacity during neuropathological processes associated with aging and disease. The value of animal models of neurological disease cannot be underestimated for their contribution towards understanding the role of the copper-ATPases in neurophysiology and disease and for identifying avenues for therapeutic initiatives.

References

Alberts, M. J., Graffagnino, C., McClenny, C., DeLong, D., Strittmatter, W., Saunders, A. M. and Roses, A. D. (1995). ApoE genotype and survival from intracerebral haemorrhage. *Lancet*, 346, p. 575.

Allen, K., Buck, N., Cheah, D., Gazeas, S., Bhathal, P. and Mercer, J. (2006). Chronological changes in tissue copper, zinc and iron in the toxic milk mouse and effects of copper loading. *Biometals*, 19, pp. 555–564.

Aronow, B. J., Lund, S. D., Brown, T. L., Harmony, J. A. and Witte, D. P. (1993). Apolipoprotein J expression at fluid-tissue interfaces: potential role in barrier cytoprotection. *Proc Natl Acad Sci USA*, 90, pp. 725–729.

Barnes, N., Tsivkovskii, R., Tsivkovskaia, N. and Lutsenko, S. (2005). The copper-transporting ATPases, Menkes and Wilson disease proteins, have distinct roles in adult and developing cerebellum. *J. Biol. Chem.*, 280, pp. 9640–9645.

Barnes, N. L., Bartee, M., Y., Braiterman, L., Gupta, A., Ustiyan, V., Zuzel, V., Kaplan, J., H., Hubbard, A., L. and Lutsenko, S. (2009). Cell-Specific Trafficking Suggests a new role for Renal ATP7B in the Intracellular Copper Storage. *Traffic*, 10, pp. 767–779.

Bartee, M. Y., Ralle, M. and Lutsenko, S. (2009). The Loop Connecting Metal-Binding Domains 3 and 4 of ATP7B Is a Target of a Kinase-Mediated Phosphorylation. *Biochemistry*, 48, pp. 5573–5581.

Bohlken, A., Cheung, B. B., Bell, J. L., Koach, J., Smith, S., Sekyere, E., Thomas, W., Norris, M., Haber, M., Lovejoy, D. B., Richardson, D. R. and Marshall, G. M. (2009). ATP7A is a novel target of retinoic acid receptor [beta]2 in neuroblastoma cells. *Br. J. Cancer*, 100, pp. 96–105.

Borjigin, J., Payne, A. S., Deng, J., Li, X., Wang, M. W., Ovodenko, B., Gitlin, J. D. and Synder, S. H. (1999). A novel pineal night-specific ATPase encoded by the Wilson disease gene. *J. Neurosci.*, 19, pp. 1018–1026.

Borm, B., Moller, L. B., Hausser, I., Emeis, M., Baerlocher, K., Horn, N. and Rossi, R. (2004). Variable clinical expression of an identical mutation in the ATP7A gene for Menkes disease/occipital horn syndrome in three affected males in a single family. *J. Pediatr.*, 145, pp. 119–121.

Buiakova, O. I., Xu, J., Lutsenko, S., Zeitlin, S., Das, K., Das, S., Ross, B. M., Mekios, C., Scheinberg, I. H. and Gilliam, T. C. (1999). Null mutation of the murine *ATP7B* (Wilson disease) gene results in intracellular copper accumulation and late-onset hepatic nodular transformation. *Hum. Mol. Genet.*, 8, pp. 1665–1671.

Bull, P. C., Thomas, G. R., Rommens, J. M., Forbes, J. R. and Cox, D. C. (1993). The Wilson disease gene is a putative copper transporting P-type ATPase similar to the Menkes gene. *Nat. Genet.*, 5, pp. 327–337 (Erratum in Nat. Genet. 1994 6:214).

Burke, R., Commons, E. and Camakaris, J. (2008). Expression and localisation of the essential copper transporter DmATP7 in Drosophila neuronal and intestinal tissues. *Int. J. Biochem. Cell Biol.*, 40, pp. 1850–1860.

Bush, A. and Curtain, C. (2008). Twenty years of metallo-neurobiology: where to now? *Eur. Biophys. J.*, 37, pp. 241–245.

Bush, A. I. and Tanzi, R. E. (2008). Therapeutics for Alzheimer's disease based on the metal hypothesis. *Neurotherapeutics*, 5, pp. 421–432.

Cater, M. A., La Fontaine, S., Deal, Y., Shield, K. and Mercer, J. F. B. (2006). ATP7B mediates vesicular sequestration of copper: insights into biliary copper excretion. *Gastroenterology*, 130, pp. 493–506.

Cater, M. A., McInnes, K. T., Li, Q.-X., Volitakis, I., La Fontaine, S., Mercer, J. F. B. and Bush, A. I. (2008). Intracellular copper deficiency increases amyloid-β secretion by diverse mechanisms *Biochem. J.*, 412, pp. 141–152.

Choi, B. S. and Zheng, W. (2009). Copper transport to the brain by the blood-brain barrier and blood-CSF barrier. *Brain Res.*, 1248, pp. 14–21.

Clancy, B., Finlay, B. L., Darlington, R. B. and Anand, K. J. (2007). Extrapolating brain development from experimental species to humans. *Neurotoxicology*, 28, pp. 931–937.

Culotta, V. C. and Gitlin, J. D. (2001). Disorders of copper transport. *In* The Metabolic and Molecular Basis of Inherited Disease. C. R. Scriver, Beaudet, A. L., Sly, W. S. and Valle, D. New York, McGraw-Hill. II: 3105–3126.

Danks, D. M. (1995). Disorders of copper transport. *In* The Metabolic and Molecular Basis of Inherited Disease. C. R. Scriver, Beaudet, A. L., Sly, W. M. and Valle, D. New York, McGraw-Hill. 1: 2211–2235.

Das, S. K. and Ray, K. (2006). Wilson's disease: an update. *Nat. Clin. Prac. Neurol.*, 2, pp. 482–493.

de Bie, P., Muller, P., Wijmenga, C. and Klomp, L. W. (2007). Molecular pathogenesis of Wilson and Menkes disease: correlation of mutations with molecular defects and disease phenotypes. *J. Med. Genet.*, 44, pp. 673–88.

DeMattos, R. B., Cirrito, J. R., Parsadanian, M., May, P. C., O'Dell, M. A., Taylor, J. W. Harmony, J. A. K., Aronow, B. J., Bales, K. R., Paul, S. M. and Holtzman, D. M. (2004). ApoE and clusterin cooperatively suppress A[beta] levels and deposition: Evidence that ApoE regulates extracellular A[beta] metabolism In Vivo. *Neuron*, 41, pp. 193–202.

Duc, H. H., Hefter, H., Stremmel, W., Castaneda-Guillot, C., Hernandez Hernandez, A., Cox, D. W. and Auburger, G. (1998). His1069Gln and six novel Wilson disease mutations: analysis of relevance for early diagnosis and phenotype. *Eur. J. Hum. Genet.*, 6, pp. 616–623.

El Meskini, R., Crabtree, K. L., Cline, L. B., Mains, R. E., Eipper, B. A. and Ronnett, G. V. (2007). ATP7A (Menkes protein) functions in axonal targeting and synaptogenesis. *Mol. Cell. Neurosci.*, 34, pp. 409–421.

El-Meskini, R., Cline, L. B., Eipper, B. A. and Ronnett, G. V. (2005). The developmentally regulated expression of Menkes protein ATP7A suggests a role in axon extension and synaptogenesis. *Dev. Neurosci.*, 27, pp. 333–348.

Emre, S., Atillasoy, E. O., Ozdemir, S., Schilsky, M., Rathna Varma, C. V., Thung, S. N., Sternlieb, I., Guy, S. R., Sheiner, P. A., Schwartz, M. E. and Miller, C. M. (2001). Orthotopic liver transplantation for Wilson's disease: a single-center experience. *Transplantation*, 72, pp. 1232–1236.

Faux, N. G., Ritchie, C. W., Gunn, A., Rembach, A., Tsatsanis, A., Bedo, J., Harrison, J., Lannfelt, L., Blennow, K., Zetterberg, H., Ingelsson, M., Masters, C. L., Tanzi, R. E., Cummings, J. L., Herd, C. M. and Bush, A. I.

(2010). PBT2 rapidly improves cognition in Alzheimer's Disease: additional phase II analyses. *J. Alzheimer's Dis.*, 20, pp. 509–516.

Ferreira, R., Heckenlively, J. R., Menkes, J. H. and Bateman, B. (1998). Menkes disease. New ocular and electroretinographic findings. *Ophthalmol.*, 105, pp. 1076–1078.

Fujii, T., Ito, M., Tsuda, H. and Mikawa, H. (1990). Biochemical study on the critical period for treatment of the mottled brindled mouse. *J. Neurochem.*, 55, pp. 885–889.

Gitlin, J. D. (2003). Wilson disease. *Gastroenterology*, 125, pp. 1868–1877.

Gouider-Khouja, N. (2009). Wilson's disease. *Parkinsonism Relat. Disord.*, 15 Suppl 3, pp. S126–9.

Gray, H. (1918). Anatomy of the human body. Philadelphia: Lea & Febiger, Bartleby.com, 2000. www.bartleby.com/107/. [Date of Printout]

Greenough, M., Pase, L., Voskoboinik, I., Petris, M. J., O'Brien, A. W. and Camakaris, J. (2004). Signals regulating trafficking of the Menkes (MNK; ATP7A) copper translocating P-type ATPase in polarized MDCK cells. *Am. J. Physiol. Cell Physiol.*, 287, pp. C1463–C1471.

Grimes, A., Hearn, C. J., Lockhart, P., Newgreen, D. F. and Mercer, J. F. B. (1997). Molecular basis of the brindled mouse mutant (Mobr): a murine model of Menkes disease. *Hum. Mol. Genet.*, 6, pp. 1037–1042.

Guo, Y., Nyasae, L., Braiterman, L. T. and Hubbard, A. L. (2005). NH2-terminal signals in ATP7B Cu-ATPase mediate its Cu-dependent anterograde traffic in polarized hepatic cells. *Am. J. Physiol. Gastrointest. Liver Physiol.*, 289, pp. 904–916.

Harold, D., Abraham, R., Hollingworth, P., Sims, R., Gerrish, A., Hamshere, M. L., Pahwa, J. S., Moskvina, V., Dowzell, K., Williams, A., Jones, N., Thomas, C., Stretton, A., Morgan, A. R., Lovestone, S., Powell, J., Proitsi, P., Lupton, M. K., Brayne, C., Rubinsztein, D. C., Gill, M., Lawlor, B., Lynch, A., Morgan, K., Brown, K. S., Passmore, P. A., Craig, D., McGuinness, B., Todd, S., Holmes, C., Mann, D., Smith, A. D., Love, S., Kehoe, P. G., Hardy, J., Mead, S., Fox, N., Rossor, M., Collinge, J., Maier, W., Jessen, F., Schurmann, B., van den Bussche, H., Heuser, I., Kornhuber, J., Wiltfang, J., Dichgans, M., Frolich, L., Hampel, H., Hull, M., Rujescu, D., Goate, A. M., Kauwe, J. S. K., Cruchaga, C., Nowotny, P., Morris, J. C., Mayo, K., Sleegers, K., Bettens, K., Engelborghs, S., De Deyn, P. P., Van Broeckhoven, C., Livingston, G., Bass, N. J., Gurling, H., McQuillin, A., Gwilliam, R., Deloukas, P., Al-Chalabi, A., Shaw, C. E.,

Tsolaki, M., Singleton, A. B., Guerreiro, R., Muhleisen, T. W., Nothen, M. M., Moebus, S., Jockel, K.-H., Klopp, N., Wichmann, H. E., Carrasquillo, M. M., Pankratz, V. S., Younkin, S. G., Holmans, P. A., O'Donovan, M., Owen, M. J. and Williams, J. (2009). Genome-wide association study identifies variants at CLU and PICALM associated with Alzheimer's disease. *Nat. Genet.*, 41, pp. 1088–1093.

Humphreys, D. T., Carver, J. A., Easterbrook-Smith, S. B. and Wilson, M. R. (1999). Clusterin has chaperone-like activity similar to that of small heat shock proteins. *J. Biol. Chem.*, 274, pp. 6875–6881.

Hung, Y. H., Bush, A. and Cherny, R. (2010). Copper in the brain and Alzheimer's disease. *J. Biol. Inorg. Chem.*, 15, pp. 61–76.

Jiang, Q., Lee, C. Y., Mandrekar, S., Wilkinson, B., Cramer, P., Zelcer, N., Mann, K., Lamb, B., Willson, T. M., Collins, J. L., Richardson, J. C., Smith, J. D., Comery, T. A., Riddell, D., Holtzman, D. M., Tontonoz, P. and Landreth, G. E. (2008). ApoE promotes the proteolytic degradation of Abeta. *Neuron*, 58, pp. 681–693.

Kaler, S. G. (1998). Diagnosis and therapy of Menkes syndrome, a genetic form of copper deficiency. *Am. J. Clin. Nutr.*, 67, pp. 1029S–1034S.

Kaler, S. G., Holmes, C. S., Goldstein, D. S., Tang, J., Godwin, S. C., Donsante, A., Liew, C. J., Sato, S. and Patronas, N. (2008). Neonatal diagnosis and treatment of Menkes disease. *N. Engl. J. Med.*, 358, pp. 605–614.

Kardos, J., Kovacs, I., Hajos, F., Kalman, M. and Simonyi, M. (1989). Nerve endings from rat brain tissue release copper upon depolarization. A possible role in regulating neuronal excitability. *Neurosci. Lett.*, 103, pp. 139–144.

Ke, B.-X., Llanos, R. M., Wright, M., Deal, Y. and Mercer, J. F. B. (2006). Alteration of copper physiology in mice overexpressing the human Menkes protein ATP7A. *Am. J. Physiol. Regul. Integr. Comp. Physiol.*, 290, pp. R1460–1467.

Kennerson, M. L., Nicholson, G. A., Kaler, S. G., Kowalski, B., Mercer, J. F. B., Tang, J., Llanos, R. M., Chu, S., Takata, R. I., Speck-Martins, C. E., Baets, J., Almeida-Souza, L., Fischer, D., Timmerman, V., Taylor, P. E., Scherer, S. S., Ferguson, T. A., Bird, T. D., De Jonghe, P., Feely, S. M. E., Shy, M. E. and Garbern, J. Y. (2010). Missense mutations in the copper transporter gene ATP7A cause X-linked distal hereditary motor neuropathy. *Am. J. Hum. Genet.*, 86, pp. 343–352.

Kim, T. J., Kim, I. O., Kim, W. S., Cheon, J. E., Moon, S. G., Kwon, J. W., Seo, J. K. and Yeon, K. M. (2006). MR imaging of the brain in Wilson disease of childhood: findings before and after treatment with clinical correlation. *Am. J. Neuroradiol.*, 27, pp. 1373–1378.

Kodama, H., Meguro, Y., Abe, T., Rayner, M. H., Suzuki, K. T., Kobayashi, S. and Nishimura, M. (1991). Genetic expression of Menkes disease in cultured astrocytes of the macular mouse. *J. Inherit. Metab. Dis.*, 14, pp. 896–901.

Kodama, H., Okabe, I., Yanagisawa, M., Nomiyama, H., Nomiyama, K., Nose, O. and Kamoshita, S. (1988). Does CSF copper level in Wilson disease reflect copper accumulation in the brain? *Pediatr. Neurol.*, 4, pp. 35–37.

Krajacic, P., Qian, Y., Hahn, P., Dentchev, T., Lukinova, N. and Dunaief, J. L. (2006). Retinal localization and copper-dependent relocalization of the Wilson and Menkes disease proteins. *Invest. Ophthalmol. Vis. Sci.*, 47, pp. 3129–3134.

Kuo, Y.-M., Gitschier, J. and Packman, S. (1997). Developmental expression of the mouse mottled and toxic milk genes suggests distinct functions for the Menkes and Wilson disease copper transporters. *Hum. Mol. Genet.*, 6, pp. 1043–1049.

La Fontaine, S., Ackland, M. L. and Mercer, J. F. B. (2010). Mammalian copper-transporting P-type ATPases, ATP7A and ATP7B: Emerging roles. *Int. J. Biochem. Cell Biol.*, 42, pp. 206–209.

La Fontaine, S., Firth, S. D., Lockhart, P. J., Brooks, H., Camakaris, J. and Mercer, J. F. B. (1999). Intracellular localization and loss of copper-responsiveness of Mnk, the murine homologue of the Menkes protein, in cells from blotchy (Mo^{blo}) and brindled (Mo^{br}) mouse mutants. *Hum. Mol. Genet.*, 8, pp. 1069–1075.

La Fontaine, S. and Mercer, J. F. B. (2007). Trafficking of the copper-ATPases, ATP7A and ATP7B: Role in copper homeostasis. *Arch. Biochem. Biophys.*, 463, pp. 149–167.

Lambert, J.-C., Heath, S., Even, G., Campion, D., Sleegers, K., Hiltunen, M., Combarros, O., Zelenika, D., Bullido, M. J., Tavernier, B., Letenneur, L., Bettens, K., Berr, C., Pasquier, F., Fievet, N., Barberger-Gateau, P., Engelborghs, S., De Deyn, P., Mateo, I., Franck, A., Helisalmi, S., Porcellini, E., Hanon, O., de Pancorbo, M. M., Lendon, C., Dufouil, C., Jaillard, C., Leveillard, T., Alvarez, V., Bosco, P., Mancuso, M., Panza, F.,

Nacmias, B., Bossu, P., Piccardi, P., Annoni, G., Seripa, D., Galimberti, D., Hannequin, D., Licastro, F., Soininen, H., Ritchie, K., Blanche, H., Dartigues, J.-F., Tzourio, C., Gut, I., Van Broeckhoven, C., Alperovitch, A., Lathrop, M. and Amouyel, P. (2009). Genome-wide association study identifies variants at CLU and CR1 associated with Alzheimer's disease. *Nat. Genet.*, 41, pp. 1094–1099.

Levinson, B., Vulpe, C., Elder, B., Martin, C., Verley, F., Packman, S. and Gitschier, J. (1994). The mottled gene is the mouse homologue of the Menkes disease gene. *Nat. Genet.*, 6, pp. 369–373.

Li, X., Chen, S., Wang, Q., Zack, D. J., Snyder, S. H. and Borjigin, J. (1998). A pineal regulatory element (PIRE) mediates transactivation by the pineal/retina-specific transcription factor CRX. *Proc. Natl. Acad. Sci. U.S.A.*, 95, pp. 1876–1881.

Linz, R., Barnes, N. L., Zimnicka, A. M., Kaplan, J. H., Eipper, B. and Lutsenko, S. (2008). Intracellular targeting of copper-transporting ATPase ATP7A in a normal and Atp7b -/- kidney. *Am. J. Physiol. Renal. Physiol.*, 294, pp. F53–61.

Llanos, R. M., Ke, B.-X., Wright, M., Deal, Y., Monty, F., Kramer, D. R. and Mercer, J. F. B. (2006). Correction of a mouse model of Menkes disease by the human Menkes gene. *Biochim. Biophys. Acta*, 1762, pp. 485–493.

Lutsenko, S., Barnes, N. L., Bartee, M. Y. and Dmitriev, O. Y. (2007). Function and Regulation of Human Copper-Transporting ATPases. *Physiol. Rev.*, 87, pp. 1011–1046.

Madsen, E. and Gitlin, J. D. (2007). Copper and Iron Disorders of the Brain. *Annu. Rev. Neurosci.*, 30, pp. 317–337.

Madsen, E. C., Morcos, P. A., Mendelsohn, B. A. and Gitlin, J. D. (2008). In vivo correction of a Menkes disease model using antisense oligonucleotides. *Proc. Natl. Acad. Sci. USA*, 105, pp. 3909–3914.

Mann, J. R., Camakaris, J., Danks, D. M. and Walliczek, E. G. (1979). Copper metabolism in mottled mouse mutants: copper therapy of brindled (Mo^{br}) mice. *Biochem. J.*, 180, pp. 605–612.

Materia, S., Cater, M. A., Klomp, L. W., Mercer, J. F., and La Fontaine, S. (2011). Clusterin (APOJ): a molecular chaperone that facilitates degradation of the copper-ATPases, ATP7A and ATP7B. *J. Biol. Chem.* 286:10073–10083

Mercer, J. F. B. (1998). Menkes syndrome and animal models. *Am. J. Clin. Nutr.*, 67 (suppl), pp. 1022S–1028S.

Mercer, J. F. B. (2001). The molecular basis of copper-transport diseases. *Trends Mol. Med.*, 7, pp. 64–69.

Mercer, J. F. B., Grimes, A., Ambrosini, L., Lockhart, P., Paynter, J. A., Dierick, H. and Glover, T. W. (1994). Mutations in the murine homologue of the Menkes disease gene in dappled and blotchy mice. *Nat. Genet.*, 6, pp. 374–378.

Michalczyk, A., Bastow, E., Greenough, M., Camakaris, J., Freestone, D., Taylor, P., Linder, M., Mercer, J. and Ackland, M. L. (2008). ATP7B expression in human breast epithelial cells is mediated by lactational hormones. *J. Histochem. Cytochem.*, 56, pp. 389–399.

Miyata, M. and Smith, J. D. (1996). Apolipoprotein E allele-specific antioxidant activity and effects on cytotoxicity by oxidative insults and [beta]-amyloid peptides. *Nat. Genet.*, 14, pp. 55–61.

Monty, J.-F., Llanos, R. M., Mercer, J. F. B. and Kramer, D. R. (2005). Copper exposure induces trafficking of the Menkes protein in intestinal epithelium of ATP7A transgenic mice. *J. Nutr.*, 135, pp. 2762–2766.

Murata, Y., Kodama, H., Abe, T., Ishida, N., Nishimura, M., Levinson, B., Gitschier, J. and Packman, S. (1997). Mutation analysis and expression of the mottled gene in the macular mouse model of Menkes disease. *Pediatr. Res.*, 42, pp. 436–442.

Murata, Y., Kodama, H., Mori, Y., Kobayashi, M. and Abe, T. (1998). Mottled gene expression and copper distribution in the macular mouse, an animal model for Menkes disease. *J. Inherit. Metab. Dis.*, 21, pp. 199–202.

Niciu, M. J., Ma, X.-M., El-Meskini, R., Ronnett, G. V., Mains, R. E. and Eipper, B. A. (2006). Deveopmental changes in the expression of ATP7A during a critical period in postnatal neurodevelopment. *Neuroscience*, 139, pp. 947–964.

Niciu, M. J., Ma, X. M., El Meskini, R., Pachter, J. S., Mains, R. E. and Eipper, B. A. (2007). Altered ATP7A expression and other compensatory responses in a murine model of Menkes disease. *Neurobiol. Dis.*, 27, pp. 278–291.

Nishihara, E., Furuyama, T., Yamashita, S. and Mori, N. (1998). Expression of copper trafficking genes in the mouse brain. *NeuroReport*, 9, pp. 3259–3263.

Norgate, M., Southon, A., Zou, S., Zhan, M., Sun, Y., Batterham, P. and Camakaris, J. (2007). Copper homeostasis gene discovery in Drosophila melanogaster. *Biometals*, 20, pp. 683–697.

Nuutinen, T., Suuronen, T., Kauppinen, A. and Salminen, A. (2009). Clusterin: A forgotten player in Alzheimer's disease. *Brain Res. Rev.*, I61, pp. 89–104.

Nyasae, L., Bustos, R., Braiterman, L., Eipper, B. and Hubbard, A. (2007). Dynamics of endogenous ATP7A (Menkes protein) in intestinal epithelial cells: copper-dependent redistribution between two intracellular sites. *Am. J. Physiol. Gastrointest. Liver Physiol.*, 292, pp. G1181–G1194.

Okabe, M., Saito, S., Saito, T., Ito, K., Kimura, S., Niioka, T. and Kurasaki, M. (1998). Histochemical localization of superoxide dismutase activity in rat brain. *Free Radic. Biol. Med.*, 24, pp. 1470–1476.

Pase, L., Voskoboinik, I., Greenough, M. and Camakaris, J. (2004). Copper stimulates trafficking of a distinct pool of the Menkes copper ATPase (ATP7A) to the plasma membrane and diverts it into a rapid recycling pool. *Biochem. J.*, 378(Pt 3), pp. 1031–1037.

Petris, M. J., Camakaris, J., Greenough, M., La Fontaine, S. and Mercer, J. F. B. (1998). A C-terminal di-leucine is required for localization of the Menkes protein in the *trans*-Golgi network. *Hum. Mol. Genet.*, 7, pp. 2063–2071.

Petris, M. J. and Mercer, J. F. B. (1999). The Menkes protein (ATP7A; MNK) cylces via the plasma membrane both in basal and elevated extracellular copper using a C-terminal di-leucine endocytic signal. *Hum. Mol. Genet.*, 8, pp. 2107–2115.

Petris, M. J., Mercer, J. F. B., Culvenor, J. G., Lockhart, P., Gleeson, P. A. and Camakaris, J. (1996). Ligand-regulated transport of the Menkes copper P-type ATPase efflux pump from the Golgi apparatus to the plasma membrane: a novel mechanism of regulated trafficking. *EMBO J.*, 15, pp. 6084–6095.

Petrukhin, K., Fischer, S. G., Pirastu, M., Tanzi, R. E., Chernov, I., Devoto, M., Brzustowicz, L. M., Cayanis, E., Vitale, E., Russo, J. J., Matseoane, D., Boukhgalter, B., Wasco, W., Figus, A. L., Loudianos, J., Cao, A., Sternlieb, I., Evgrafov, O., Parano, E., Pavone, L., Warburton, D., Ott, J., Penchaszadeh, G. K., Scheinberg, I. H. and Gilliam, T. C. (1993). Mapping, cloning ang genetic characterization of the region containing the Wilson disease gene. *Nature Genetics*, 5, pp. 338–343.

Pfeiffer, R. F. (2007). Wilson's Disease. *Semin. Neurol.*, 27, pp. 123–132.

Phillips, M., Camakaris, J. and Danks, D. M. (1986). Comparisons of copper deficiency states in the murine mutants blotchy and brindled. *Biochem. J.*, 238, pp. 177–183.

Phinney, A., Drisaldi, B., SD Schmidt, A., Lugowski, S., Coronado, V., Liang, Y., Horne, P., Yang, J., Sekoulidis, J., Coomaraswamy, J., Chishti, M., Cox, D., Mathews, P., Nixon, R., Carlson, G., George-Hyslop, P. S. and Westaway, D. (2003). In vivo reduction of amyloid-beta by a mutant copper transporter. *Proc. Natl. Acad. Sci. USA*, 100, pp. 14193–14198.

Pilankatta, R., Lewis, D., Adams, C. M. and Inesi, G. (2009). High yield heterologous expression of wild-type and mutant Cu+-ATPase (ATP7B, Wilson disease protein) for functional characterization of catalytic activity and serine residues undergoing copper-dependent phosphorylation. *J. Biol. Chem.*, 284, pp. 21307–21316.

Poon, S., Easterbrook-Smith, S. B., Rybchyn, M. S., Carver, J. A. and Wilson, M. R. (2000). Clusterin is an ATP-independent chaperone with very broad substrate specificity that stabilizes stressed proteins in a folding-competent state. *Biochemistry*, 39, pp. 15953–15960.

Prohaska, J. R. and Gybina, A. A. (2004). Intracellular copper transport in mammals. *J. Nutr.*, 134, pp. 1003–1006.

Qian, Y., Tiffany-Castiglioni, E., Welsh, J. and Harris, E. D. (1998). Copper efflux from murine microvascular cells requires expression of the Menkes disease Cu-ATPase. *J. Nutr.*, 128, pp. 1276–1282.

Quinn, J. F., Crane, S., Harris, C. and Wadsworth, T. L. (2009). Copper in Alzheimer's disease: too much or too little? *Expert Rev. Neurother.*, 9, pp. 631–637.

Roelofsen, H., Wolters, H., Luyn, M. J. A. V., Miura, N., Kuipers, F. and Vonk, R. J. (2000). Copper-induced apical trafficking of ATP7B in polarized hepatoma cells provides a mechanism for biliary copper excretion. *Gastroenterology*, 119, pp. 782–793.

Royce, P. M., Camakaris, J., Mann, J. R. and Danks, D. M. (1982). Copper metabolism in mottled mouse mutants: the effect of copper therapy on lysyl oxidase activity in brindled (Mo^{br}) mice. *Biochem. J.*, 202, pp. 369–371.

Saito, T., Nagao, T., Okabe, M. and Saito, K. (1996). Neurochemical and histochemical evidence for an abnormal catecholamine metabolism in the cerebral cortex of the Long-Evans Cinnamon rat before excessive copper accumulation in the brain. *Neurosci. Lett.*, 216, pp. 195–198.

Saito, T., Okabe, M., Hosokawa, T., Kurasaki, M., Hata, A., Endo, F., Nagano, K., Matsuda, I., Urakami, K. and Saito, K. (1999). Immunohistochemical determination of the Wilson copper-transporting P-type ATPase in the brain tissues of the rat. *Neurosci. Lett.*, 266, pp. 13–16.

Schiefermeier, M., Kollegger, H., Madl, C., Polli, C., Oder, W., Kuhn, H., Berr, F. and Ferenci, P. (2000). The impact of apolipoprotein E genotypes on age at onset of symptoms and phenotypic expression in Wilson's disease. *Brain*, 123 Pt 3, pp. 585–590.

Schleif, M. L., Craig, A. M. and Gitlin, J. D. (2005). NMDA receptor activation mediates copper homeostasis in hippocampal neurons. *J. Neurosci.*, 25, pp. 239–246.

Schlief, M. L., West, T., Craig, A. M., Holtzman, D. M. and Gitlin, J. D. (2006). Role of the Menkes copper-transporting ATPase in NMDA receptor-mediated neuronal toxicity. *Proc. Natl. Acad. Sci. USA*, 103, pp. 14919–14924.

Schumacher, G., Platz, K. P., Mueller, A. R., Neuhaus, R., Luck, W., Langrehr, J. M., Settmacher, U., Steinmueller, T., Becker, M. and Neuhaus, P. (2001). Liver transplantation in neurologic Wilson's disease. *Transplant. Proc.*, 33, pp. 1518–1519.

Stephenson, S. E. M., Dubach, D., Lim, C. M., Mercer, J. F. B. and La Fontaine, S. (2005). A single PDZ domain protein interacts with the Menkes copper ATPase, ATP7A: A new protein implicated in copper homeostasis. *J. Biol. Chem.*, 280, pp. 33270–33279.

Steveson, T. C., Ciccotosto, D. D., Ma, X.-M., Mueller, G. P., Mains, R. E. and Eipper, B. A. (2003). Menkes protein contirbutes to the function of peptidylglycine α-amidating monooxygenase. *Endocrinology*, 144, pp. 188–200.

Strand, S., Hofmann, W. J., Grambihler, A., Hug, H., Volkmann, M., Otto, G., Wesch, H., Mariani, S. M., Hack, V., Stremmel, W., Krammer, P. H. and Galle, P. R. (1998). Hepatic failure and liver cell damage in acute Wilson's disease involve CD95 (APO-1/Fas) mediated apoptosis. *Nat. Med.*, 4, pp. 588–593.

Tang, J., Donsante, A., Desai, V., Patronas, N. and Kaler, S. G. (2008). Clinical outcomes in Menkes disease patients with a copper-responsive ATP7A mutation, G727R. *Mol. Genet. Metab.*, 95, pp. 174–181.

Tanzi, R. E., Petrukhin, K., Chernov, I., Pellequer, J. L., Wasco, W., Ross, B., Romano, D. M., Parano, E., Pavone, L., Brzustowicz, L. M., Devoto,

M., Peppercorn, J., Bush, A. I., Sternlieb, I., Pirastu, M., Gusella, J. F., Evgrafov, O., Penchaszadeh, G. K., Honig, B., Edelman, I. S., Soares, M. B., Scheinberg, I. H. and Gilliam, T. C. (1993). The Wilson disease gene is a copper transporting ATPase with homology to the Menkes disease gene. *Nature Genet.*, 5, pp. 344–350.

Teasdale, G. M., Nicoll, J. A., Murray, G. and Fiddes, M. (1997). Association of apolipoprotein E polymorphism with outcome after head injury. *Lancet*, 350, pp. 1069–1071.

Terada, K., Aiba, N., Yang, X.-L., Iida, M., Nakai, M., Miura, N. and Sugiyama, T. (1999). Biliary excretion of copper in LEC rat after introduction of copper transporting P-type ATPase, ATP7B. *FEBS Lett.*, 448, pp. 53–56.

Terada, K., Nakako, T., Yang, X.-L., Iida, M., Aiba, N., Minamiya, Y., Nakai, M., Sakaki, T., Miura, N. and Sugiyama, T. (1998). Restoration of holoceruloplasmin synthesis in LEC rat after infusion of recombinant adenovirus bearing WND cDNA. *J. Biol. Chem.*, 273, pp. 1815–1820.

Trougakos, I. P. and Gonos, E. S. (2006). Regulation of clusterin/apolipoprotein J, a functional homologue to the small heat shock proteins, by oxidative stress in ageing and age-related diseases. *Free Radic. Res.*, 40, pp. 1324–1334.

Tumer, Z. and Moller, L. B. (2009). Menkes disease. *Eur. J. Hum. Genet.*, 18, pp. 511–518.

van Es, M. A., and van den Berg, L. H. (2009). Alzheimer's disease beyond APOE. *Nature Genetics*, 41(10), 1047–1048

Veldhuis, N., Gaeth, A., Pearson, R., Gabriel, K. and Camakaris, J. (2009a). The multi-layered regulation of copper translocating P-type ATPases. *Biometals*, 22, pp. 177–190.

Veldhuis, N. A., Valova, V. A., Gaeth, A. P., Palstra, N., Hannan, K. M., Michell, B. J., Kelly, L. E., Jennings, I., Kemp, B. E., Pearson, R. B., Robinson, P. J. and Camakaris, J. (2009b). Phosphorylation regulates copper-responsive trafficking of the Menkes copper transporting P-type ATPase. *Int. J. Biochem. Cell Biol.*, 41, pp. 2403–2412.

Vulpe, C., Levinson, B., Whitney, S., Packman, S. and Gitschier, J. (1993). Isolation of a candidate gene for Menkes disease and evidence that it encodes a copper-transporting ATPase. *Nat. Genet.*, 3, pp. 7–13.

Weisner, B., Hartard, C. and Dieu, C. (1987). CSF copper concentration: a new parameter for diagnosis and monitoring therapy of Wilson's disease with cerebral manifestation. *J. Neurol. Sci.*, 79, pp. 229–237.

Wilson, M. R., Yerbury, J. J. and Poon, S. (2008). Potential roles of abundant extracellular chaperones in the control of amyloid formation and toxicity. *Mol. Biosyst.*, 4, pp. 42–52.

Yerbury, J. J., Poon, S., Meehan, S., Thompson, B., Kumita, J. R., Dobson, C. M. and Wilson, M. R. (2007). The extracellular chaperone clusterin influences amyloid formation and toxicity by interacting with prefibrillar structures. *FASEB J.*, 21, pp. 2312–2322.

Yoshimura, N. (1994). Histochemical localization of copper in various organs of brindled mice. *Pathol. Int.*, 44, pp. 14–19.

Chapter 6

Role of the Amyloid Precursor Protein and Copper in Alzheimer's Disease

Loredana Spoerri,[a,b] Kevin J. Barnham,[a,b,c] Gerd Multhaup,[d] and Roberto Cappai[a,b]

[a]*Department of Pathology, The University of Melbourne, Parkville, Victoria 3010, Australia*

[b]*Bio21 Molecular Science and Biotechnology Institute, The University of Melbourne, Parkville, Victoria 3010, Australia*

[c]*Mental Health Research Institute, Parkville, Victoria 3052, Australia*

[d]*Freie Universitaet Berlin, Institut fuer Chemie/Biochemie, Thielallee 63, D-14195 Berlin, Germany*

The Amyloid Precursor Protein is a highly studied protein due to its assumed central role in the pathogenic pathway that leads to Alzheimer's disease. It is expressed into a number of isoforms and undergoes complex proteolytic processing. Structural, biochemical and cellular data identify the Amyloid Precursor Protein as copper binding protein. This interaction modulates Amyloid Precursor Protein metabolism as well as copper homeostasis. This interaction could also have a direct impact upon the pathogenic pathway that causes Alzheimer's diseases.

Brain Diseases and Metalloproteins
Edited by David R. Brown

ISBN 978-981-4316-01-9 (Hardcover), 978-981-4364-07-2 (eBook)
www.panstanford.com

6.1 The Amyloid Precursor Protein

6.1.1 *Introduction*

The Amyloid Precursor Protein (APP) is a ubiquitously expressed type-1 membrane protein with a long extracellular domain and a short cytoplasmic tail. The APP molecule occupies a central role in Alzheimer's Disease (AD) as the source of the amyloid β peptide (Aβ) There are three main isoforms of human APP, APP_{770}, APP_{751} and APP_{696} (subscript indicating the number of amino acid residues) which are generated by alternative splicing of two domains, the Kunitz-type protease inhibitor (KPI) domain and the OX2 domain (reviewed in the section *"APP domains"*) (Ponte *et al.*, 1988).

The abundance of each isoform is related to the cell type, thus while APP_{695} is highly expressed in neurons, APP_{770} and APP_{751} are more abundant in non-neuronal cells (Haass *et al.*, 1991; Konig *et al.*, 1992; McGeer *et al.*, 1992; Sandbrink *et al.*, 1994; Rohan de Silva *et al.*, 1997; Sandbrink *et al.*, 1997). However, all three isoforms are expressed in brain (Neve *et al.*, 1988).

In addition to the human APP gene, two other paralogs, APLP1 and APLP2 (Wasco *et al.*, 1992; Wasco *et al.*, 1993), and several orthologs such as *Caenorhabditis elegans* (Daigle and Li, 1993), *Drosophila melanogaste* (Rosen *et al.*, 1989), *Xenopus laevis,* (Okado and Okamoto, 1995), electric ray (*Narke japonica*) (Iijima *et al.*, 1998), and puffer fish (*Fugu rubripes and Tetraodon fluviatilis*) (Villard *et al.*, 1998) comprise the APP-gene family.

APP is proteolytically processed by two different pathways, named non-amyloidogenic and amyloidogenic, the latter producing the toxic Aβ peptide (Busciglio *et al.*, 1993; Haass *et al.*, 1993). During amyloidogenic processing, APP is first cleaved at its β-site (after Met597 APP_{695}) generating the extracellular soluble sAPPβ (Seubert *et al.*, 1993) and the membrane bound C-terminal domain (C99) species (Gabuzda *et al.*, 1994). C99 can subsequently be processed by γ-secretase giving rise to Aβ and the APP intracellular domain (AICD) (Anderson *et al.*, 1992). The γ-cleavage occurs at two different sites, after Val637 and Ala639, generating Aβ40 and Aβ42 respectively. Before the γ-cleavage, C99 is thought to be cut at two other sites named ε and ζ, after Leu646 and Val643 respectively (Yu *et al.*, 2001; Weidemann *et al.*, 2002; Zhao *et al.*, 2004). As suggested by its name, the non-amyloidogenic pathway prevents Aβ formation by cleaving APP at the α-site located within the Aβ sequence (after Lys612) (Roberts *et al.,* 1994; Lammich *et al.,* 1999; Allinson *et al.,* 2003). This event results in the extracellular shedding of sAPPα and the generation of the 83 amino acid C-terminal fragment (C83) (Weidemann *et al.,* 1989) which can be further processed by ε-, ζ- and γ-secretases in a similar way as for C99, giving rise to the analogous AICD and a small p3 (Haass *et al.,* 1993) fragment instead of Aβ. Despite numerous investigations, the proteolyic processing of APP by α-secretase is still not fully elucidated. A zinc metalloproteinase cleaving APP at the Lys612-Leu613 bond has been identified as an α-secretase (Roberts *et al.,* 1994). Several candidates have been proposed to carry out this task, all belonging to the ADAM (Disintegrin And Metalloproteinase) family, the most plausible being ADAM9, ADAM10 and ADAM17 (Allinson *et al.,* 2003). β-secretase is an aspartyl protease that cleaves the bond between Met597 and Asp598 and is encoded by the BACE (BACE1) gene (Hussain *et al.,* 1999; Sinha *et al.,* 1999; Vassar *et al.,* 1999; Yan *et al.,* 1999). The γ-secretase is an aspartyl protease that cleaves APP within its transmembrane domain and exhibits low sequence specificity. Four essential subunits form the γ-secretase multiprotein complex: APH-1 (anterior pharyx defective 1), nicastrin, PS1 (presenilin-1) or PS2 (presenilin-2), and PEN-2 (presenilin enhancer-2) (Yu *et al.,* 2000; Francis *et al.,* 2002; Goutte *et al.,* 2002; Edbauer *et al.,* 2003). For an extensive review on APP processing and secretases involvement in this process see (Chow *et al.,* 2009; Xu 2009).

The precise function of APP hasn't been identified yet, but the involvement of the full length protein and its proteolytic fragments in different activities such as cell adhesion, neurite outgrowth, synaptogenesis, neuronal survival, apoptosis, modulation of synaptoplasticity and axonal transport has been demonstrated (reviewed in (Jacobsen and Iverfeldt, 2009)). APP can be subdivided in several smaller domains with their own structure and specific activity (see Fig. 6.1).

6.1.2 *APP Domains*

6.1.2.1 *The growth factor domain (GFD) - APP_{28-123}*

The N-terminus of APP starts with a signal peptide sequence which is removed during translocation into the ER by signal peptidase. This is

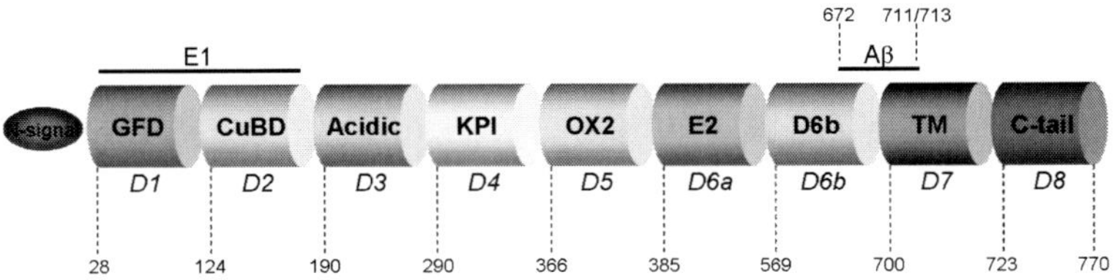

Figure 6.1 Cartoon representing the arrangement of APP domains with the respective names and amino acids numbers. In italics is the alternative APP domains nomenclature. From N-terminus to C-terminus is the Growth Factor Domain (GFD), the Copper Binding Domain (CuBD), the Acidic domain (Acidic), the Kunitz-type Protease Inhibitor Domain, the OX2 Domain (OX2), the E2 Domain (E2), the D6a Domain (D6a), the Transmembrane Domain (TM) and the Cytoplasmic Domain (C-tail) (APP_{695}). See also Colour Insert.

followed by a cysteine-rich region containing a growth factor domain (GFD, also called D1), encompassed by residues 28–123. The GFD binds heparin and can promote neuronal development (Small *et al.*, 1994; Ohsawa *et al.*, 1997; Rossjohn *et al.*, 1999). The presence of heparan sulphate proteoglycan (HSPG) is indispensible for neuronal growth to occur and APP residues Lys-99, Arg-100 and Arg-102 are necessary for heparin binding (Small *et al.*, 1994). X-ray analysis of the GFD structure revealed it shared structural characteristics with other cysteine-rich growth factors such as epidermal growth factor, tumor necrosis factor, nerve growth factor, midkine, hepatocytes growth factor and vascular endothelial growth factor, supporting the hypothesis that APP N-terminal may function as a growth factor *in vivo* (Rossjohn *et al.*, 1999).

6.1.2.2 *The copper binding domain (CuBD) – $APP_{124–189}$*

The GFD is followed by a copper-binding domain (CuBD) (Hesse *et al.*, 1994; Multhaup *et al.*, 1996; Barnham *et al.*, 2003) encompassed by residues 133–189 (Barnham *et al.*, 2003) whose characteristics will be discussed in detail later on in this chapter. GFD and CuBD are also named as E1 domain which form a folding unit as recently analyzed by Dahms and colleagues, 2010 (Dahms *et al.*, 2010).

6.1.2.3 *The acidic domain – $APP_{190–289}$*

Next to the CuBD is an acidic region (D3) rich in glutamic and aspartic acid residues (~ 45% of the total residues). This domain is released when APP undergoes processing by caspases, leading to programmed cell death in rat glial cultured cells. It has been proposed that this apoptotic activity is mediated via inhibition of cell adhesion and induced expression of Fas and iNOS, genes playing a central role in apoptosis (Sun *et al.*, 2004).

6.1.2.4 *The Kunitz-type protease inhibitor (KPI) domain – $APP_{290–365}$*

Following D3 is the Kunitz-type protease inhibitor (KPI or D4) domain and the OX2 domain (D5) (Kitaguchi *et al.*, 1988; Ponte *et al.*, 1988; Tanaka *et al.*, 1988; Tanzi *et al.*, 1988). These two domains can be alternatively spliced to give rise to the three major APP isoforms: APP_{770}, APP_{751} and APP_{695}, the latter two being the most abundant forms (Palmert *et al.*, 1988; Sandbrink *et al.*, 1994). A possible role in influencing cell adhesion has been proposed for the KPI domain (Hynes *et al.*, 1990; Gillian *et al.*, 1997), which is present in both APP_{770} and APP_{751} isoforms. Other investigations identified a different neuroprotective mechanism between KPI positive (APP_{770} and APP_{751}) and KPI negative (APP_{695}) forms, according to the neuronal stress type the cell is experiencing (Willoughby *et al.*, 1992; Mucke *et al.*, 1994; Rosa *et al.*, 2005). The link between APP isoform expression and AD remains to be clarified since some studies observed an increase in KPI containing isoform in patients compared to healthy controls (Moir *et al.*, 1998) while others couldn't prove any differential expression of the three isoforms between these two groups (Panegyres *et al.*, 2000).

6.1.2.5 *The OX2 domain – $APP_{366–384}$*

The 19 residue OX2 domain, present only in the APP_{770} isoform, is 43% homologous to the MRC OX-2 antigen (Kang and Muller-Hill 1989; Weidemann *et al.*, 1989), a surface glycoprotein expressed in rat that belongs to the immunoglobulin superfamily (McCaughan *et al.*, 1987). However, the putative APP OX-2 domain shares only 8 identical amino acids with the MRC OX-2 antigen and lacks an important residue for the immunoglobulin fold (Cys21). Moreover, the homology between the two proteins is decreased when sequences from the same species are compared, suggesting the function of this domain is still speculative (Richards *et al.*, 1995).

6.1.2.6 *The E2 domain – $APP_{385-568}$*

Linked to the acidic domain (in APP_{695}), KPI (in APP_{751}) or OX2 domain (in APP_{770}) is a 275 amino acid carbohydrate region consisting of the highly helical E2 domain (also known as D6a or CAPPD) followed by an unstructured domain (D6b). The E2 domain bears two distinct coiled-coil substructures connected through a continuous helix and is found as an antiparallel dimer in solution. This peculiar dimerization could suggest a role for membrane bound APP in cell-cell adhesion via interaction of APP on two neighbouring cells (Wang and Ha 2004; Soba *et al.,* 2005). The E2 region bears a second APP heparin binding site (the first is located in the GFD domain as explained above) containing a RERMS motif thought to be important for APP cell growth promotion and differentiation (Narindrasorasak *et al.,* 1991; Ninomiya *et al.,* 1993; Jin *et al.,* 1994; Clarris *et al.,* 1997). However, crystallography studies indicated the RERMS was not located in a convenient location for the proposed role, since the sequence is part of an α-helical structure and resides at the APP dimeric interface. Therefore only a major conformational change could fully expose the RERMS motif for interaction (Wang and Ha, 2004). In contrast to the APP-N-terminal analogous GFD, the E2 trophic activity of this sequence doesn't appear to be mediated through heparin binding (Ninomiya *et al.,* 1994). E2 has also been shown to interact with the CR-cluster domain of sorLA, a protein involved in modulating APP transport in the Golgi. APP binding to sorLA results in sequestration of APP in the Golgi and in decreased Aβ generation (Andersen *et al.,* 2005; Andersen *et al.,* 2006).

6.1.2.7 *The D6b domain – $APP_{569-670}$*

No structure or activity has been attributed to this domain (Kong *et al.,* 2008) and it may serve as a linker region between the transmembrane domain and the highly structured E2/D6a domain. This region contains APP exon 15 which can be spliced off to give rise to an alternative APP isoform, APP_{677}, also called L-APP. This isoform was first discovered in peripheral mononuclear leukocytes and activated microglial cells (Konig *et al.,* 1992). L-APP mRNA constitutes between 25 and 72% of total APP transcripts in peripheral tissues and lower than 10% in the central nervous system. The splicing out of exon 15 fuses exon 14 with exon 16 and generates a new post-translational modification motif ENEGSG where xylosyltransferase can attach a chondroitin sulphate glycosaminoglycan moiety (Pangalos *et al.*, 1995a; Pangalos *et al.*, 1995b). The modified L-APP is denoted as appican and its function remains unclear. However it may be involved in wound healing, cells adhesion, neurite outgrowth and axon guidance (Brittis *et al.*, 1992; Margolis and Margolis 1993; Okamoto *et al.*, 1994). This alternative splicing decreased the generation of Aβ relative to p3 (Hartmann *et al.*, 1996), an effect probably due to exon 15 vicinity to Aβ cleavages site.

6.1.2.8 *The juxtamembrane and transmembrane domains – $APP_{671-723}$*

Following the D6b domain is the juxtamembrane area and the transmembrane domain: these two portions contain the Aβ, a 40 or 42 amino acids peptide (most common forms) released during APP amyloidogenic processing and plays an important role in AD due to its neuronal and synaptic toxicity. However, in contrast with its toxic characteristics, Aβ has also proven to have neurotophic activity at sub nanomolar concentrations (Yankner *et al.*, 1990).

The transmembrane domain contains an amino acid motif GxxxG that has a regulatory impact on the length of Aβ species generated from APP processing (Munter *et al.*, 2007). This topic will be reviewed in detail in section 6.4.3 (APP dimerization) in this chapter.

6.1.2.9 *The cytoplasmic domain – $APP_{723-770}$*

The final domain is the APP intracellular cytoplasmic domain (AICD) which contains several regions capable of interacting with a multitude of intracellular proteins. Within the AICD domain, resides an evolutionary conserved region (from *C. elegans* to humans) that is of particular importance: the YENPTY motif. Similar sequences in the iron ion delivery protein Transferrin and the low-density lipoprotein (LDL) receptor are responsible for protein internalization (Chen *et al.*, 1990; Collawn *et al.*, 1990), suggesting the YENPTY motif may have an analogous function in APP. This hypothesis was verified via the creation of APP C-terminal tail deletion mutants which exhibited decreased Aβ secretion correlating with a reduction in cell surface APP internalization (Koo and Squazzo, 1994; LeBlanc and Gambetti

1994; Essalmani *et al.*, 1996) and mutagenesis experiments identifying the YENP sequence as the key motif for this event (Perez *et al.*, 1999).

The AICD exhibits a disordered structure in solution (Ramelot *et al.*, 2000) which builds up into two α-helices upon interaction with Fe65 (Radzimanowski *et al.*, 2008). Binding of APP to Fe65 has been shown to regulate APP transcription (von Rotz *et al.*, 2004) and Fe65 overexpression influences Aβ generation (Guenette *et al.*, 1999; Chang *et al.*, 2003).

6.2 Copper Physiology

6.2.1 *Copper Modulation with Age*

There is considerable evidence that copper (Cu) levels in the brain, plasma and CFS of different animal models and humans rise with aging (Adlard and Bush, 2006). Studies in rats showed a rapid increase in copper levels in 3 to 14 day old brains with no significant change in animals older than 3 weeks (Tarohda *et al.*, 2004). In the same animal model, plasma copper levels sharply rise 10 days after birth (Martinez Lista *et al.*, 1993). Similarly, a 46% increase in brain copper content (whole brain, excluding olfactory bulb, cerebellum and brain stem) was observed in 3 to 18 month old mice (Maynard *et al.*, 2002).

A study on human males aged 8–89 years revealed a positive correlation between plasma copper concentration and aging which was emphasized by an acute increase in subjects above 75 years of age (Madaric *et al.*, 1994). Aging related elevation of copper is well supported by several others studies (McMaster *et al.*, 1992; Iskra *et al.*, 1993; Menditto *et al.*, 1993; Milne and Johnson, 1993). The same metal content alteration could be reflected in the brain since an age-related increase in ceruloplasmin (Connor *et al.*, 1993) and an elevated presence of copper clusters has been observed in this organ (Wender *et al.*, 1992). It is possible that this increase in metal concentration is region-specific since it has been demonstrated that copper content varies independently across the different brain areas (Rajan *et al.*, 1997; Dobrowolska *et al.*, 2008). It's important to emphasize that the cause of brain metal alterations are probably due to disrupted homeostasis rather than increased metal exposure since the brain is well isolated and protected by the blood brain barrier (BBB) which strictly regulates ion uptake (Bush and Tanzi, 2008).

6.2.2 *Copper Modulation with Alzheimer's Disease*

Several findings demonstrate that besides aging, copper levels also correlate with AD. Copper content was raised in the cerebro-spinal fluid (CSF) (Basun *et al.*, 1991) and serum (Squitti *et al.*, 2002) of AD patients compared to healthy controls. These data are supported by an interesting study of a pair of elderly monozygotic female twins discordant for AD where the twin with greater cognitive impairment diagnosed for AD was found to bear a 44% elevation in copper serum (Squitti *et al.*, 2004). However, some other studies challenge this hypothesis since they couldn't demonstrate any changes in the metal content in AD affected patients' CSF (Molina *et al.*, 1998; Gerhardsson *et al.*, 2008), plasma (Gerhardsson *et al.*, 2008) or serum (Molina *et al.*, 1998). Further discussion on the clinical association between copper and AD is in section 6.3.

Data on brain copper content are discordant. The discrepancies could be due to the complexity of this organ and its zone-specific variability in copper content as mentioned above. Bulk techniques such as neutron activation analysis (NAA) on multiple samples from different regions of brain (anterior frontal lobe, middle frontal lobe, motor cortex, anterior temporal lobe, middle temporal lobe, hippocampus, superior parietal lobe, and occipital lobe) (Tandon, 1994) and on hippocampus and cerebral cortex (Ward, 1987) reveal unchanged copper levels in AD patients compared to health controls. However, a decrease in copper levels in AD patients was detected using the same technique in hippocampus and amygdala (Deibel *et al.*, 1996). Conversely, a marginally significant ($p = 0.056$) increase in copper levels was observed in neuropil from AD patients, compared to healthy controls, with enrichment in metal concentration in the rim of senile plaques (Lovell *et al.*, 1998a).

6.3 Copper and Alzheimer's Disease

If the aforementioned modulation in copper distribution truly represents a hallmark in AD, the next fundamental step is to define and understand the mechanisms involved in this event in order to identify it as cause or effect. The literature on possible AD related effects due to copper levels and distribution modulation is broad and mainly comprises the involvement of the metal in initiation and propagation of inflammatory response (Campbell, 2006) and generation of reactive oxygen species (Adlard and Bush, 2006).

6.3.1 *Inflammatory Events*

It is known that in AD patients there is an upregulation of brain inflammatory response compared to healthy controls, characterized by increased microglia activation and cytokine secretion (McGeer and McGeer, 2003). Findings showing decreased incidence of AD in rheumatoid arthritis patients treated with anti-inflammatory drugs support the possibility that these inflammatory events are present in early stages of the illness and promote its advancement (McGeer, 1990). The way neuroinflammation may contribute to cognitive impairment is unclear, but it could potentially happen through hindered neurogenesis and altered synaptic plasticity, events triggered by the secreted bacterial toxin LPS (lipopolysaccharide) and well known to provoke strong immune response in animals (Monje *et al.*, 2003; Hellstrom *et al.*, 2005). Since studies demonstrated that copper elicits secretion and activation of several cytokines, expression of genes involved in immune response and production of blood neutrophil granulocytes (Kennedy *et al.*, 1998; Schmalz *et al.*, 1998; Rice *et al.*, 2001; Suska *et al.*, 2003; Suska *et al.*, 2005), it may be reasonable to attribute it a role in AD brain inflammatory events upregulation and consequent neurodamaging.

6.3.2 *Oxidative Stress*

Free copper ions can contribute to the formation of reactive oxygen species through the Haber–Weiss reaction (Pourahmad and O'Brien, 2000; Gaetke and Chow 2003; Letelier *et al.*, 2005; Moriwaki *et al.*, 2008). In this process, cupric ions are reduced by superoxide, or another reducing agent, leading to cuprous species, which in turn are able to catalyze the generation of a hydroxyl radical from a hydrogen peroxide molecule. The produced radical has a highly powerful oxidative potential and the ability to react with a multitude of biological molecules resulting in irreversible modifications. Evidence of increased oxidative stress in AD patients' brain is supported by the presence of typical hallmarks such as lipid peroxidation and protein and DNA oxidation.

Metals are the main sources of physiological ROS, therefore copper, with its strong redox ability and its relatively abundant presence in the human body, may constitute a threat to the health of the organism if not tightly regulated. Copper defective homeostasis, could lead to oxidative stress and consequent lipid, protein and DNA damage, potentially resulting in typical AD brain injuries as neuronal and synaptic loss.

6.3.2.1 *Lipid peroxidation*

Brain membrane phospholipids are composed of polyunsaturated fatty acids (PUFA), especially arachidonic and docosalhexaenoic acids (AA and DA respectively), which are very susceptible to peroxidation since extraction of hydrogen atoms from side-chain methylene carbons is facilitated by the presence of double bonds. Lipid peroxidation is characterized by decreased levels of lipids with concomitant elevated presence of their oxidation products as aldehydes and isoprostanes. Decreased levels of AA, DA and other easy oxidable lipids were observed in AD in several studies (Nitsch *et al.*, 1992; Prasad *et al.*, 1998).

PUFA peroxidation leads to the formation of neurotoxic aldehydes as acrolein (2-propenal) and HNE (4-hydroxy-2-transnonenal) that can react with other proteins and modify their structure and function.

Elevated brain levels of protein-bound acrolein were detected in AD patients (Lovell *et al.*, 2001) and alterations in membrane proteins structure were established in gerbil synaptosomes (Pocernich *et al.* 2001), in particular protein carbonylation (Mello *et al.*, 2007). These findings are complementary to the neurotoxicity observed in rat hippocampal neurons upon acrolein incubation (Lovell *et al.*, 2001). The presence of acrolein-modified protein was also detected in AD NFTs (neurofibrillary tangles) (Calingasan *et al.*, 1999) and tau phosphorylation was shown to be raised in human neuroblastoma cells and in primary cultures of mouse embryo cortical neurons exposed to acrolein (Gomez-Ramos *et al.*, 2003).

HNE levels are elevated in AD brain and CSF compared to healthy controls (Sayre *et al,*. 1997; Markesbery and Lovell 1998; Volkel *et al.*, 2006) and these aldehydes were shown to be toxic to rat hippocampal and cortical primary neurons (Mark *et al.*, 1997; Long *et al.*, 2008), especially in striatal synaptosomes (Lopachin *et al.*, 2009). Incubation of HNE with gerbil synaptosomal membrane led to changes in protein conformation and bilayer fluidity (Subramaniam *et al.*, 1997), events that may be involved in the aforementioned toxicity. Finally, HNE brain injections cause a decrease in cholineacteyltransferase activity, a process that is remarkably inhibited in AD patients (Katzman and Saitoh, 1991).

Other altered lipid peroxidation markers in AD are raised levels of isoprostanes (products of non-enzymatic oxidation of AA and DA) and decreased activity of gluthatione S-transferase (enzyme involved in HNE clearing) (Lovell *et al.*, 1998b; Roberts *et al.*, 1998) but their mechanisms are less well understood.

6.3.2.2 *Protein oxidation*

ROS can oxidize proteins through reactions with amino acids side-chains which generate carbonyl groups. Protein carbonylation can occur through direct attack from ROS or from the interaction with one of the products resulting from lipid peroxidation or glycation, such as acrolein and HNE (Berlett and Stadtman, 1997; Butterfield, 1997; Dean *et al.*, 1997).

Specific oxidation of some proteins correlates to AD. High levels of oxidized creatine kinase (CK), β-actin and glutamine synthetase were detected in AD patient's brain (Butterfield, 1997; Aksenov *et al.*, 2000; Aksenov *et al.*, 2001). CK is an enzyme involved in the energetics of Ca^{2+} homeostasis and its oxidation and subsequent decreased activity observed in patients (Hensley *et al.*, 1995; Aksenov *et al.*, 2000) may result in the accumulation or depletion of Ca^{2+} with associated toxicity. Glutamine synthetase, whose activity is known to be reduced in AD brains (Hensley *et al.*, 1995), is an enzyme responsible for the conversion of glutamate to glutamine and its impaired function may lead to glutamate accumulation and consequents toxic effects as NMDA receptors over activation and excess neuronal calcium. Finally, activity levels of the important cytoskeletal structural molecule β-actin were found to be decreased in AD brains (Bajo *et al.*, 2001).

6.3.2.3 *DNA oxidation*

There is increasing evidence that oxidative DNA stress may contribute to neuronal death in AD. ROS can attack DNA causing strand breaks, DNA-DNA and DNA-protein cross-linking, base release and mutations. The most copious oxidized base product seems to be 8-hydroxy-guanine (OH8dG) which represents the ideal biomarker for DNA oxidative stress as it can be sensitively detected in urine, serum and tissue. Two separate studies showed a threefold (Mecocci *et al.,* 1994) and a tenfold (Wang *et al.,* 2005) increase (respectively) in OH8dG levels in AD brain in comparison to healthy controls, a strong increment that was related to mitochondrial DNA since nuclear DNA exhibited only a slight (but significant) increase in this oxidation product. Even if less sensitive, other oxidized bases (such as 8-hydroxy-adenine, 5-hydroxy-cytosine, 5-hydroxy-uracil, fapyadenine and fapyguanine) can be used as biomarker and have been shown to be raised in AD brain (Gabbita *et al.,* 1998; Wang *et al.,* 2005). Lovell and colleagues (Lovell *et al.,* 1999) made a step further in the analysis of DNA damage biomarkers and demonstrated that OH8dG levels are raised also in AD CSF. Furthermore, they found a reduction in free OH8dG levels, a product of DNA repairing, suggesting that DNA restoration may be hampered in disease.

There are several other studies demonstrating that DNA repair mechanisms are altered in AD patients compared to healthy controls. de la Monte and colleagues (de la Monte *et al.,* 2000) found elevated DNA nicking and fragmentation in AD temporal lobe and Lovell and colleagues (Lovell *et al.,* 2000) showed lowered activity of 8-oxoguanineglycosylase, an enzyme responsible for the removal of oxidized bases, concomitant with a raised helicase activity, an enzyme also responsible of DNA repair, in several brain regions (except in inferior parietal lobule). Poly (ADP-ribose) polymerase, which aids to fix DNA strands, exhibited elevated activity in frontal and temporal cortex of AD (Love *et al.,* 1999). The decrease in DNA repair activity may be interpreted as a cause of oxidative stress, conversely, elevated activity levels may correspond to an up regulation of the enzymes. All these findings taken together indicate significant DNA damage in AD patients due to oxidative stress which together with deficient DNA repairing machinery could play an important role in development and progression of the illness.

6.3.3 *Gene expression*

Beyond inflammatory events and oxidative stress, another event appears to link AD and Cu: modulated APP expression and processing. Copper has been shown to influence expression of several genes (van De Sluis *et al.*, 2002; Klomp *et al.*, 2003; van Bakel *et al.*, 2005; Auclair *et al.*, 2006; Muller *et al.*, 2007) and a variety of genes are known to be altered in the AD brain compared to healthy controls (Ginsberg *et al.*, 2000; Loring *et al.*, 2001; Colangelo *et al.*, 2002), however the only gene whose expression is changed under either conditions (Cu modulation and AD) seems to be the APP gene.

6.4 APP, Copper and Alzheimer's Disease

6.4.1 *APP Expression*

Several studies report APP gene deregulations in Alzheimer's disease, contributing to the theory that APP is a key protein in AD. The data on APP gene regulation in AD show some discordancies but most studies support altered expression in patients. In brain, mRNA APP is either increased (Cohen *et al.*, 1988; Vitek, 1989) or decreased (Preece *et al.*, 2004) and some studies show concomitant increment and diminution of APP expression according to the specific cell type (Higgins *et al.*, 1988; Clark and Parhad, 1989) or correlating to the levels of illness markers (i.e., NTS and NP) (Lewis *et al.*, 1988). In peripheral mononuclear blood cell (PMBC) preparations, which include lymphocytes, monocytes and macrophages APP expression was increased in AD (Jiang *et al.*, 2003) but a more specific analysis of lymphocytes alone didn't detect any difference in APP expression levels (Ebstein *et al.*, 1996).

Altered APP expression has been observed under copper modulation in different systems by several groups. Bellingham and colleagues demonstrated that copper depleted human fibroblasts (overexpressing the Menkes protein) exhibit reduced APP protein and mRNA levels suggesting a down-regulation in gene expression (Bellingham *et al.*, 2004b). To support these findings they were able to identify an APP copper-regulatory region by using promoter deletion constructs. The positive correlation between Cu levels and APP expression was demonstrated by other groups. A study on 25 healthy postmenopausal women adopting a diet with different copper and zinc (Zn) contents for 190 days demonstrated decreased APP levels in platelets upon low Cu intake accompanied with high Zn supplementation (Davis *et al.*, 2000). However, the small sample size ($n = 4$) for this condition (low Cu and high Zn diet) limits the significance of these results which needs further studies to be confirmed. Conversely, Armendariz and colleagues observed increased APP expression under chronic Cu overload in mutant fibroblasts accumulating excess metal (Armendariz *et al.*, 2004).

All these results taken together could appear incongruous, but it is necessary to take into account the variability between studies (Davis *et al.*, 2000; Armendariz *et al.*, 2004; Bellingham *et al.*, 2004b). For example, the type of cells used, the treatment conditions and other factors such as the APP isoform which is examined, may influence APP expression levels. This is highlighted by Ebstein *et al.* who demonstrated that total mRNA levels remained unchanged but a closer examination revealed a lower APP_{751}/APP_{770} ratio in AD patients compared to controls (Ebstein *et al.*, 1996).

6.4.2 *APP Processing*

Aβ accumulation in hippocampus and cerebral cortex of AD patients represents a primary hallmark of the illness, suggesting an overproduction of this neurotoxic peptide or/and its impaired clearance. There are several studies demonstrating increased β-secretase enzymatic activity in AD brain (Fukumoto *et al.*, 2002; Tyler *et al.*, 2002; Yang *et al.*, 2003; Li *et al.*, 2004). Impaired α-secretase has also been observed (Tyler *et al.*, 2002) and may contribute to increased Aβ generation since α- and the β-secretases appears to compete for their common substrate APP.

Cu modulation influences the processing of the APP protein. There seems to be agreement in the literature regarding a negative correlation between copper levels and the AD amyloidogenic processing of APP in AD. Elevated extracellular copper concentrations in CHO cells resulted in an increase in sAPPα and p3 levels, products released from the α-secretory pathway, with a concomitant decrease in β-secretory pathway-generated Aβ and p3.5, the latter being the product of an alternative β-secretase activity (Simons *et al.*, 1996). At the same time cell-associated full-length APP appeared to increase.

The authors concluded the existence of two different copper dependent regulating mechanisms, one acting on Aβ production and the other on APP synthesis (Borchardt *et al.*, 1999). This study was confirmed using another cell model (human fibroblasts) which showed that conversely to Cu enrichment, Cu depletion stimulates the amyloidogenic pathway. Increased Aβ was observed in neuroblastoma cells (N2a) but differently from human fibroblasts, where this effect was explained as a consequence of altered APP cleavage, it was attributed to an inhibition in Aβ degradation (Cater *et al.*, 2008). Another study employing Cu^{II}(gtsm) (glyoxalbis(*N*(4)-methylthiosemicarbazonato), a Cu(II)btsc complex susceptible to intracellular reduction and subsequent release of Cu, showed that treatment of CHO cells overexpressing APP with this compound resulted in elevated intracellular copper levels accompanied by a reduction in Aβ levels. Aβ depletion in the culture media was associated with an increased breakdown due to an up-regulation of metalloproteases (Donnelly *et al.*, 2008).

In vivo studies are consistent with the Cu-Aβ relationship observed *in vitro*. Phinney *et al.* raised Cu levels in TgCRND8 AD transgenic mice (mice model expressing APP with a combined Swedish and Indiana familial mutation and exhibiting Aβ burden) by crossing it with toxic-milk (tx^J) mice carrying a mutated Cu-ATPase7b transporter. They detected a reduction in Aβ plaques and Aβ in CNS and plasma (Phinney *et al.*, 2003). Analogously to this latter study, a copper enriched diet also lowered CNS Aβ levels and Aβ plaques in the APP23 transgenic mice AD model, through increased brain Cu (Bayer *et al.*, 2003).

6.4.3 *APP Dimerization*

Cu can influence APP dimerization. The existence of APP dimers is supported by increasing evidence from structural and biochemical data on full length APP and APP domains (Beher *et al.*, 1996; Rossjohn *et al.*, 1999; Scheuermann *et al.*, 2001; Wang and Ha, 2004; Soba *et al.*, 2005; Gralle *et al.*, 2006; Munter *et al.*, 2007; Kaden *et al.*, 2008; Eggert *et al.*, 2009; Gralle *et al.*, 2009; Kaden *et al.*, 2009). One of the first studies testing the effect of Cu on APP showed that binding of the metal inhibited APP binding to a so called homophilic peptide encompassing residues 448–478 (Hesse *et al.*, 1994). The masking of the APP homophilic binding site was hypothesized to be the cause of the hampered binding and since a protein conformational change was less likely to happen this effect was attributed to Cu-induced APP aggregation indicating possible dimerization (Hesse *et al.*, 1994). Growing evidence supports the idea that APP dimerization represents an important event in AD, particularly by affecting the amyloidogenic pathway. If so, another potential link can be drawn between Cu and AD.

There is evidence for APP dimerization playing an important role in the generation of the toxic Aβ peptide. Munter *et al.*, (Munter *et al.*, 2007) illustrated the existence of dimeric APP in cells and generated stabilized APP dimers in SY5Y cells by expressing a mutant APP (L613C). An increased Aβ42 production was detected in this system and attributed to the mutationally induced dimerization of cellular APP. Effects of APP dimerization on processing were also analyzed by another group, but results on Aβ production were discordant (Eggert *et al.*, 2009). The controlled APP dimerization was achieved through generation of an APP-chimeric molecule resulting in decreased Aβ. This effect was attributed to a modulation of γ-secretase cleavage due to APP dimerization (Eggert *et al.*, 2009).

The APP transmembrane sequence (TMS) has been shown to dimerize via the GxxxG motif with point mutations in this area resulting in weakened dimerization strength, reduced generation of Aβ42, increased Aβ38 and shorter species but unchanged Aβ40 levels in cells. These findings indicate a crucial importance for the strength of interaction between two APP molecules (mediated by the TMS residues G29 and G33, in the Aβ sequence) for the γ-secretase cleavages (Munter *et al.*, 2007). Finally, the APP hairpin loop encompassing residues 91 to 111 has been shown to be a ligand for APP and reduce Aβ production by disrupting APP dimers (Kaden *et al.*, 2008).

The molecular association between APP and its mammalian homologs was also explored. In systematically addressing this issue, live cell imaging of transiently transfected HEK293 cells showed that APLP1 mainly localizes to the cell surface, whereas APP and APLP2 are mostly found in intracellular compartments (Kaden *et al.*, 2009). Homo- and heterotypic cis interactions of APP family members could be detected by FRET and co-immunoprecipitation analysis and occur in a modular mode. Only APLP1 formed trans interactions in the HEK293 cells, supporting the argument for a putative specific role of APLP1 in cell adhesion. Deletion mutants of APP family members revealed two highly conserved regions as important for the protein crosstalk. In particular, the N-terminal half of the ectodomain was crucial for APP and APLP2 interactions. By contrast, multimerization of APLP1 was only partially dependent on this domain but strongly on the C-terminal half of the ectodomain, e.g., the E2 domain The coexpression of APP with APLP1 or APLP2 leads to diminished generation of Aβ42 (Kaden *et al.,* 2009). Despite some discrepancies, these data taken together suggest an important role for APP dimerization in the generation of the toxic Aβ peptide.

Further studies are required to elucidate this effect and to understand if and how AD risk factors such as high cholesterol levels, lipid homeostasis, oxidative stress and metal unbalances may affect APP dimer stability.

6.4.4 *APP and Copper Homeostasis*

A substantial amount of literature suggests copper dishomeostasis as a cause of AD, since it triggers toxic effects such as inflammatory events, oxidative stress, protein expression and metabolism modulation and protein aggregation. There is also the possibility that altered metal levels are actually the result (and not the cause) of metabolic changes such as APP up/down-regulation (White *et al.*, 1999b; Maynard *et al.*, 2002; Phinney *et al.*, 2003; Bellingham *et al.*, 2004a; Suazo *et al.*, 2009). The most likely cause appears to be APP expression since several studies demonstrated its involvement in Cu homeostasis regulation. A specific increase in Cu levels was observed in cerebral cortex and liver from $APP^{-/-}$ mice suggesting a role for this protein in Cu homeostasis (White *et al.*, 1999b). Cell culture and *in vivo* studies are concordant since $APP^{-/-}$ mice primary cortical neurons showed higher Cu accumulation under enriched metal conditions compared to their WT counterparts but only when APLP2 (APP paralogue) was concomitantly knocked-out. In fact, Cu levels remained similar in $APP^{-/-}$/$APLP2^{+/+}$ compared to $APP^{+/+}$/$APLP2^{+/+}$ (WT) neurons indicating an exchangeable role in Cu homeostasis for APP and its paralogue. Moreover, $APP^{+/-}$/$APLP2^{+/+}$ presented a less remarkable metal accumulation compared to $APP^{-/-}$/

APLP2$^{+/+}$ suggesting a gene-dosage effect of APP on Cu regulation (Bellingham *et al.*, 2004a). Similar results were obtained in mice embryonic fibroblasts with the exception that in this model the APP single-allele expression appeared to be sufficient to maintain normal Cu cellular levels since no difference could be detected between APP$^{+/-}$/APLP2$^{+/+}$ and APP$^{+/+}$/APLP2$^{+/+}$ (Bellingham *et al.*, 2004a).

These findings indicate an inverse correlation between Cu levels and APP expression in brain, however a recent study demonstrated that Cu accumulates in HEK293 (Human Embryonic Kidney) cells overexpressing APP (Suazo *et al.*, 2009). This discrepancy raises speculation of a different APP function dependent on the cell type in which it is expressed, the protein may have a role in Cu efflux in certain cells (e.g., neuronal cells and connective tissue cells) whereas in other tissues, like the kidney, it may serve as a chaperone, storage or even have a role in copper uptake and promote metal cellular accumulation.

The fact that the increase in metal levels was detected in the cerebral cortex human brain, an area commonly affected in AD, and not in other brain regions (i.e., cerebellum) supports APP involvement in AD related Cu imbalances. Conversely, in Tg2576 mice, an AD mouse model which overexpresses APP containing the Swedish mutation (K595N/M596L), copper levels are significantly reduced in brain (Maynard *et al.*, 2002). Decreased copper levels were also observed in TgCRND8 mice exhibiting Aβ burden (Phinney *et al.*, 2003). These two studies don't provide a direct relationship between APP and Cu, however increased amounts of Aβ, which are related to APP expression and processing, appear to play a crucial role in Cu homeostasis. These observations suggested that restoring brain Cu homeostasis might have a beneficial influence on the progression of AD biomarkers. Therefore, in a clinical trial potential beneficial effects of oral intake of Cu-(II)-orotate-dihydrate (8 mg Cu daily) in AD patients were investigated. The efficacy of oral Cu supplementation in the treatment of AD in a prospective, randomized, double-blind, placebo-controlled phase 2 clinical trial in patients with mild AD for 12 months was evaluated. Sixty-eight subjects were randomized. Patients with mild AD received either Cu-(II)-orotate-dihydrate (verum group; 8 mg Cu daily) or placebo (placebo group). CSF was collected at beginning and at the end of the study after 12 months.

The treatment was well-tolerated. There were however no significant differences in primary outcome measures (Alzheimer's Disease Assessment Scale, Cognitive subscale, Mini Mental Status Examination) between the verum [Cu-(II)-orotate-dihydrate; 8 mg Cu daily] and the placebo group. The primary outcome measures in CSF were Aβ42, Tau and Phospho-Tau. The clinical trial demonstrated that long-term oral intake of 8 mg Cu can be excluded as a risk factor for AD based on CSF biomarker analysis. Cu intake had no effect on the progression of Tau and Phospho-Tau levels in CSF. While Aβ 42 levels declined by 30% in the placebo group ($p = 0.001$), they decreased only by 10% ($p = 0.04$) in the verum group. Since decreased CSF Aβ42 is a diagnostic marker for AD, this observation indicated that Cu treatment had a positive effect on a relevant AD biomarker. Using Mini-Mental State Examination (MMSE) and ADAS-Cog (Alzheimer Disease Assessment Scale-cognitive subscale) it was previously demonstrated that there are no Cu treatment effects on cognitive performance. Finally, CSF Aβ42 levels declined significantly in both groups within 12 months supporting the notion that CSF Aβ42 may be valid not only for diagnostic but also for prognostic purposes in AD (Kessler *et al.*, 2008b).

Short-term high Cu intake has been reported not to affect Cu status or functions related to Cu status, only long-term high Cu intake can result in increases in some parameters in young men (Turnlund *et al.*, 2004). The plasma Cu levels declined only in the placebo group during the 12 months-period however the placebo group had higher Cu levels at the beginning of the study by accident. Cu levels in the verum group were unchanged, which seems to be paradoxical. Significantly lower levels of plasma Cu has been measured in AD patients who fulfilled the criteria of CSF diagnosis for AD (Kessler *et al.*, 2006). Moreover, reduced plasma Cu levels were found in patients with higher ADAScog scores (making more mistakes in this neuropsychological test) (Pajonk *et al.*, 2005). One may speculate that Cu treatment normalized Cu levels in plasma in the verum group by enhanced uptake and transport and improved tissue homeostasis. However, Cu treatment had no beneficial effect on cognitive abilities tested by MMSE and ADAScog in AD patients of the present clinical phase II pilot study (Kessler *et al.*, 2008a).

In conclusion, despite a number of findings supporting the hypothesis of environmental Cu modulating AD, the results demonstrated that oral Cu intake has neither a detrimental nor a promoting effect on the progression of AD.

6.5 APP Copper Binding Domain (CuBD)

We can conclude that, even if not fully defined, a functional relationship exists between APP and Cu. As reviewed above, the literature on this topic is extensive, covering several aspects of this interaction. Narrowing down the focus to the APP domain responsible for these events, the CuBD, may help to shed some light on the mechanisms involved. Structural elucidations recently obtained may explain the activities of this domain at a molecular level and its putative functions.

6.5.1 *Cu Binding to APP CuBD*

Hesse and colleagues were the first to discover that APP is able to bind copper (dissociation constant = 10nM). They localised a type II copper binding domain to the APP area encompassing residues 135–155 of APP695 and identified His147, His149 and His151 as the putative amino acids responsible for copper binding. Conservation of one of the two His-Xxx-His motifs was enough to maintain copper binding properties (Hesse *et al.,* 1994). The ability of APP CuBD to bind copper is now supported from affinity and structural studies (Multhaup *et al.,* 1996; White *et al.* 2002; Barnham *et al.,* 2003; Kong *et al.,* 2007a; Kong *et al.,* 2007b).

6.5.2 *CuBD Structure*

The three dimensional structure of the CuBD domain of APP (residues 124–189 expressed in yeast) was solved by NMR spectroscopy (Barnham *et al.,* 2003). The tertiary structure is composed by an α-helix (147–159) packed against three β-sheets (133–139, 162–167, 181–188) and stabilized by three disulfide bridges (C133–C187, C144–C174 and C158–C186) and a small hydrophobic core (Leu-136, Trp-150, Val-153, Ala-154, Leu-165, Met-170, Val-182, and Val-185). Strand β3 is connected to strand β1 and to the helix via C133–C187 and C144–C174 bonds, respectively, C158–C186 bond links together two loop located at the end of the molecule.

His147, His151, Tyr168 and Met170 are arranged in a tetrahedral coordination sphere and were identified as the probable ligands responsible for copper binding. Broadening of resonance signals of the four residues upon Cu^{2+} addition confirmed their vicinity to the metal, as expected for nuclei close to paramagnetic centres such as Cu^{2+}. Interestingly, after 48h hours Cu^{2+} incubation, Met170 exhibited a new resonance indicating a residue modification. The observed changes in chemical shifts suggested the presence of an oxidized sulphur atom supporting the hypothesis of Met170 being oxidized during Cu^{2+} reduction.

EPR analysis was used to analyse the coordination sphere around the metal and resulted in a distorted square planar conformation of the four putative ligands. Since cupric ions prefer square planar coordination and cuprous ions a tetrahedral configuration, the hybrid conformation observed for the APP CuBD appears to allow the binding of the oxidised metal but may consequently favour its reduction to Cu^{1+} in order to obtain a better fit into the binding site.

These putative ligands and their coordination constitute a novel binding domain that nevertheless evokes the one belonging to blue copper proteins which exhibit the same geometry with similar amino acids as ligands (2His, 1Met and 1 Cys). The tertiary structure is reminiscent of copper chaperones such as the Menkes copper-transporting ATPase fragment, the metallochaperone Atx1 and copper chaperone for SOD1 (CCS) which all display an α-helix packed next to a triple strand β-sheet topology. Consistent with the putative chaperone function, the binding site was found to be located at the surface of the structure, its placement allowing metal sequestration on binding of the chaperone to its target. These data, together with findings which demonstrate a role of APP in neuronal homeostasis, suggest a function for APP as a neuronal metallotransporter (Barnham *et al.,* 2003).

Kong and colleagues determined the structure of apo-, Cu^{2+}-bound- and Cu^{+}-bound-CuBD through X-ray crystallography (Kong *et al.,* 2007b). The apo form data were obtained from three crystal forms owing different crystal lattices but all displaying identical Cu-binding sites. The X-ray structures tightly superimposed with those derived from NMR studies and confirmed an α-helix packed against a three-strand β-sheet topology. The main differences between the crystal and the NMR structures, in terms of orientation, were observed for the His147 and for some loops involved in crystal contacts which exhibited a slightly difference position.

The Cu^{2+}-bound structure revealed His147, His151 and Tyr168 as ligands for copper binding together with two water molecules, one equatorially located slightly above the amino acids plane and the other in an axial position. The coordination sphere recalled the one classically observed for five-coordinated cupric ions where two or three histidine nitrogen atoms and oxygen atoms act as ligands in a square pyramidal geometry.

EPR and EXFAS analysis further confirmed this geometry (named model A). Moreover, two other potential, however less relevant, models (models B and C) were also generated through the latter technique. Model B displayed a six-coordinated structure similar to model A but contained an extra axial ligand modelled as an oxygen atom. Model C proposed a five-coordinated sphere where the equatorial water molecule had been replaced by a sulphur atom in model A.

The Cu^{+}-bound structure appeared to be very similar to the cupric ion model except for the lack of the axial ligand (water molecule). Interestingly, the square pyramidal geometry was maintained despite four-coordinated Cu^{+} favouring tetrahedral conformations. The authors suggested that this unstable interaction may allow prompt transfer of the metal to putative partners.

Despite His147, His 151 and Tyr 168 being identified as residues responsible for Cu^{1+} binding, the crystallography data contradicted the involvement of Met170 (suggested by NMR data) in this event due to its distance from the bound metal. Based on the Cu-titration NMR findings, that demonstrated Met170 proximity to the paramagnetic centre (Cu^{2+}) and its consequential oxidation, Met170 was proposed to act as an electron donor either via its possible ability to come in contact with the metal ion through conformational changes or by exploiting neighbouring ligands for electron transfer.

Bringing these findings to a physiologically relevant level, the water ligands were hypothesized to represent a docking site and to be displaced during binding of the putative partner. Initially, potential ligands were proposed to be an APP binding partners (e.g., copper transport proteins), another CuBD ligand or one of the APP adjacent domains. However, APP was subsequently excluded as a putative partner since no oligomerization was observed upon Cu binding in this study. A Cu-triggered interaction may be the cause of altered APP processing previously observed (Borchardt *et al.,* 1999), since any evident conformational change in CuBD structure could not be detected in these experiments (Kong *et al.,* 2007a; Kong *et al.,* 2007b).

In summary, APP His147, His151 and Tyr168 appear to be the putative binding ligands for Cu since all of them have been confirmed by NMR and crystallography investigations. The role of Met170 remains unclear but crystallography data exclude its involvement in Cu binding due to its physical emplacement.

6.5.3 *Copper Reduction*

The binding of Cu doesn't appear to constitute the sole and principal function of the CuBD but rather a stage that allows a second (in chronological terms) activity, the reduction of the cupric ion. Earlier studies, employing a synthetic peptide encompassing only a little portion of the CuBD domain and including a single cysteine residue Cys144 (Multhaup *et al.,* 1996; Ruiz *et al.,* 1999) demonstrated that CuBD can reduce Cu^{2+} to Cu(I) and proposed a role for Cys144. However, in the more physiologically relevant intact CuBD (APP 133–189) there are no free cysteines and Cys144 forms a disulfide bridge with C174 therefore its oxidation is less likely. It may be that the absence of Met170, combined with the presence of a non-oxidized Cys144, favours the behaviour of this latter amino acid as an electron donor during the reduction of copper, which would explain its previously observed oxidation.

The main source of electrons, indispensable for Cu reduction, appears to come from an oxidable residue in the CuBD itself (its absence hampering the redox activity). Structural studies using the entire CuBD (APP 124–189) expressed in yeast (no free cysteines) suggest Met170 as the amino acid for this role (Barnham *et al.,* 2003; Kong *et al.,* 2007b). However, in its absence, another oxidable residue in the vicinity of the binding site, as Cys144 when not involved in disulfide bridging, carries out this task.

The considerable distance of Met170 from the binding site may argue against this hypothesis, but a conformational change or the employment of its neighbouring residues for electron transfer, as proposed by Kong and colleagues (Kong *et al.,* 2007a; Kong *et al.,* 2007b), may help to overcome this hindrance. However, the oxidation of Met170 *in vivo* appears less likely since this event would alter the characteristics of the binding site. The source of electrons for Cu reduction could instead come from an APP binding partner interacting with the binding site via displacement of the water molecules as suggested by Kong *et al.* (Kong *et al.,* 2007b). In this case the presence of a readily renewable pool of reducing species would be necessary to ensure the turnover and the supply of reducing agents for every cupric ion. The lack of putative binding partners and/or reducing agents from the environment used during previous investigations may explain the employment of other available oxidable entities such as Met170 or Cys144.

6.5.4 *Copper Homeostasis*

Cu binding and its reduction alone may not constitute the whole activity of CuBD but rather being part of a chain of steps leading to the ultimate function, Cu homeostasis. Three features make the CuBD an eligible candidate for this role, its structural homology to other Cu chaperones, the binding site surface location and the unfavourable ligand amino acids conformation around the reduced cuprous ion (Barnham *et al.,* 2003; Kong *et al.,* 2007b). A role for the CuBD as a Cu chaperone is supported by the surface location of the binding domain and its babb topology which is a common architecture shared amongst other Cu chaperones such as Atx1 (yeast metallochaperone), HAH1 (human metallochaperone), CCS (copper chaperone for SOD1) and the Menkes copper-transporting ATPase fragment. In fact, since Cu exposure to the surrounding environment could lead to undesired reactions (e.g., Fenton reaction), a binding site surface location is unusual, except for chaperones, where the transfer of the metal is facilitated by an easily accessible location. Finally, according to crystallography data, once Cu reduction occurs, the ion is coordinated by four ligands in a distorted square planar conformation, an unfavourable geometry for cuprous ions which are known to prefer a tetrahedral arrangement. This unstable state could help the extrusion of the metal and its prompt transfer to a putative binding partner (Barnham *et al.* 2003; Kong *et al.,* 2007b).

The majority of the literature (White *et al.,* 1999b; Maynard *et al.,* 2002; Bellingham *et al.,* 2004a) indicates the participation of

APP in cellular Cu export since its presence or absence appears to decrease or raise Cu levels, respectively. White and colleague used atomic absorption spectroscopy to measure the copper content of several neuronal (cerebral cortex and cerebellum) and non-neuronal (liver, spleen and serum) tissues of WT and APP knockout mice and demonstrated increased metal levels in cerebral cortex and liver of animals deprived of APP expression (White *et al.,* 1999b). Primary cortical neurons and embryonic fibroblasts cultures obtained from similar animal models were employed by Bellingham and colleague to analyze cellular copper content. Again, the lack of APP (together with the lack of APLP2) resulted in an accumulation of cellular copper in both cells cultures (Bellingham *et al.,* 2004a). Conversely, the overexpression of APP containing the Swedish mutation in Tg2576 mice caused a decrease in brain copper levels (Maynard *et al.,* 2002). Hypothetically, APP could bind Cu inside the cell and export it to the plasma membrane, by exposure of its ectodomain in the extracellular environment. The cleavage of the N-terminal domain (where the CuBD resides) could finalize this process by removing the metal from the surrounding cellular environment. As a type-1 transmembrane protein, APP never exposes its CuBD (N-terminal located) into the cytosolic environment limiting its ability to bind copper only to the lumen of the various organelles it travels through (ER, Golgi, secretory vesicles). Therefore this hypothetical model implies that the proteins from where APP obtains the metal from (almost all cellular Cu is complexed (Rae *et al.,* 1999)) are lumenal proteins, or transmembrane proteins with an ectodomain inside these organelles. Participation of APP in cellular Cu uptake is also plausible (Cerpa *et al.,* 2004; Suazo *et al.,* 2009).

Human embryonic kidney cell cultures (HEK293) overexpressing APP exhibited increased cellular copper accumulation compared to the corresponding non-transfected cells. The effect of synthetic peptides encompassing the CuBD (APP 135–155) was also tested. The exogenous peptides proved to have a similar influence on copper content as the endogenously overexpressed APP. Mutation of Cys144 abolished this effect (Suazo *et al.,* 2009). Cerpa and colleague analysed copper content of hippocampus region isolated from rats to which $^{64}CuSO_4$ and CuBD synthetic peptides (APP 135–155) had been injected (in this brain region) 14 hours beforehand. $^{64}CuSO_4$ co-injection with the WT CuBD peptide resulted in metal accumulation compared to when $^{64}CuSO_4$ was injected alone. Again Cys144 mutation reduced this effect to half (Cerpa *et al.*, 2004). In this case, plasma membrane APP may bind extracellular cupric ions, reduce them to cuprous species and import them inside the cell through endocytosis or via metal transfer to Ctr1 and Ctr2 integral membrane proteins responsible for Cu^{1+} uptake from the extracellular fluid into the cytoplasm. A role for APP as a Cu importer is supported by the observation that mutations in the CuBD (Cys144) result in decreased reducing activity plus lowered Cu uptake through the plasma membrane into the cell (Cerpa *et al.*, 2004; Suazo *et al.*, 2009). In fact, mutation of this crucially involved residue in the structural stability of the CuBD is likely to cause unfolding of the domain and consequent loss of function.

Studies in animals have reported that normalized or elevated Cu levels can inhibit or even remove Alzheimer's disease-related pathological plaques and exert a desirable amyloid-modifying effect. Thus, engineered nanocarriers composed of diverse core-shell architectures were tested to modulate Cu levels under physiological conditions through bypassing the cellular Cu uptake systems (Treiber *et al.*, 2009). Two different nanocarrier systems were able to transport Cu across the plasma membrane of yeast or higher eukaryotic cells, CS-NPs (core-shell nanoparticles) and CMS-NPs (core-multishell nanoparticles). Intracellular Cu levels could be increased up to 3-fold above normal with a sublethal dose of carriers. Both types of carriers released their bound guest molecules into the cytosolic compartment where they were accessible for the Cu-dependent enzyme SOD1. In particular, CS-NPs reduced Aβ levels and targeted intracellular organelles more efficiently than CMS-NPs. Fluorescently labeled CMS-NPs unraveled a cellular uptake mechanism, which depended on clathrin-mediated endocytosis in an energy-dependent manner. In contrast, the transport of CS-NPs was most likely driven by a concentration gradient. Overall, nanocarriers depending on the nature of the surrounding shell functioned by mediating import of Cu across cellular membranes, increased levels of bioavailable Cu, and affected Aβ turnover. Taken together, this illustrate that Cu-charged nanocarriers can achieve a reasonable metal ion specificity and represent an alternative to metal-complexing agents. Conclusively, carrier strategies have potential for the treatment of metal ion deficiency disorders.

In the case of raised intracellular Cu levels, a contribution of APP to cellular Cu accumulation, rather than Cu import, appears less probable. Its interaction with the metal seems to be more transient rather than a prolonged event due to the immediate Cu reduction and the consequent unfavourable coordination geometry of the binding ligands around the metal.

6.5.5 *APP-Mediated Cu Toxicity*

Two schools of thought appear to arise regarding the role of APP in Cu-toxicity. One hypothesis is that APP functions as a mediator of Cu-toxicity, demonstrated by studies showing increased Cu–APP binding and reduction and elevated protein and lipid oxidation following ROS generation (White *et al.*, 1999a; White *et al.*, 2002). APP WT ($APP^{+/+}$) and APP knock out ($APP^{-/-}$) primary neurons were used to assess the effect of endogenous APP on Cu-toxicity. $APP^{+/+}$ neurons were more susceptible to Cu toxicity than $APP^{-/-}$ neurons but no difference in Cu uptake were detected (probably due to the presence of APLP2). The exacerbated toxicity correlated with increased levels of lipid peroxidation.

Despite less physiologically relevant, investigations employing synthetic peptides encompassing a portion of the CuBD (rather than the entire domain) generated interesting results. Exogenous synthetic peptide encompassing APP 142–166 was able to mimic the detrimental consequences of endogenous APP, elevating the toxic Cu effect in $APP^{-/-}$ neurons. This effect was abrogated when His147, His149 and His151 were substituted in this peptide (White *et al.*, 1999a). Moreover neurons incubation with full length APP or sAPPα resulted (in both cases) in enhanced lipid peroxidation and Cu toxicity promotion (White *et al.*, 2002).

The testing of synthetic peptides encompassing APP CuBD paralogs and orthologs generated some interesting data on the evolution of APP behaviour regarding Cu mediated toxicity. The human APP, *Xenopus* APP and human APLP2 CuBDs promoted toxicity, on the other hand APL-1 (*C.elegans*) inhibited it (White *et al.*, 2002). The key point behind the potentiation of Cu toxicity appears to be the conservation of histidines at positions 147 and 151 (human APP numbering) which, upon mutation to Tyr and Lys respectively, result in protection against Cu toxicity. Substitution of these two residues with other amino acids resulted in an inert phenotype regarding Cu toxicity. The toxic and protective CuBD phenotypes were associated with differences in Cu binding and reduction (White *et al.*, 2002).

The other school of thought attributes to APP a protective effect against Cu-toxicity. An increase in Cu reduction has been observed in these studies and ROS generation has been accepted as a consequence of this event (Cerpa *et al.*, 2004; Suazo *et al.*, 2009). Therefore other mechanisms are thought to intervene to overcome the oxidative stress in order to bring beneficial outcomes. In previous studies, sAPP is shown to have neuroprotective (Mattson *et al.*, 1993; Smith-Swintosky *et al.*, 1994; Furukawa *et al.*, 1996; Morimoto *et al.*, 1998) and neurotrophic (Milward *et al.*, 1992; Small *et al.*, 1994; Ohsawa *et al.*, 1997; Rossjohn *et al.*, 1999) characteristics, and induction ofmetallothionein proteins (metal chelator and antioxidant function) has been observed with the concomitant rescue of APP Cu toxicity (Suazo *et al.*, 2009). This difference in promoting or decreasing Cu-toxicity may be related to the cellular Cu uptake which was observed to be increased when APP exhibited protective features (Cerpa *et al.*, 2004; Suazo *et al.*, 2009) but remained unchanged when toxicity was promoted (White *et al.*, 1999a), suggesting a possible detrimental accumulation of cuprous ions in the extracellular environment.

6.6 Conclusions

The literature strongly suggests that copper plays a role in AD. Moreover, the metabolism of APP can be affected by copper and APP, in turn, can modulate copper homeostasis. Therefore, the interaction between copper and APP could directly influence the molecular pathways in AD. A better understanding of the function of the APP CuBD will provide important insights into the pathobiology of AD. Despite APP possesses the characteristics of a copper chaperone, the regulation of this activity on copper homeostasis *in vivo* remains unclear. The same applies to the role of APP in Cu-mediated toxicity, with findings supporting a role for APP in either promoting Cu-toxicity or in protecting against Cu-toxicity. APP modulation of Cu homeostasis and Cu-mediated toxicity represent complex events possibly involving several other molecules. One such molecular target is the proteoglycan molecule glypican-1. Both APP:Cu or APLP2:Cu complexes can regulate the metabolism of the heparin sulfate sidechains in glypican-1 in vitro and in vivo (Cappai *et al.*, 2005) this indicates glypican-1 is a molecular target for APP:Cu and APLP2:Cu. Therefore, future research should focus on elucidating the APP:copper interaction in vivo and to clarify the consequences of these interactions on cellular function, protein metabolism, copper homeostasis and AD pathogenesis.

References

Adlard, P. A. and Bush, A. I. (2006). Metals and Alzheimer's disease. *J Alzheimers Dis*, 10, pp. 145–63.

Aksenov, M., Aksenova, M., Butterfield, D. A. and Markesbery, W. R. (2000). Oxidative modification of creatine kinase BB in Alzheimer's disease brain. *J Neurochem*, 74, pp. 2520–7.

Aksenov, M. Y., Aksenova, M. V., Butterfield, D. A., Geddes, J. W. and Markesbery, W. R. (2001). Protein oxidation in the brain in Alzheimer's disease. *Neuroscience*, 103, pp. 373–83.

Allinson, T. M., Parkin, E. T., Turner, A. J. and Hooper, N. M. (2003). ADAMs family members as amyloid precursor protein alpha-secretases. *J Neurosci Res*, 74, pp. 342–52.

Andersen, O. M., Reiche, J., Schmidt, V., Gotthardt, M., Spoelgen, R., Behlke, J., von Arnim, C. A., Breiderhoff, T., Jansen, P., Wu, X., Bales, K. R., Cappai, R., Masters, C. L., Gliemann, J., Mufson, E. J., Hyman, B. T., Paul, S. M., Nykjaer, A. and Willnow, T. E. (2005). Neuronal sorting protein-related receptor sorLA/LR11 regulates processing of the amyloid precursor protein. *Proc Natl Acad Sci USA*, 102, pp. 13461–6.

Andersen, O. M., Schmidt, V., Spoelgen, R., Gliemann, J., Behlke, J., Galatis, D., McKinstry, W. J., Parker, M. W., Masters, C. L., Hyman, B. T., Cappai, R. and Willnow, T. E. (2006). Molecular dissection of the interaction between amyloid precursor protein and its neuronal trafficking receptor SorLA/LR11. *Biochemistry*, 45, pp. 2618–28.

Anderson, J. P., Chen, Y., Kim, K. S. and Robakis, N. K. (1992). An alternative secretase cleavage produces soluble Alzheimer amyloid precursor protein containing a potentially amyloidogenic sequence. *J Neurochem*, 59, pp. 2328–31.

Armendariz, A. D., Gonzalez, M., Loguinov, A. V. and Vulpe, C. D. (2004). Gene expression profiling in chronic copper overload reveals upregulation of Prnp and App. *Physiol Genomics*, 20, pp. 45–54.

Auclair, S., Feillet-Coudray, C., Coudray, C., Schneider, S., Muckenthaler, M. U. and Mazur, A. (2006). Mild copper deficiency alters gene expression of proteins involved in iron metabolism. *Blood Cells Mol Dis*, 36, pp. 15–20.

Bajo, M., Yoo, B. C., Cairns, N., Gratzer, M. and Lubec, G. (2001). Neurofilament proteins NF-L, NF-M and NF-H in brain of patients with Down syndrome and Alzheimer's disease. *Amino Acids*, 21, pp. 293–301.

Barnham, K. J., McKinstry, W. J., Multhaup, G., Galatis, D., Morton, C. J., Curtain, C. C., Williamson, N. A., White, A. R., Hinds, M. G., Norton, R. S., Beyreuther, K., Masters, C. L., Parker, M. W. and Cappai, R. (2003). Structure of the Alzheimer's disease amyloid precursor protein copper binding domain. A regulator of neuronal copper homeostasis. *J Biol Chem*, 278, pp. 17401–7.

Basun, H., Forssell, L. G., Wetterberg, L. and Winblad, B. (1991). Metals and trace elements in plasma and cerebrospinal fluid in normal aging and Alzheimer's disease. *J Neural Transm Park Dis Dement Sect*, 3, pp. 231–58.

Bayer, T. A., Schafer, S., Simons, A., Kemmling, A., Kamer, T., Tepest, R., Eckert, A., Schussel, K., Eikenberg, O., Sturchler-Pierrat, C., Abramowski, D., Staufenbiel, M. and Multhaup, G. (2003). Dietary Cu stabilizes brain superoxide dismutase 1 activity and reduces amyloid Abeta production in APP23 transgenic mice. *Proc Natl Acad Sci USA*, 100, pp. 14187–92.

Beher, D., Hesse, L., Masters, C. L. and Multhaup, G. (1996). Regulation of amyloid protein precursor (APP) binding to collagen and mapping of the binding sites on APP and collagen type I. *J Biol Chem*, 271, pp. 1613–20.

Bellingham, S. A., Ciccotosto, G. D., Needham, B. E., Fodero, L. R., White, A. R., Masters, C. L., Cappai, R. and Camakaris, J. (2004a). Gene knockout of amyloid precursor protein and amyloid precursor-like protein-2 increases cellular copper levels in primary mouse cortical neurons and embryonic fibroblasts. *J Neurochem*, 91, pp. 423–8.

Bellingham, S. A., Lahiri, D. K., Maloney, B., La Fontaine, S., Multhaup, G. and Camakaris, J. (2004b). Copper depletion down-regulates expression of the Alzheimer's disease amyloid-beta precursor protein gene. *J Biol Chem*, 279, pp. 20378–86.

Berlett, B. S. and Stadtman, E. R. (1997). Protein oxidation in aging, disease, and oxidative stress. *J Biol Chem*, 272, pp. 20313–6.

Borchardt, T., Camakaris, J., Cappai, R., Masters, C. L., Beyreuther, K. and Multhaup, G. (1999). Copper inhibits beta-amyloid production and stimulates the non-amyloidogenic pathway of amyloid-precursor-protein secretion. *Biochem J*, 344 Pt 2, pp. 461–7.

Brittis, P. A., Canning, D. R. and Silver, J. (1992). Chondroitin sulfate as a regulator of neuronal patterning in the retina. *Science*, 255, pp. 733–6.

Busciglio, J., Gabuzda, D. H., Matsudaira, P. and Yankner, B. A. (1993). Generation of beta-amyloid in the secretory pathway in neuronal and nonneuronal cells. *Proc Natl Acad Sci USA*, 90, pp. 2092–6.

Bush, A. I. and Tanzi, R. E. (2008). Therapeutics for Alzheimer's disease based on the metal hypothesis. *Neurotherapeutics*, 5, pp. 421–32.

Butterfield, D. A. (1997). Protein oxidation processes in aging brain. *Cell Aging Gerontol*, 2, pp. 161–191.

Calingasan, N. Y., Uchida, K. and Gibson, G. E. (1999). Protein-bound acrolein: a novel marker of oxidative stress in Alzheimer's disease. *J Neurochem*, 72, pp. 751–6.

Campbell, A. (2006). The role of aluminum and copper on neuroinflammation and Alzheimer's disease. *J Alzheimers Dis*, 10, pp. 165–72.

Cappai, R., Cheng, F., Ciccotosto, G. D., Needham, B. E., Masters, C. L., Multhaup, G., Fransson, L. A. and Mani, K. (2005). The amyloid precursor protein (APP) of Alzheimer disease and its paralog, APLP2, modulate the Cu/Zn-Nitric Oxide-catalyzed degradation of glypican-1 heparan sulfate in vivo. *J Biol Chem*, 280, pp. 13913–20.

Cater, M. A., McInnes, K. T., Li, Q. X., Volitakis, I., La Fontaine, S., Mercer, J. F. and Bush, A. I. (2008). Intracellular copper deficiency increases amyloid-beta secretion by diverse mechanisms. *Biochem J*, 412, pp. 141–52.

Cerpa, W. F., Barria, M. I., Chacon, M. A., Suazo, M., Gonzalez, M., Opazo, C., Bush, A. I. and Inestrosa, N. C. (2004). The N-terminal copper-binding domain of the amyloid precursor protein protects against Cu^{2+} neurotoxicity in vivo. *FASEB J*, 18, pp. 1701–3.

Chang, Y., Tesco, G., Jeong, W. J., Lindsley, L., Eckman, E. A., Eckman, C. B., Tanzi, R. E. and Guenette, S. Y. (2003). Generation of the beta-amyloid peptide and the amyloid precursor protein C-terminal fragment gamma are potentiated by FE65L1. *J Biol Chem*, 278, pp. 51100–7.

Chen, W. J., Goldstein, J. L. and Brown, M. S. (1990). NPXY, a sequence often found in cytoplasmic tails, is required for coated pit-mediated internalization of the low density lipoprotein receptor. *J Biol Chem*, 265, pp. 3116–23.

Chow, V. W., Mattson, M. P., Wong, P. C. and Gleichmann, M. (2009). An Overview of APP Processing Enzymes and Products. *Neuromolecular Med*, pp. 1–12.

Clark, A. W. and Parhad, I. M. (1989). Expression of neuronal mRNAs in Alzheimer type degeneration of the nervous system. *Can J Neurol Sci*, 16, pp. 477–82.

Clarris, H. J., Cappai, R., Heffernan, D., Beyreuther, K., Masters, C. L. and Small, D. H. (1997). Identification of heparin-binding domains in the amyloid precursor protein of Alzheimer's disease by deletion mutagenesis and peptide mapping. *J Neurochem*, 68, pp. 1164–72.

Cohen, M. L., Golde, T. E., Usiak, M. F., Younkin, L. H. and Younkin, S. G. (1988). In situ hybridization of nucleus basalis neurons shows increased beta-amyloid mRNA in Alzheimer disease. *Proc Natl Acad Sci USA*, 85, pp. 1227–31.

Colangelo, V., Schurr, J., Ball, M. J., Pelaez, R. P., Bazan, N. G. and Lukiw, W. J. (2002). Gene expression profiling of 12633 genes in Alzheimer hippocampal CA1: transcription and neurotrophic factor down-regulation and up-regulation of apoptotic and pro-inflammatory signaling. *J Neurosci Res*, 70, pp. 462–73.

Collawn, J. F., Stangel, M., Kuhn, L. A., Esekogwu, V., Jing, S. Q., Trowbridge, I. S. and Tainer, J. A. (1990). Transferrin receptor internalization sequence YXRF implicates a tight turn as the structural recognition motif for endocytosis. *Cell*, 63, pp. 1061–72.

Connor, J. R., Tucker, P., Johnson, M. and Snyder, B. (1993). Ceruloplasmin levels in the human superior temporal gyrus in aging and Alzheimer's disease. *Neurosci Lett*, 159, pp. 88–90.

Dahms, S. O., Hoefgen, S., Roeser, D., Schlott, B., Guhrs, K. H. and Than, M. E. Structure and biochemical analysis of the heparin-induced E1 dimer of the amyloid precursor protein. *Proc Natl Acad Sci USA*, pp. 108(6), pp. 2629.

Daigle, I. and Li, C. (1993). apl-1, a Caenorhabditis elegans gene encoding a protein related to the human beta-amyloid protein precursor. *Proc Natl Acad Sci USA*, 90, pp. 12045–9.

Davis, C. D., Milne, D. B. and Nielsen, F. H. (2000). Changes in dietary zinc and copper affect zinc-status indicators of postmenopausal women, notably, extracellular superoxide dismutase and amyloid precursor proteins. *Am J Clin Nutr*, 71, pp. 781–8.

de la Monte, S. M., Luong, T., Neely, T. R., Robinson, D. and Wands, J. R. (2000). Mitochondrial DNA damage as a mechanism of cell loss in Alzheimer's disease. *Lab Invest*, 80, pp. 1323–35.

Dean, R. T., Fu, S., Stocker, R. and Davies, M. J. (1997). Biochemistry and pathology of radical-mediated protein oxidation. *Biochem J*, 324 (Pt 1), pp. 1–18.

Deibel, M. A., Ehmann, W. D. and Markesbery, W. R. (1996). Copper, iron, and zinc imbalances in severely degenerated brain regions in Alzheimer's disease: possible relation to oxidative stress. *J Neurol Sci*, 143, pp. 137–42.

Dobrowolska, J., Dehnhardt, M., Matusch, A., Zoriy, M., Palomero-Gallagher, N., Koscielniak, P., Zilles, K. and Becker, J. S. (2008). Quantitative imaging of zinc, copper and lead in three distinct regions of the human brain by laser ablation inductively coupled plasma mass spectrometry. *Talanta*, 74, pp. 717–23.

Donnelly, P. S., Caragounis, A., Du, T., Laughton, K. M., Volitakis, I., Cherny, R. A., Sharples, R. A., Hill, A. F., Li, Q. X., Masters, C. L., Barnham, K. J. and White, A. R. (2008). Selective intracellular release of copper and zinc ions from bis(thiosemicarbazonato) complexes reduces levels of Alzheimer disease amyloid-beta peptide. *J Biol Chem*, 283, pp. 4568–77.

Ebstein, R. P., Nemanov, L., Lubarski, G., Dano, M., Trevis, T. and Korczyn, A. D. (1996). Changes in expression of lymphocyte amyloid precursor protein mRNA isoforms in normal aging and Alzheimer's disease. *Brain Res Mol Brain Res*, 35, pp. 260–8.

Edbauer, D., Winkler, E., Regula, J. T., Pesold, B., Steiner, H. and Haass, C. (2003). Reconstitution of gamma-secretase activity. *Nat Cell Biol*, 5, pp. 486–8.

Eggert, S., Midthune, B., Cottrell, B. and Koo, E. H. (2009). Induced dimerization of the amyloid precursor protein (APP) leads to decreased amyloid-beta protein (Abeta) production. *J Biol Chem*, pp. 28943–28952.

Essalmani, R., Macq, A. F., Mercken, L. and Octave, J. N. (1996). Missense mutations associated with familial Alzheimer's disease in Sweden lead to the production of the amyloid peptide without internalization of its precursor. *Biochem Biophys Res Commun*, 218, pp. 89–96.

Francis, R., McGrath, G., Zhang, J., Ruddy, D. A., Sym, M., Apfeld, J., Nicoll, M., Maxwell, M., Hai, B., Ellis, M. C., Parks, A. L., Xu, W., Li, J., Gurney, M., Myers, R. L., Himes, C. S., Hiebsch, R., Ruble, C., Nye, J. S. and Curtis, D. (2002). aph-1 and pen-2 are required for Notch pathway signaling, gamma-secretase cleavage of betaAPP, and presenilin protein accumulation. *Dev Cell*, 3, pp. 85–97.

Fukumoto, H., Cheung, B. S., Hyman, B. T. and Irizarry, M. C. (2002). Beta-secretase protein and activity are increased in the neocortex in Alzheimer disease. *Arch Neurol*, 59, pp. 1381–9.

Furukawa, K., Sopher, B. L., Rydel, R. E., Begley, J. G., Pham, D. G., Martin, G. M., Fox, M. and Mattson, M. P. (1996). Increased activity-regulating and neuroprotective efficacy of alpha-secretase-derived secreted amyloid precursor protein conferred by a C-terminal heparin-binding domain. *J Neurochem*, 67, pp. 1882–96.

Gabbita, S. P., Lovell, M. A. and Markesbery, W. R. (1998). Increased nuclear DNA oxidation in the brain in Alzheimer's disease. *J Neurochem*, 71, pp. 2034–40.

Gabuzda, D., Busciglio, J., Chen, L. B., Matsudaira, P. and Yankner, B. A. (1994). Inhibition of energy metabolism alters the processing of amyloid precursor protein and induces a potentially amyloidogenic derivative. *J Biol Chem*, 269, pp. 13623–8.

Gaetke, L. M. and Chow, C. K. (2003). Copper toxicity, oxidative stress, and antioxidant nutrients. *Toxicology*, 189, pp. 147–63.

Gerhardsson, L., Lundh, T., Minthon, L. and Londos, E. (2008). Metal concentrations in plasma and cerebrospinal fluid in patients with Alzheimer's disease. *Dement Geriatr Cogn Disord*, 25, pp. 508–15.

Gillian, A. M., McFarlane, I., Lucy, F. M., Overly, C., McConlogue, L. and Breen, K. C. (1997). Individual isoforms of the amyloid beta precursor protein demonstrate differential adhesive potentials to constituents of the extracellular matrix. *J Neurosci Res*, 49, pp. 154–60.

Ginsberg, S. D., Hemby, S. E., Lee, V. M., Eberwine, J. H. and Trojanowski, J. Q. (2000). Expression profile of transcripts in Alzheimer's disease tangle-bearing CA1 neurons. *Ann Neurol*, 48, pp. 77–87.

Gomez-Ramos, A., Diaz-Nido, J., Smith, M. A., Perry, G. and Avila, J. (2003). Effect of the lipid peroxidation product acrolein on tau phosphorylation in neural cells. *J Neurosci Res*, 71, pp. 863–70.

Goutte, C., Tsunozaki, M., Hale, V. A. and Priess, J. R. (2002). APH-1 is a multipass membrane protein essential for the Notch signaling pathway in Caenorhabditis elegans embryos. *Proc Natl Acad Sci USA*, 99, pp. 775–9.

Gralle, M., Botelho, M. G. and Wouters, F. S. (2009). Neuroprotective secreted amyloid precursor protein acts by disrupting amyloid precursor protein dimers. *J Biol Chem*, pp. 15016–15025.

Gralle, M., Oliveira, C. L., Guerreiro, L. H., McKinstry, W. J., Galatis, D., Masters, C. L., Cappai, R., Parker, M. W., Ramos, C. H., Torriani, I. and Ferreira, S. T. (2006). Solution conformation and heparin-induced dimerization of the full-length extracellular domain of the human amyloid precursor protein. *J Mol Biol*, 357, pp. 493–508.

Guenette, S. Y., Chen, J., Ferland, A., Haass, C., Capell, A. and Tanzi, R. E. (1999). hFE65L influences amyloid precursor protein maturation and secretion. *J Neurochem*, 73, pp. 985–93.

Haass, C., Hung, A. Y., Schlossmacher, M. G., Teplow, D. B. and Selkoe, D. J. (1993). beta-Amyloid peptide and a 3-kDa fragment are derived by distinct cellular mechanisms. *J Biol Chem*, 268, pp. 3021–4.

Haass, C., Hung, A. Y. and Selkoe, D. J. (1991). Processing of beta-amyloid precursor protein in microglia and astrocytes favors an internal localization over constitutive secretion. *J Neurosci*, 11, pp. 3783–93.

Hartmann, T., Bergsdorf, C., Sandbrink, R., Tienari, P. J., Multhaup, G., Ida, N., Bieger, S., Dyrks, T., Weidemann, A., Masters, C. L. and Beyreuther, K. (1996). Alzheimer's disease betaA4 protein release and amyloid precursor protein sorting are regulated by alternative splicing. *J Biol Chem*, 271, pp. 13208–14.

Hellstrom, I. C., Danik, M., Luheshi, G. N. and Williams, S. (2005). Chronic LPS exposure produces changes in intrinsic membrane properties and a sustained IL-beta-dependent increase in GABAergic inhibition in hippocampal CA1 pyramidal neurons. *Hippocampus*, 15, pp. 656–64.

Hensley, K., Hall, N., Subramaniam, R., Cole, P., Harris, M., Aksenov, M., Aksenova, M., Gabbita, S. P., Wu, J. F., Carney, J. M. and *et al.* (1995). Brain regional correspondence between Alzheimer's disease histopathology and biomarkers of protein oxidation. *J Neurochem*, 65, pp. 2146–56.

Hesse, L., Beher, D., Masters, C. L. and Multhaup, G. (1994). The beta A4 amyloid precursor protein binding to copper. *FEBS Lett*, 349, pp. 109–16.

Higgins, G. A., Lewis, D. A., Bahmanyar, S., Goldgaber, D., Gajdusek, D. C., Young, W. G., Morrison, J. H. and Wilson, M. C. (1988). Differential regulation of amyloid-beta-protein mRNA expression within hippocampal neuronal subpopulations in Alzheimer disease. *Proc Natl Acad Sci USA*, 85, pp. 1297–301.

Hussain, I., Powell, D., Howlett, D. R., Tew, D. G., Meek, T. D., Chapman, C., Gloger, I. S., Murphy, K. E., Southan, C. D., Ryan, D. M., Smith, T. S., Simmons, D. L., Walsh, F. S., Dingwall, C. and Christie, G. (1999). Identification of a novel aspartic protease (Asp 2) as beta-secretase. *Mol Cell Neurosci*, 14, pp. 419–27.

Hynes, T. R., Randal, M., Kennedy, L. A., Eigenbrot, C. and Kossiakoff, A. A. (1990). X-ray crystal structure of the protease inhibitor domain of Alzheimer's amyloid beta-protein precursor. *Biochemistry*, 29, pp. 10018–22.

Iijima, K., Lee, D. S., Okutsu, J., Tomita, S., Hirashima, N., Kirino, Y. and Suzuki, T. (1998). cDNA isolation of Alzheimer's amyloid precursor protein from cholinergic nerve terminals of the electric organ of the electric ray. *Biochem J*, 330 (Pt 1), pp. 29–33.

Iskra, M., Patelski, J. and Majewski, W. (1993). Concentrations of calcium, magnesium, zinc and copper in relation to free fatty acids and cholesterol in serum of atherosclerotic men. *J Trace Elem Electrolytes Health Dis*, 7, pp. 185–8.

Jacobsen, K. T. and Iverfeldt, K. (2009). Amyloid precursor protein and its homologues: a family of proteolysis-dependent receptors. *Cell Mol Life Sci*, pp. 2299–318.

Jiang, S., Zhang, M., Ren, D., Tang, G., Lin, S., Qian, Y., Zhang, Y., Jiang, K., Li, F. and Wang, D. (2003). Enhanced production of amyloid precursor protein mRNA by peripheral mononuclear blood cell in Alzheimer's disease. *Am J Med Genet B Neuropsychiatr Genet*, 118B, pp. 99–102.

Jin, L. W., Ninomiya, H., Roch, J. M., Schubert, D., Masliah, E., Otero, D. A. and Saitoh, T. (1994). Peptides containing the RERMS sequence of amyloid beta/A4 protein precursor bind cell surface and promote neurite extension. *J Neurosci*, 14, pp. 5461–70.

Kaden, D., Munter, L. M., Joshi, M., Treiber, C., Weise, C., Bethge, T., Voigt, P., Schaefer, M., Beyermann, M., Reif, B. and Multhaup, G. (2008). Homophilic interactions of the amyloid precursor protein (APP) ectodomain are regulated by the loop region and affect beta-secretase cleavage of APP. *J Biol Chem*, 283, pp. 7271–9.

Kaden, D., Voigt, P., Munter, L. M., Bobowski, K. D., Schaefer, M. and Multhaup, G. (2009). Subcellular localization and dimerization of APLP1 are strikingly different from APP and APLP2. *J Cell Sci*, 122, pp. 368–77.

Kang, J. and Muller-Hill, B. (1989). The sequence of the two extra exons in rat preA4. *Nucleic Acids Res*, 17, pp. 2130.

Katzman, R. and Saitoh, T. (1991). Advances in Alzheimer's disease. *Faseb J*, 5, pp. 278–86.

Kennedy, T., Ghio, A. J., Reed, W., Samet, J., Zagorski, J., Quay, J., Carter, J., Dailey, L., Hoidal, J. R. and Devlin, R. B. (1998). Copper-dependent inflammation and nuclear factor-kappaB activation by particulate air pollution. *Am J Respir Cell Mol Biol*, 19, pp. 366–78.

Kessler, H., Bayer, T. A., Bach, D., Schneider-Axmann, T., Supprian, T., Herrmann, W., Haber, M., Multhaup, G., Falkai, P. and Pajonk, F. G. (2008a). Intake of copper has no effect on cognition in patients with mild Alzheimer's disease: a pilot phase 2 clinical trial. *J Neural Transm*, 115, pp. 1181–7.

Kessler, H., Pajonk, F. G., Bach, D., Schneider-Axmann, T., Falkai, P., Herrmann, W., Multhaup, G., Wiltfang, J., Schafer, S., Wirths, O. and Bayer, T. A. (2008b). Effect of copper intake on CSF parameters in patients with mild Alzheimer's disease: a pilot phase 2 clinical trial. *J Neural Transm*, 115, pp. 1651–9.

Kessler, H., Pajonk, F. G., Meisser, P., Schneider-Axmann, T., Hoffmann, K. H., Supprian, T., Herrmann, W., Obeid, R., Multhaup, G., Falkai, P. and Bayer, T. A. (2006). Cerebrospinal fluid diagnostic markers correlate with lower plasma copper and ceruloplasmin in patients with Alzheimer's disease. *J Neural Transm*, 113, pp. 1763–9.

Kitaguchi, N., Takahashi, Y., Tokushima, Y., Shiojiri, S. and Ito, H. (1988). Novel precursor of Alzheimer's disease amyloid protein shows protease inhibitory activity. *Nature*, 331, pp. 530–2.

Klomp, A. E., van de Sluis, B., Klomp, L. W. and Wijmenga, C. (2003). The ubiquitously expressed MURR1 protein is absent in canine copper toxicosis. *J Hepatol*, 39, pp. 703–9.

Kong, G. K., Adams, J. J., Cappai, R. and Parker, M. W. (2007a). Structure of Alzheimer's disease amyloid precursor protein copper-binding domain at atomic resolution. *Acta Crystallogr Sect F Struct Biol Cryst Commun*, 63, pp. 819–24.

Kong, G. K., Adams, J. J., Harris, H. H., Boas, J. F., Curtain, C. C., Galatis, D., Masters, C. L., Barnham, K. J., McKinstry, W. J., Cappai, R. and Parker, M. W. (2007b). Structural studies of the Alzheimer's amyloid precursor protein copper-binding domain reveal how it binds copper ions. *J Mol Biol*, 367, pp. 148–61.

Kong, G. K., Miles, L. A., Crespi, G. A., Morton, C. J., Ng, H. L., Barnham, K. J., McKinstry, W. J., Cappai, R. and Parker, M. W. (2008). Copper binding to the Alzheimer's disease amyloid precursor protein. *Eur Biophys J*, 37, pp. 269–79.

Konig, G., Monning, U., Czech, C., Prior, R., Banati, R., Schreiter-Gasser, U., Bauer, J., Masters, C. L. and Beyreuther, K. (1992). Identification and differential expression of a novel alternative splice isoform of the beta A4 amyloid precursor protein (APP) mRNA in leukocytes and brain microglial cells. *J Biol Chem*, 267, pp. 10804–9.

Koo, E. H. and Squazzo, S. L. (1994). Evidence that production and release of amyloid beta-protein involves the endocytic pathway. *J Biol Chem*, 269, pp. 17386–9.

Lammich, S., Kojro, E., Postina, R., Gilbert, S., Pfeiffer, R., Jasionowski, M., Haass, C. and Fahrenholz, F. (1999). Constitutive and regulated alpha-secretase cleavage of Alzheimer's amyloid precursor protein by a disintegrin metalloprotease. *Proc Natl Acad Sci USA*, 96, pp. 3922–7.

LeBlanc, A. C. and Gambetti, P. (1994). Production of Alzheimer 4kDa beta-amyloid peptide requires the C-terminal cytosolic domain of the amyloid precursor protein. *Biochem Biophys Res Commun*, 204, pp. 1371–80.

Letelier, M. E., Lepe, A. M., Faundez, M., Salazar, J., Marin, R., Aracena, P. and Speisky, H. (2005). Possible mechanisms underlying copper-induced damage in biological membranes leading to cellular toxicity. *Chem Biol Interact*, 151, pp. 71–82.

Lewis, D. A., Higgins, G. A., Young, W. G., Goldgaber, D., Gajdusek, D. C., Wilson, M. C. and Morrison, J. H. (1988). Distribution of precursor amyloid-beta-protein messenger RNA in human cerebral cortex: relationship to neurofibrillary tangles and neuritic plaques. *Proc Natl Acad Sci USA*, 85, pp. 1691–5.

Li, R., Lindholm, K., Yang, L. B., Yue, X., Citron, M., Yan, R., Beach, T., Sue, L., Sabbagh, M., Cai, H., Wong, P., Price, D. and Shen, Y. (2004). Amyloid beta peptide load is correlated with increased beta-secretase activity in sporadic Alzheimer's disease patients. *Proc Natl Acad Sci USA*, 101, pp. 3632–7.

Long, E. K., Murphy, T. C., Leiphon, L. J., Watt, J., Morrow, J. D., Milne, G. L., Howard, J. R. and Picklo, M. J., Sr. (2008). Trans-4-hydroxy-2-hexenal is a neurotoxic product of docosahexaenoic (22:6; n-3) acid oxidation. *J Neurochem*, 105, pp. 714–24.

Lopachin, R. M., Geohagen, B. C. and Gavin, T. (2009). Synaptosomal toxicity and nucleophilic targets of 4-hydroxy-2-nonenal. *Toxicol Sci*, 107, pp. 171–81.

Loring, J. F., Wen, X., Lee, J. M., Seilhamer, J. and Somogyi, R. (2001). A gene expression profile of Alzheimer's disease. *DNA Cell Biol*, 20, pp. 683–95.

Love, S., Barber, R. and Wilcock, G. K. (1999). Increased poly (ADP-ribosyl) ation of nuclear proteins in Alzheimer's disease. *Brain*, 122 (Pt 2), pp. 247–53.

Lovell, M. A., Gabbita, S. P. and Markesbery, W. R. (1999). Increased DNA oxidation and decreased levels of repair products in Alzheimer's disease ventricular CSF. *J Neurochem*, 72, pp. 771–6.

Lovell, M. A., Robertson, J. D., Teesdale, W. J., Campbell, J. L. and Markesbery, W. R. (1998a). Copper, iron and zinc in Alzheimer's disease senile plaques. *J Neurol Sci*, 158, pp. 47–52.

Lovell, M. A., Xie, C. and Markesbery, W. R. (1998b). Decreased glutathione transferase activity in brain and ventricular fluid in Alzheimer's disease. *Neurology*, 51, pp. 1562–6.

Lovell, M. A., Xie, C. and Markesbery, W. R. (2000). Decreased base excision repair and increased helicase activity in Alzheimer's disease brain. *Brain Res*, 855, pp. 116–23.

Lovell, M. A., Xie, C. and Markesbery, W. R. (2001). Acrolein is increased in Alzheimer's disease brain and is toxic to primary hippocampal cultures. *Neurobiol Aging*, 22, pp. 187–94.

Madaric, A., Ginter, E. and Kadrabova, J. (1994). Serum copper, zinc and copper/zinc ratio in males: influence of aging. *Physiol Res*, 43, pp. 107–11.

Margolis, R. K. and Margolis, R. U. (1993). Nervous tissue proteoglycans. *Experientia*, 49, pp. 429–46.

Mark, R. J., Lovell, M. A., Markesbery, W. R., Uchida, K. and Mattson, M. P. (1997). A role for 4-hydroxynonenal, an aldehydic product of lipid peroxidation, in disruption of ion homeostasis and neuronal death induced by amyloid beta-peptide. *J Neurochem*, 68, pp. 255–64.

Markesbery, W. R. and Lovell, M. A. (1998). Four-hydroxynonenal, a product of lipid peroxidation, is increased in the brain in Alzheimer's disease. *Neurobiol Aging*, 19, pp. 33–6.

Martinez Lista, E., Sole, J., Arola, L. and Mas, A. (1993). Changes in plasma copper and zinc during rat development. *Biol Neonate*, 64, pp. 47–52.

Mattson, M. P., Cheng, B., Culwell, A. R., Esch, F. S., Lieberburg, I. and Rydel, R. E. (1993). Evidence for excitoprotective and intraneuronal calcium-regulating roles for secreted forms of the beta-amyloid precursor protein. *Neuron*, 10, pp. 243–54.

Maynard, C. J., Cappai, R., Volitakis, I., Cherny, R. A., White, A. R., Beyreuther, K., Masters, C. L., Bush, A. I. and Li, Q. X. (2002). Overexpression of Alzheimer's disease amyloid-beta opposes the age-dependent elevations of brain copper and iron. *J Biol Chem*, 277, pp. 44670–6.

McCaughan, G. W., Clark, M. J. and Barclay, A. N. (1987). Characterization of the human homolog of the rat MRC OX-2 membrane glycoprotein. *Immunogenetics*, 25, pp. 329–35.

McGeer, E. G. and McGeer, P. L. (2003). Inflammatory processes in Alzheimer's disease. *Prog Neuropsychopharmacol Biol Psychiatry*, 27, pp. 741–9.

McGeer, P. L. (1990). Antiinflammatory drugs and Alzheimer's disease. *Lancet*, 42, pp. 447–449.

McGeer, P. L., Akiyama, H., Kawamata, T., Yamada, T., Walker, D. G. and Ishii, T. (1992). Immunohistochemical localization of beta-amyloid precursor protein sequences in Alzheimer and normal brain tissue by light and electron microscopy. *J Neurosci Res*, 31, pp. 428–42.

McMaster, D., McCrum, E., Patterson, C. C., Kerr, M. M., O'Reilly, D., Evans, A. E. and Love, A. H. (1992). Serum copper and zinc in random samples of the population of Northern Ireland. *Am J Clin Nutr*, 56, pp. 440–6.

Mecocci, P., MacGarvey, U. and Beal, M. F. (1994). Oxidative damage to mitochondrial DNA is increased in Alzheimer's disease. *Ann Neurol*, 36, pp. 747–51.

Mello, C. F., Sultana, R., Piroddi, M., Cai, J., Pierce, W. M., Klein, J. B. and Butterfield, D. A. (2007). Acrolein induces selective protein carbonylation in synaptosomes. *Neuroscience*, 147, pp. 674–9.

Menditto, A., Morisi, G., Alimonti, A., Caroli, S., Petrucci, F., Spagnolo, A. and Menotti, A. (1993). Association of serum copper and zinc with serum electrolytes and with selected risk factors for cardiovascular disease in men aged 55–75 years. NFR Study Group. *J Trace Elem Electrolytes Health Dis*, 7, pp. 251–3.

Milne, D. B. and Johnson, P. E. (1993). Assessment of copper status: effect of age and gender on reference ranges in healthy adults. *Clin Chem*, 39, pp. 883–7.

Milward, E. A., Papadopoulos, R., Fuller, S. J., Moir, R. D., Small, D., Beyreuther, K. and Masters, C. L. (1992). The amyloid protein precursor of Alzheimer's disease is a mediator of the effects of nerve growth factor on neurite outgrowth. *Neuron*, 9, pp. 129–37.

Moir, R. D., Lynch, T., Bush, A. I., Whyte, S., Henry, A., Portbury, S., Multhaup, G., Small, D. H., Tanzi, R. E., Beyreuther, K. and Masters, C. L. (1998). Relative increase in Alzheimer's disease of soluble forms of cerebral Abeta amyloid protein precursor containing the Kunitz protease inhibitory domain. *J Biol Chem*, 273, pp. 5013–9.

Molina, J. A., Jimenez-Jimenez, F. J., Aguilar, M. V., Meseguer, I., Mateos-Vega, C. J., Gonzalez-Munoz, M. J., de Bustos, F., Porta, J., Orti-Pareja, M., Zurdo, M., Barrios, E. and Martinez-Para, M. C. (1998). Cerebrospinal fluid levels of transition metals in patients with Alzheimer's disease. *J Neural Transm*, 105, pp. 479–88.

Monje, M. L., Toda, H. and Palmer, T. D. (2003). Inflammatory blockade restores adult hippocampal neurogenesis. *Science*, 302, pp. 1760–5.

Morimoto, T., Ohsawa, I., Takamura, C., Ishiguro, M., Nakamura, Y. and Kohsaka, S. (1998). Novel domain-specific actions of amyloid precursor protein on developing synapses. *J Neurosci*, 18, pp. 9386–93.

Moriwaki, H., Osborne, M. R. and Phillips, D. H. (2008). Effects of mixing metal ions on oxidative DNA damage mediated by a Fenton-type reduction. *Toxicol In Vitro*, 22, pp. 36–44.

Mucke, L., Masliah, E., Johnson, W. B., Ruppe, M. D., Alford, M., Rockenstein, E. M., Forss-Petter, S., Pietropaolo, M., Mallory, M. and Abraham, C. R. (1994). Synaptotrophic effects of human amyloid beta protein precursors in the cortex of transgenic mice. *Brain Res*, 666, pp. 151–67.

Muller, P., van Bakel, H., van de Sluis, B., Holstege, F., Wijmenga, C. and Klomp, L. W. (2007). Gene expression profiling of liver cells after copper overload *in vivo* and *in vitro* reveals new copper-regulated genes. *J Biol Inorg Chem*, 12, pp. 495–507.

Multhaup, G., Schlicksupp, A., Hesse, L., Beher, D., Ruppert, T., Masters, C. L. and Beyreuther, K. (1996). The amyloid precursor protein of Alzheimer's disease in the reduction of copper(II) to copper(I). *Science*, 271, pp. 1406–9.

Munter, L. M., Voigt, P., Harmeier, A., Kaden, D., Gottschalk, K. E., Weise, C., Pipkorn, R., Schaefer, M., Langosch, D. and Multhaup, G. (2007). GxxxG motifs within the amyloid precursor protein transmembrane sequence are critical for the etiology of Abeta42. *Embo J*, 26, pp. 1702–12.

Narindrasorasak, S., Lowery, D., Gonzalez-DeWhitt, P., Poorman, R. A., Greenberg, B. and Kisilevsky, R. (1991). High affinity interactions between the Alzheimer's beta-amyloid precursor proteins and the basement membrane form of heparan sulfate proteoglycan. *J Biol Chem*, 266, pp. 12878–83.

Neve, R. L., Finch, E. A. and Dawes, L. R. (1988). Expression of the Alzheimer amyloid precursor gene transcripts in the human brain. *Neuron*, 1, pp. 669–77.

Ninomiya, H., Roch, J. M., Jin, L. W. and Saitoh, T. (1994). Secreted form of amyloid beta/A4 protein precursor (APP) binds to two distinct APP binding sites on rat B103 neuron-like cells through two different domains, but only one site is involved in neuritotropic activity. *J Neurochem*, 63, pp. 495–500.

Ninomiya, H., Roch, J. M., Sundsmo, M. P., Otero, D. A. and Saitoh, T. (1993). Amino acid sequence RERMS represents the active domain of amyloid beta/A4 protein precursor that promotes fibroblast growth. *J Cell Biol*, 121, pp. 879–86.

Nitsch, R. M., Blusztajn, J. K., Pittas, A. G., Slack, B. E., Growdon, J. H. and Wurtman, R. J. (1992). Evidence for a membrane defect in Alzheimer disease brain. *Proc Natl Acad Sci USA*, 89, pp. 1671–5.

Ohsawa, I., Takamura, C. and Kohsaka, S. (1997). The amino-terminal region of amyloid precursor protein is responsible for neurite outgrowth in rat neocortical explant culture. *Biochem Biophys Res Commun*, 236, pp. 59–65.

Okado, H. and Okamoto, H. (1995). Developmental regulation of Xenopus beta-amyloid precursor protein gene expression. *Gerontology*, 41 Suppl 1, pp. 7–12.

Okamoto, M., Mori, S., Ichimura, M. and Endo, H. (1994). Chondroitin sulfate proteoglycans protect cultured rat's cortical and hippocampal neurons from delayed cell death induced by excitatory amino acids. *Neurosci Lett*, 172, pp. 51–4.

Pajonk, F. G., Kessler, H., Supprian, T., Hamzei, P., Bach, D., Schweickhardt, J., Herrmann, W., Obeid, R., Simons, A., Falkai, P., Multhaup, G. and Bayer, T. A. (2005). Cognitive decline correlates with low plasma concentrations of copper in patients with mild to moderate Alzheimer's disease. *J Alzheimers Dis*, 8, pp. 23–7.

Palmert, M. R., Golde, T. E., Cohen, M. L., Kovacs, D. M., Tanzi, R. E., Gusella, J. F., Usiak, M. F., Younkin, L. H. and Younkin, S. G. (1988). Amyloid protein precursor messenger RNAs: differential expression in Alzheimer's disease. *Science*, 241, pp. 1080–4.

Panegyres, P. K., Zafiris-Toufexis, K. and Kakulas, B. A. (2000). Amyloid precursor protein gene isoforms in Alzheimer's disease and other neurodegenerative disorders. *J Neurol Sci*, 173, pp. 81–92.

Pangalos, M. N., Efthimiopoulos, S., Shioi, J. and Robakis, N. K. (1995a). The chondroitin sulfate attachment site of appican is formed by splicing out exon 15 of the amyloid precursor gene. *J Biol Chem*, 270, pp. 10388–91.

Pangalos, M. N., Shioi, J. and Robakis, N. K. (1995b). Expression of the chondroitin sulfate proteoglycans of amyloid precursor (appican) and amyloid precursor-like protein 2. *J Neurochem*, 65, pp. 762–9.

Perez, R. G., Soriano, S., Hayes, J. D., Ostaszewski, B., Xia, W., Selkoe, D. J., Chen, X., Stokin, G. B. and Koo, E. H. (1999). Mutagenesis identifies new signals for beta-amyloid precursor protein endocytosis, turnover, and the generation of secreted fragments, including Abeta42. *J Biol Chem*, 274, pp. 18851–6.

Phinney, A. L., Drisaldi, B., Schmidt, S. D., Lugowski, S., Coronado, V., Liang, Y., Horne, P., Yang, J., Sekoulidis, J., Coomaraswamy, J., Chishti, M.

A., Cox, D. W., Mathews, P. M., Nixon, R. A., Carlson, G. A., St George-Hyslop, P. and Westaway, D. (2003). *In vivo* reduction of amyloid-beta by a mutant copper transporter. *Proc Natl Acad Sci USA*, 100, pp. 14193–8.

Pocernich, C. B., Cardin, A. L., Racine, C. L., Lauderback, C. M. and Butterfield, D. A. (2001). Glutathione elevation and its protective role in acrolein-induced protein damage in synaptosomal membranes: relevance to brain lipid peroxidation in neurodegenerative disease. *Neurochem Int*, 39, pp. 141–9.

Ponte, P., Gonzalez-DeWhitt, P., Schilling, J., Miller, J., Hsu, D., Greenberg, B., Davis, K., Wallace, W., Lieberburg, I. and Fuller, F. (1988). A new A4 amyloid mRNA contains a domain homologous to serine proteinase inhibitors. *Nature*, 331, pp. 525–7.

Pourahmad, J. and O'Brien, P. J. (2000). A comparison of hepatocyte cytotoxic mechanisms for Cu2+ and Cd2+. *Toxicology*, 143, pp. 263–73.

Prasad, M. R., Lovell, M. A., Yatin, M., Dhillon, H. and Markesbery, W. R. (1998). Regional membrane phospholipid alterations in Alzheimer's disease. *Neurochem Res*, 23, pp. 81–8.

Preece, P., Virley, D. J., Costandi, M., Coombes, R., Moss, S. J., Mudge, A. W., Jazin, E. and Cairns, N. J. (2004). Amyloid precursor protein mRNA levels in Alzheimer's disease brain. *Brain Res Mol Brain Res*, 122, pp. 1–9.

Radzimanowski, J., Simon, B., Sattler, M., Beyreuther, K., Sinning, I. and Wild, K. (2008). Structure of the intracellular domain of the amyloid precursor protein in complex with Fe65–PTB2. *EMBO Rep*, 9, pp. 1134–40.

Rae, T. D., Schmidt, P. J., Pufahl, R. A., Culotta, V. C. and O'Halloran, T. V. (1999). Undetectable intracellular free copper: the requirement of a copper chaperone for superoxide dismutase. *Science*, 284, pp. 805–8.

Rajan, M. T., Jagannatha Rao, K. S., Mamatha, B. M., Rao, R. V., Shanmugavelu, P., Menon, R. B. and Pavithran, M. V. (1997). Quantification of trace elements in normal human brain by inductively coupled plasma atomic emission spectrometry. *J Neurol Sci*, 146, pp. 153–66.

Ramelot, T. A., Gentile, L. N. and Nicholson, L. K. (2000). Transient structure of the amyloid precursor protein cytoplasmic tail indicates preordering of structure for binding to cytosolic factors. *Biochemistry*, 39, pp. 2714–25.

Rice, T. M., Clarke, R. W., Godleski, J. J., Al-Mutairi, E., Jiang, N. F., Hauser, R. and Paulauskis, J. D. (2001). Differential ability of transition metals to induce pulmonary inflammation. *Toxicol Appl Pharmacol*, 177, pp. 46–53.

Richards, S. J., Hodgman, C. and Sharpe, M. (1995). Reported sequence homology between Alzheimer amyloid770 and the MRC OX-2 antigen does not predict function. *Brain Res Bull*, 38, pp. 305–6.

Roberts, L. J., 2nd, Montine, T. J., Markesbery, W. R., Tapper, A. R., Hardy, P., Chemtob, S., Dettbarn, W. D. and Morrow, J. D. (1998). Formation of isoprostane-like compounds (neuroprostanes) *in vivo* from docosahexaenoic acid. *J Biol Chem*, 273, pp. 13605–12.

Roberts, S. B., Ripellino, J. A., Ingalls, K. M., Robakis, N. K. and Felsenstein, K. M. (1994). Non-amyloidogenic cleavage of the beta-amyloid precursor protein by an integral membrane metalloendopeptidase. *J Biol Chem*, 269, pp. 3111–6.

Rohan de Silva, H. A., Jen, A., Wickenden, C., Jen, L. S., Wilkinson, S. L. and Patel, A. J. (1997). Cell-specific expression of beta-amyloid precursor protein isoform mRNAs and proteins in neurons and astrocytes. *Brain Res Mol Brain Res*, 47, pp. 147–56.

Rosa, M. L., Guimaraes, F. S., de Oliveira, R. M., Padovan, C. M., Pearson, R. C. and Del Bel, E. A. (2005). Restraint stress induces beta-amyloid precursor protein mRNA expression in the rat basolateral amygdala. *Brain Res Bull*, 65, pp. 69–75.

Rosen, D. R., Martin-Morris, L., Luo, L. Q. and White, K. (1989). A Drosophila gene encoding a protein resembling the human beta-amyloid protein precursor. *Proc Natl Acad Sci USA*, 86, pp. 2478–82.

Rossjohn, J., Cappai, R., Feil, S. C., Henry, A., McKinstry, W. J., Galatis, D., Hesse, L., Multhaup, G., Beyreuther, K., Masters, C. L. and Parker, M. W. (1999). Crystal structure of the N-terminal, growth factor-like domain of Alzheimer amyloid precursor protein. *Nat Struct Biol*, 6, pp. 327–31.

Ruiz, F. H., Gonzalez, M., Bodini, M., Opazo, C. and Inestrosa, N. C. (1999). Cysteine 144 is a key residue in the copper reduction by the beta-amyloid precursor protein. *J Neurochem*, 73, pp. 1288–92.

Sandbrink, R., Masters, C. L. and Beyreuther, K. (1994). Beta A4-amyloid protein precursor mRNA isoforms without exon 15 are ubiquitously expressed in rat tissues including brain, but not in neurons. *J Biol Chem*, 269, pp. 1510–7.

Sandbrink, R., Monning, U., Masters, C. L. and Beyreuther, K. (1997). Expression of the APP gene family in brain cells, brain development and aging. *Gerontology*, 43, pp. 119–31.

Sayre, L. M., Zelasko, D. A., Harris, P. L., Perry, G., Salomon, R. G. and Smith, M. A. (1997). 4-Hydroxynonenal-derived advanced lipid peroxidation end products are increased in Alzheimer's disease. *J Neurochem*, 68, pp. 2092–7.

Scheuermann, S., Hambsch, B., Hesse, L., Stumm, J., Schmidt, C., Beher, D., Bayer, T. A., Beyreuther, K. and Multhaup, G. (2001). Homodimerization of amyloid precursor protein and its implication in the amyloidogenic pathway of Alzheimer's disease. *J Biol Chem*, 276, pp. 33923–9.

Schmalz, G., Schuster, U. and Schweikl, H. (1998). Influence of metals on IL-6 release in vitro. *Biomaterials*, 19, pp. 1689–94.

Seubert, P., Oltersdorf, T., Lee, M. G., Barbour, R., Blomquist, C., Davis, D. L., Bryant, K., Fritz, L. C., Galasko, D., Thal, L. J. and *et al.* (1993). Secretion of beta-amyloid precursor protein cleaved at the amino terminus of the beta-amyloid peptide. *Nature*, 361, pp. 260–3.

Simons, M., de Strooper, B., Multhaup, G., Tienari, P. J., Dotti, C. G. and Beyreuther, K. (1996). Amyloidogenic processing of the human amyloid precursor protein in primary cultures of rat hippocampal neurons. *J Neurosci*, 16, pp. 899–908.

Sinha, S., Anderson, J. P., Barbour, R., Basi, G. S., Caccavello, R., Davis, D., Doan, M., Dovey, H. F., Frigon, N., Hong, J., Jacobson-Croak, K., Jewett, N., Keim, P., Knops, J., Lieberburg, I., Power, M., Tan, H., Tatsuno, G., Tung, J., Schenk, D., Seubert, P., Suomensaari, S. M., Wang, S., Walker, D., Zhao, J., McConlogue, L. and John, V. (1999). Purification and cloning of amyloid precursor protein beta-secretase from human brain. *Nature*, 402, pp. 537–40.

Small, D. H., Nurcombe, V., Reed, G., Clarris, H., Moir, R., Beyreuther, K. and Masters, C. L. (1994). A heparin-binding domain in the amyloid protein precursor of Alzheimer's disease is involved in the regulation of neurite outgrowth. *J Neurosci*, 14, pp. 2117–27.

Smith-Swintosky, V. L., Pettigrew, L. C., Craddock, S. D., Culwell, A. R., Rydel, R. E. and Mattson, M. P. (1994). Secreted forms of beta-amyloid precursor protein protect against ischemic brain injury. *J Neurochem*, 63, pp. 781–4.

Soba, P., Eggert, S., Wagner, K., Zentgraf, H., Siehl, K., Kreger, S., Lower, A., Langer, A., Merdes, G., Paro, R., Masters, C. L., Muller, U., Kins, S. and Beyreuther, K. (2005). Homo- and heterodimerization of APP family members promotes intercellular adhesion. *Embo J*, 24, pp. 3624–34.

Squitti, R., Cassetta, E., Dal Forno, G., Lupoi, D., Lippolis, G., Pauri, F., Vernieri, F., Cappa, A. and Rossini, P. M. (2004). Copper perturbation in 2 monozygotic twins discordant for degree of cognitive impairment. *Arch Neurol*, 61, pp. 738–43.

Squitti, R., Lupoi, D., Pasqualetti, P., Dal Forno, G., Vernieri, F., Chiovenda, P., Rossi, L., Cortesi, M., Cassetta, E. and Rossini, P. M. (2002). Elevation of serum copper levels in Alzheimer's disease. *Neurology*, 59, pp. 1153–61.

Suazo, M., Hodar, C., Morgan, C., Cerpa, W., Cambiazo, V., Inestrosa, N. C. and Gonzalez, M. (2009). Overexpression of amyloid precursor protein increases copper content in HEK293 cells. *Biochem Biophys Res Commun*, 382(4), pp. 740–744.

Subramaniam, R., Roediger, F., Jordan, B., Mattson, M. P., Keller, J. N., Waeg, G. and Butterfield, D. A. (1997). The lipid peroxidation product, 4-hydroxy-2-trans-nonenal, alters the conformation of cortical synaptosomal membrane proteins. *J Neurochem*, 69, pp. 1161–9.

Sun, K. H., Sun, G. H., Su, Y., Chang, C. I., Chuang, M. J., Wu, W. L., Chu, C. Y. and Tang, S. J. (2004). Acidic-rich region of amyloid precursor protein induces glial cell apoptosis. *Apoptosis*, 9, pp. 833–41.

Suska, F., Esposito, M., Gretzer, C., Kalltorp, M., Tengvall, P. and Thomsen, P. (2003). IL-1alpha, IL-1beta and TNF-alpha secretion during *in vivo/ex vivo* cellular interactions with titanium and copper. *Biomaterials*, 24, pp. 461–8.

Suska, F., Gretzer, C., Esposito, M., Emanuelsson, L., Wennerberg, A., Tengvall, P. and Thomsen, P. (2005). *In vivo* cytokine secretion and NF-kappaB activation around titanium and copper implants. *Biomaterials*, 26, pp. 519–27.

Tanaka, S., Nakamura, S., Ueda, K., Kameyama, M., Shiojiri, S., Takahashi, Y., Kitaguchi, N. and Ito, H. (1988). Three types of amyloid protein precursor mRNA in human brain: their differential expression in Alzheimer's disease. *Biochem Biophys Res Commun*, 157, pp. 472–9.

Tandon, L. (1994). RNAA for arsenic, cadmium, copper, and molybdenum in CNS tissues from subjects with age-related neurodegenerative diseases. *Journal of Radioanalytical and Nuclear Chemistry*, 179, pp. 331–339.

Tanzi, R. E., McClatchey, A. I., Lamperti, E. D., Villa-Komaroff, L., Gusella, J. F. and Neve, R. L. (1988). Protease inhibitor domain encoded by an amyloid protein precursor mRNA associated with Alzheimer's disease. *Nature*, 331, pp. 528–30.

Tarohda, T., Yamamoto, M. and Amamo, R. (2004). Regional distribution of manganese, iron, copper, and zinc in the rat brain during development. *Anal Bioanal Chem*, 380, pp. 240–6.

Treiber, C., Quadir, M. A., Voigt, P., Radowski, M., Xu, S., Munter, L. M., Bayer, T. A., Schaefer, M., Haag, R. and Multhaup, G. (2009). Cellular copper import by nanocarrier systems, intracellular availability, and effects on amyloid beta peptide secretion. *Biochemistry*, 48, pp. 4273–84.

Turnlund, J. R., Jacob, R. A., Keen, C. L., Strain, J. J., Kelley, D. S., Domek, J. M., Keyes, W. R., Ensunsa, J. L., Lykkesfeldt, J. and Coulter, J. (2004). Long-term high copper intake: effects on indexes of copper status, antioxidant status, and immune function in young men. *Am J Clin Nutr*, 79, pp. 1037–44.

Tyler, S. J., Dawbarn, D., Wilcock, G. K. and Allen, S. J. (2002). alpha- and beta-secretase: profound changes in Alzheimer's disease. *Biochem Biophys Res Commun*, 299, pp. 373–6.

van Bakel, H., Strengman, E., Wijmenga, C. and Holstege, F. C. (2005). Gene expression profiling and phenotype analyses of S. cerevisiae in response to changing copper reveals six genes with new roles in copper and iron metabolism. *Physiol Genomics*, 22, pp. 356–67.

van De Sluis, B., Rothuizen, J., Pearson, P. L., van Oost, B. A. and Wijmenga, C. (2002). Identification of a new copper metabolism gene by positional cloning in a purebred dog population. *Hum Mol Genet*, 11, pp. 165–73.

Vassar, R., Bennett, B. D., Babu-Khan, S., Kahn, S., Mendiaz, E. A., Denis, P., Teplow, D. B., Ross, S., Amarante, P., Loeloff, R., Luo, Y., Fisher, S., Fuller, J., Edenson, S., Lile, J., Jarosinski, M. A., Biere, A. L., Curran, E., Burgess, T., Louis, J. C., Collins, F., Treanor, J., Rogers, G. and Citron, M. (1999). Beta-secretase cleavage of Alzheimer's amyloid precursor protein by the transmembrane aspartic protease BACE. *Science*, 286, pp. 735–41.

Villard, L., Tassone, F., Crnogorac-Jurcevic, T., Clancy, K. and Gardiner, K. (1998). Analysis of pufferfish homologues of the AT-rich human APP gene. *Gene*, 210, pp. 17–24.

Vitek, M. P. (1989). Increasing amyloid peptide precursor production and its impact on Alzheimer's disease. *Neurobiol Aging*, 10, pp. 471–3; discussion 477–8.

Volkel, W., Sicilia, T., Pahler, A., Gsell, W., Tatschner, T., Jellinger, K., Leblhuber, F., Riederer, P., Lutz, W. K. and Gotz, M. E. (2006). Increased brain levels of 4-hydroxy-2-nonenal glutathione conjugates in severe Alzheimer's disease. *Neurochem Int*, 48, pp. 679–86.

von Rotz, R. C., Kohli, B. M., Bosset, J., Meier, M., Suzuki, T., Nitsch, R. M. and Konietzko, U. (2004). The APP intracellular domain forms nuclear multiprotein complexes and regulates the transcription of its own precursor. *J Cell Sci*, 117, pp. 4435–48.

Wang, J., Xiong, S., Xie, C., Markesbery, W. R. and Lovell, M. A. (2005). Increased oxidative damage in nuclear and mitochondrial DNA in Alzheimer's disease. *J Neurochem*, 93, pp. 953–62.

Wang, Y. and Ha, Y. (2004). The X-ray structure of an antiparallel dimer of the human amyloid precursor protein E2 domain. *Mol Cell*, 15, pp. 343–53.

Ward, N. I. (1987). Neutron activation analysis techniques for identifying elemental status in Alzheimer's disease *J. Radioanal. Nucl. Chem.*, 113, pp. 515–526.

Wasco, W., Bupp, K., Magendantz, M., Gusella, J. F., Tanzi, R. E. and Solomon, F. (1992). Identification of a mouse brain cDNA that encodes a protein related to the Alzheimer disease-associated amyloid beta protein precursor. *Proc Natl Acad Sci USA*, 89, pp. 10758–62.

Wasco, W., Gurubhagavatula, S., Paradis, M. D., Romano, D. M., Sisodia, S. S., Hyman, B. T., Neve, R. L. and Tanzi, R. E. (1993). Isolation and characterization of APLP2 encoding a homologue of the Alzheimer's associated amyloid beta protein precursor. *Nat Genet*, 5, pp. 95–100.

Weidemann, A., Eggert, S., Reinhard, F. B., Vogel, M., Paliga, K., Baier, G., Masters, C. L., Beyreuther, K. and Evin, G. (2002). A novel epsilon-cleavage within the transmembrane domain of the Alzheimer amyloid precursor protein demonstrates homology with Notch processing. *Biochemistry*, 41, pp. 2825–35.

Weidemann, A., Konig, G., Bunke, D., Fischer, P., Salbaum, J. M., Masters, C. L. and Beyreuther, K. (1989). Identification, biogenesis, and localization of precursors of Alzheimer's disease A4 amyloid protein. *Cell*, 57, pp. 115–26.

Wender, M., Szczech, J., Hoffmann, S. and Hilczer, W. (1992). Electron paramagnetic resonance analysis of heavy metals in the aging human brain. *Neuropatol Pol*, 30, pp. 65–72.

White, A. R., Multhaup, G., Galatis, D., McKinstry, W. J., Parker, M. W., Pipkorn, R., Beyreuther, K., Masters, C. L. and Cappai, R. (2002). Contrasting, species-dependent modulation of copper-mediated neurotoxicity by the Alzheimer's disease amyloid precursor protein. *J Neurosci*, 22, pp. 365–76.

White, A. R., Multhaup, G., Maher, F., Bellingham, S., Camakaris, J., Zheng, H., Bush, A. I., Beyreuther, K., Masters, C. L. and Cappai, R. (1999a). The Alzheimer's disease amyloid precursor protein modulates copper-induced toxicity and oxidative stress in primary neuronal cultures. *J Neurosci*, 19, pp. 9170–9.

White, A. R., Reyes, R., Mercer, J. F., Camakaris, J., Zheng, H., Bush, A. I., Multhaup, G., Beyreuther, K., Masters, C. L. and Cappai, R. (1999b). Copper levels are increased in the cerebral cortex and liver of APP and APLP2 knockout mice. *Brain Res*, 842, pp. 439–44.

Willoughby, D. A., Johnson, S. A., Pasinetti, G. M., Tocco, G., Najm, I., Baudry, M. and Finch, C. E. (1992). Amyloid precursor protein mRNA encoding the Kunitz protease inhibitor domain is increased by kainic acid-induced seizures in rat hippocampus. *Exp Neurol*, 118, pp. 332–9.

Xu, X. (2009). Gamma-secretase catalyzes sequential cleavages of the AbetaPP transmembrane domain. *J Alzheimers Dis*, 16, pp. 211–24.

Yan, R., Bienkowski, M. J., Shuck, M. E., Miao, H., Tory, M. C., Pauley, A. M., Brashier, J. R., Stratman, N. C., Mathews, W. R., Buhl, A. E., Carter, D. B., Tomasselli, A. G., Parodi, L. A., Heinrikson, R. L. and Gurney, M. E. (1999). Membrane-anchored aspartyl protease with Alzheimer's disease beta-secretase activity. *Nature*, 402, pp. 533–7.

Yang, L. B., Lindholm, K., Yan, R., Citron, M., Xia, W., Yang, X. L., Beach, T., Sue, L., Wong, P., Price, D., Li, R. and Shen, Y. (2003). Elevated beta-secretase expression and enzymatic activity detected in sporadic Alzheimer's disease. *Nat Med*, 9, pp. 3–4.

Yankner, B. A., Duffy, L. K. and Kirschner, D. A. (1990). Neurotrophic and neurotoxic effects of amyloid beta protein: reversal by tachykinin neuropeptides. *Science*, 250, pp. 279–82.

Yu, C., Kim, S. H., Ikeuchi, T., Xu, H., Gasparini, L., Wang, R. and Sisodia, S. S. (2001). Characterization of a presenilin-mediated amyloid precursor protein carboxyl-terminal fragment gamma. Evidence for distinct mechanisms involved in gamma -secretase processing of the APP and Notch1 transmembrane domains. *J Biol Chem*, 276, pp. 43756–60.

Yu, G., Nishimura, M., Arawaka, S., Levitan, D., Zhang, L., Tandon, A., Song, Y. Q., Rogaeva, E., Chen, F., Kawarai, T., Supala, A., Levesque, L., Yu, H., Yang, D. S., Holmes, E., Milman, P., Liang, Y., Zhang, D. M., Xu, D. H., Sato, C., Rogaev, E., Smith, M., Janus, C., Zhang, Y., Aebersold, R., Farrer, L. S., Sorbi, S., Bruni, A., Fraser, P. and St George-Hyslop, P. (2000). Nicastrin modulates presenilin-mediated notch/glp-1 signal transduction and betaAPP processing. *Nature*, 407, pp. 48–54.

Zhao, G., Mao, G., Tan, J., Dong, Y., Cui, M. Z., Kim, S. H. and Xu, X. (2004). Identification of a new presenilin-dependent zeta-cleavage site within the transmembrane domain of amyloid precursor protein. *J Biol Chem*, 279, pp. 50647–50.

Chapter 7

Role of Aluminum and Other Metal Ions in the Pathogenesis of Alzheimer's Disease

Silvia Bolognin and Paolo Zatta
CNR-Institute for Biomedical Technologies, Padua "Metalloproteins" Unit, Department of Biology, University of Padua, Viale G. Colombo 3-35121, Padua, Italy

Alzheimer's Disease (AD) is the most common cause of dementia in the elderly and with the increase of life expectancy in the developing countries it is becoming a problem with a relevant health and social impact. The etiology of the disease is still elusive but a huge number of reports indicate that, among putative aggravating factors, several metal ions (aluminum, zinc, copper, iron) could detrimentally impair the aggregation of β-amyloid (Aβ), a key protein apparently involved in the pathology. While studying the molecular basis of AD, it has become clear that the protein conformation plays a critical role in the pathogenic process. In this chapter, we will focus on the role of metal ions, specifically aluminum, in affecting Aβ-amyloid aggregation examining also the toxicity of these complexes in cell culture models.

Brain Diseases and Metalloproteins
Edited by David R. Brown

ISBN 978-981-4316-01-9 (Hardcover), 978-981-4364-07-2 (eBook)
www.panstanford.com

7.1 Alzheimer's Disease

Alzheimer's Disease (AD) is one of the most common age-related neurodegenerative disorders (ND). Neurologically, it is initially characterized by a series of mild cognitive impairments, deficits

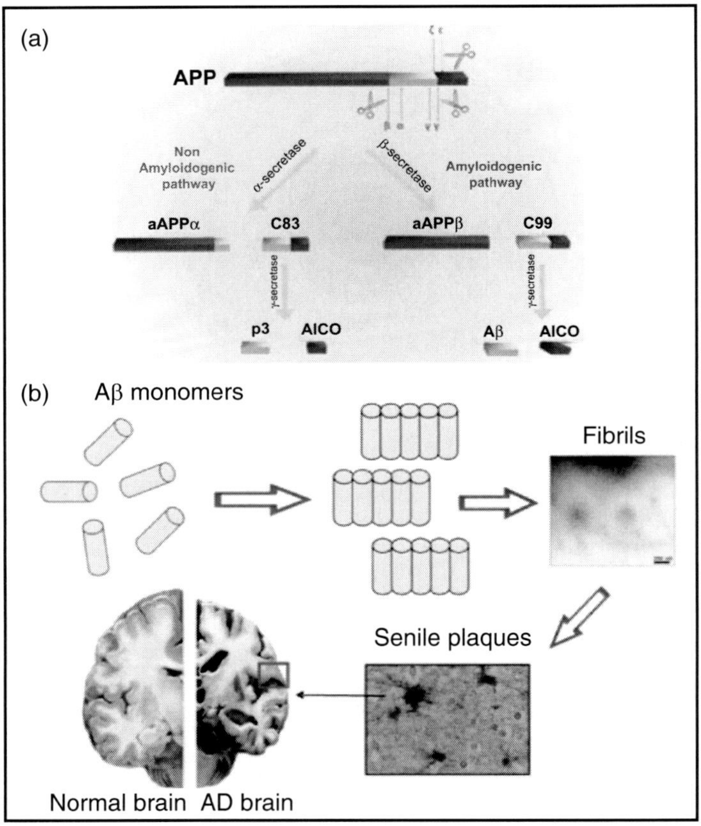

Figure 7.1 A) APP processing. The amyloidogenic pathway involves the sequential cleavage by β-secretase and γ-secretase with the production of Aβ and APP-intracellular domain (AICD). The non-amyloidogenic pathway involves the cleavage by α-secretase with the release of sAPPα from the cell surface. B) Aβ aggregation follows a sequence by which the accumulation of soluble Aβ, is followed by the appearance of low molecular weight oligomers that rapidly associate in higher-order aggregates and finally precipitate to form senile plaques. Aβ aggregation is greatly influenced by all the metal ions (e.g., Al, Cu, Fe, and Zn) that are found in both the core and rim of the AD SP. See also Colour Insert.

in short-term memory, loss of spatial memory, and emotional imbalances. As the disease progresses, these symptoms become more severe, and ultimately results in the total loss of executive functions. Histologically, hallmarks of the disease are the loss of neurons in the cerebral cortex, the intracellular deposition of neurofibrillary tangles containing hyperphosphorylated τ protein and the the presence of extraneuronal senile plaques (SP), whose core is mainly constituted by a peptide mixture of 39–43 residues called β-amyloid (Aβ) (Haass and Selkoe, 2007; Iqbal *et al.,* 2009).

This peptide derives from a large transmembrane precursor protein (APP) which can be proteolytically processed into two different pathways (Fig. 7.1). The amyloidogenic pathway is characterized by the cleavage of β-secretase which results with the secretion of the large ectodomain called sAPPβ. The remaining C-terminal fragment is further processed by γ-secretase, generating Aβ peptide. The non-amyloidogenic pathway starts with APP cleavage by α-secretase, generating the N-terminal ectodomain sAPPα and a 83 amino acid fragment which is further processes by γ-secretase.

7.2 The Amyloid Cascade and Oligomer Hypothesis

In the early nineties Hardy & Higgins (1992) proposed that the key event in AD etiopathogenesis was the increased production and decreased clearance of Aβ peptides, which resulted in the extracellular deposition of insoluble proteinaceus deposits into SP. This detrimental deposition was suggested to trigger a cascade of events, finally promoting neuronal death (amyloid cascade hypothesis). Over time, the initial idea that SP were the only culprit initiating neuronal loss has fallen out of favour in the light of more recent evidence. As a matter of fact, the pathological relevance of SP has been questioned by many scientists as fibrillar deposits have been detected also in non-demented individuals, suggesting that the plaque load does not always represent a good parameter to identify the degree of dementia in humans. Moreover, it has been reported that many AD patients with severely impaired memory showed no SP at *post-mortem* analysis (Nordberg, 2008). In addition, mouse models of AD display memory deficits before the observation of SP in the brain (Lesne *et al.,* 2006).

These data have given rise to new speculations supporting the need of an amyloid cascade revision. Current studies, investigating the importance of the various soluble Aβ assembly, indicate that early-stage Aβ aggregates, oligomers, could be more relevant to AD etiology and correlate better than insoluble deposits with the severity of dementia (Naslund *et al.,* 2000). Significantly, elevated levels of soluble oligomers were indeed found in AD brain compared to controls (Tomic *et al.,* 2009). The focus of the research moved then toward Aβ oligomerization pathway which finally ends with fibril formation. Thus, soluble oligomers seem to act a paramount role in triggering the early events causing the disease while SP, despite contributing to neuronal injury (Tsai *et al.,* 2004), are more likely a *reservoir* of toxicity or even a protection (Caughey & Lansbury, 2003).

Furthermore, it is emerging that Aβ accumulates also intracellularly in mouse models of AD and in human AD brains and this could represent a previously unknown detrimental contribution of the peptide to the disease progression (LaFerla *et al.,* 2007).

The *in vitro* investigations using synthetic peptides have shown a complex pattern of Aβ aggregation, with differences between $A\beta_{1-40}$ and $A\beta_{1-42}$ oligomerization (Ricchelli *et al.,* 2005). Nevertheless, it is almost unanimously recognized that Aβ exists in a 'natively unfolded' conformation which undergoes nucleation-dependent polymerization (Roychaudhuri *et al.,* 2009). On the contrary, there is not a general consensus in discriminating which is the most toxic species of Aβ. According to few studies the highest neurotoxicity was associated only with the Aβ dimers and not with the higher oligomers (Shankar *et al.*, 2008). Other groups showed that much larger oligomers, termed Aβ*56, rather than the smaller dimeric/trimeric aggregates could be more detrimental (Lesne *et al.,* 2006; Cheng *et al.,* 2007). These oligomers possibly composed of 12 monomers, despite being observed in a transgenic mouse model, have not already been isolated in AD brains.

Clarification of this issue seems now crucial for uncovering of the role played by Aβ oligomerization in the pathogenesis of AD.

7.3 Metal Dysmetabolism in AD

Many studies have suggested a potential involvement of several metal ions, such as copper (Cu), zinc (Zn), iron (Fe) and aluminum (Al), in the development and/or progression of AD (Crichton *et al.*, 2008; Hureau & Faller, 2009; Zatta *et al.*, 2009).

The above metals, apart from Al, are fundamental for the correct brain functioning, however at the same time they need to be strictly regulated to avoid the triggering of detrimental cell processes; indeed, depletion and accumulation of these metals can lead both to abnormal interactions with proteins or nucleic acids and to consequent cell damage. The brain therefore strictly regulates the metal ion fluxes as there is no passive transport of metal ions across the blood brain barrier (BBB). Thus, for AD but also for the major ND, the described metal imbalance is not simply and solely due to an increased exposure to metals but, rather, to a more complicated impairment in relevant homeostatic mechanisms.

Thus, the detailed significance of brain metal dyshomeostasis in AD is still a matter of intense debate and it primarily depends on the obtainment of reliable and unambiguous analytical data on brain tissues. Unfortunately, so far, the availability of adequate sets of analytical and/or chemical speciation data for brain metals is disappointingly scarce, in that most studies report only fragmentary and sporadic data which are sometimes contradictory. The collection of reliable analytical data for brain metals is complicated further by the extreme structural, functional, topographical and architectural complexity of the brain itself (Speziali & Orvini, 2003). Thus, the assessment of metal dyshomeostais is currently founded on several independent, often unrelated, qualitative indications rather than on conclusive quantitative data. It follows that new and systematic analytical data on brain metal concentration distribution and speciation are strongly required.

Generally, the amygdale and hippocampus were sites where the concentrations of the various metals significantly differed in AD brain compared to controls (Speziali & Orvini, 2003). Moreover, increased metal concentrations have been found within the SP compared with the surrounding tissues in human AD brains (Lovell *et al.*, 1998) as well as in transgenic mice (Rajendran *et al.*, 2009).

It is also important to bear in mind that the dyshomeostasis of a single metal ion would upset the whole metal homeostatic pool, resulting in a significant imbalance of the elemental levels in the body (serum, CSF and brain) as a result of what we called "domino effect" (Bolognin *et al.*, 2009).

7.3.1 *Copper*

The potential and controversial involvement of Cu in the etiopathogenesis of AD is a matter of intense debate as the analytical data regarding Cu concentration in brain tissues are often unclear. For example, Deibel *et al.*, (1996) reported a general decrease of approximately 20% in total Cu brain level in AD brain compared to controls, but other groups did not confirm this data (Loeffler *et al.*, 1996).

A possible explanation of the proposed association between AD and Cu metabolism could lie in the control exerted by Cu on Aβ levels. Cu dietary supplementation in AD animal models has been found to determine a decreased of secreted Aβ (Bayer *et al.*, 2003) suggesting that Cu supplementation, or better proper delivery into the brain, could be beneficial (Bayer *et al.*, 2006). At this regard, Crouch *et al.*, (2009) demonstrated that increased intracellular Cu availability inhibited the accumulation of Aβ oligomers and τ phosphorylation.

The effect of an oral Cu supplementation was evaluated also in the cerebrospinal fluid (CSF) of AD patients with the result of a stabilizing effect in terms of contrasting the decrease of $A\beta_{1-42}$ (Kessler *et al.*, 2008), which is generally reported in AD patients compared to controls (Lewczuk *et al.*, 2004). Nevertheless, this effect did not correspond to any sign of cognitive improvement (Bolognin *et al.*, unpublished data).

Undoubtedly, further studies are necessary to better clarify the correlation between Cu and Aβ, bearing in mind that none of the above studies have demonstrated the existence of a Cu-deficient condition in the brain of AD patients and other ND (Zatta & Frank, 2007).

7.3.2 *Zinc*

The finding that Zn was enriched in the SP and its concentration was increased in the neuropil of AD patients as compared to controls (Lovell *et al.*, 1998) strengthened the hypothesis of a potential role played by the metal in the pathogenesis of AD. Later, other studies confirmed these results in brain tissues (Stoltenberg *et al.*, 2005; Miller *et al.*, 2006). On the contrary, the attempt to quantify Zn in AD serum (Rulon *et al.*, 2000; Dong *et al.*, 2008) or CSF (Molina *et al.*, 1998; Gerhardsson *et al.*, 2008) did not lead to conclusive results.

Despite these controversial reports, studies on the role of Zn in AD have shown that Zn binding to Aβ could reduce Zn availability at the synaptic cleft leading to deleterious effects considering the role of Zn in neuronal signaling and synaptic plasticity (see Frederickson *et al.*, 2005). Moreover, alteration in the concentration of Zn in the brain could enhance Aβ precipitation, giving rise to protease-resistant, and thus possibly/potential neurotoxic, amorphous aggregates. As a second observation, it has been hypothesized that vesicular Zn released in the synaptic cleft during neurotransmission may be one contributing factor for the recruitment of Aβ oligomers to synaptic terminals (Deshpande *et al.*, 2009).

Furthermore, when transgenic mice lacking Zn transporter 3 (ZnT3) were crossed with Aβ-producing mice, a marked decrease of the plaque load was observed (Lee *et al.*, 2002), suggesting that synaptic Zn may play a role in enhancing Aβ aggregation and SP deposition.

In vitro studies have reported an interaction also between Zn and τ protein. Low micromolar concentrations of Zn accelerated fibril formation of τ fragment 244–372 (Mo *et al.*, 2009) thus detrimentally favoring the formation of β-sheet structures, which can undergo filament nucleation and elongation, eventually forming neurofibrillary tangles.

Behavioral studies on transgenic mice examining the effect of Zn supplementation reported an increased impairment of spatial memory, but with a concomitant unexpected reduction of Aβ deposits (Linkous *et al.*, 2009). In contrast, compounds affecting Zn homeostasis have been shown to decrease Aβ brain deposition (Lee *et al.*, 2004; Adlard *et al.*, 2008).

Cuajungco and Faget (2003) reviewed several controversial data proposing a paradoxical role for Zn in AD. It is likely that Zn could be released following oxidative/nitrosative stress factors implicated in AD etiology and in turn, this Zn increase can trigger neuronal death giving rise to a vicious cycle.

Moreover, it is worth noting that MT I and II were immunohistochemically found to be dramatically increased in astrocytes of AD brains compared with those of controls (Zambenedetti *et al.*, 1998). Considering that Zn is a potent inducer of this family of proteins, a correlation between Zn levels and the pathology should be further investigated.

7.3.3 *Iron*

The increase of Fe levels and Fe-binding proteins in the AD brain have been extensively reported in the literature (Lovell *et al.*, 1998; Cahill *et al.*, 2009) and several hypotheses have been made in the attempt of verifying this correlation (see Altamura & Muckenthaler, 2009). Smith *et al.*, (2007) stated this could be a secondary effect caused by, for example, increased heme oxygenase activity in response to oxidative stress (Schipper, 2004). This could be further strengthened by the fact that, despite being found to interact with Aβ *in vitro* (Hu *et al.*, 2006), Fe does not co-purify with Aβ extracted from plaques (Opazo *et al.*, 2002). Most recently, other pathways have been explored; in particular, it has been assumed that Fe could directly influence Aβ production through the modulation of furin, a ubiquitous enzyme, whose proteolytic activity is required for many cellular processes, including α- and γ-secretase processing. According to Silvestri and Camaschella (2008) high cellular Fe levels lower furin activity, which in turn reduces α-secretase activity favouring β- and γ-secretase activity with the consequent enhancement of Aβ production. In accordance, the furin mRNA level was reduced in AD brain patients (Hwang *et al.*, 2006).

A further correlation between Fe and AD is based on the observation that oxidative stress markers are highly expressed in AD-affected brain regions (e.g., Zambenedetti *et al.*, 1998) and this matches with the redox-active nature of Fe. Fe-dependent ROS production is indeed able to increase Fe cellular uptake (Pantopoulos & Hentze, 1998) which, in turn, could increase oxidative damage giving rise to a vicious cycle. Smith *et al.*, (2010) provided evidence that Fe and Fe-mediated redox activity were significantly elevated at the earliest preclinical stage of AD.

Another recent study reported that total Fe concentration in the CSF did not significantly differ in AD, mild cognitive impairment (MCI) and control patients, while redox active Fe was higher in the MCI cases compared to those with a definite diagnosis of AD (Lavados *et al.*, 2008). This supports the early contribution of Fe dyshomeostasis and redox activity in AD pathogenesis.

In summary, although the mechanism of Fe accumulation in AD is still unknown, it is necessary to consider Fe as an important cofactor. It is also undeniable that diseases directly related to increased levels of this metal (e.g., haemochromatosis) are not characterized by enhanced deposition of SP, proving that it could be one of many other contributing factors.

7.3.4 *Aluminum*

A wide number of papers has been published dealing with the potential exposure to Al as a contributing factor for AD pathogenesis. Significantly raised levels of Al were indeed reported in the parietal cortex of the AD brain as compared with controls (Yumoto *et al.*, 2009). Signs of Al dyshomeostasis were also recently found in a triple transgenic AD mouse, the 3xTg-AD, that expresses mutant APP, PS1, and tau, and is therefore considered to recapitulate the hallmarks of AD pathology (Oddo *et al.*, 2003). In that study, experiments employing mass spectrometry indicate that, when compared with the distribution of other AD-relevant metals (Zn, Cu, and Fe), Al is the only metal ion that was significantly increased in the cortex of 14 month old 3xTg-AD mice (Drago *et al.*, 2008b). Although it is evident that AD animal models cannot fully represent the complexity of AD, it is worth of noticing that also in this important model there was a significant alteration of Al brain concentration.

The so called "Aluminum affair" in AD began when Alfrey *et al.* (1976) described for the first time a neurological condition resembling AD dementia which was called dialysis encephalopathy (DE). DE consists of an abnormal accumulation of Al in the brain of uremic patients with renal failure undergoing chronic dialysis, which occurs when tap water, without any further purification, was used in the dialysis process (Zatta *et al.*, 2004; Zatta, 2006). This effect was reversible as once Al was removed from the "dialysis bath" the DE practically disappeared. These findings have given rise to a widespread speculation as to whether AD and Al could be linked, but no conclusive results were established (Reusche, 2003) and this is still a rather open issue. As a matter of fact, it is still elusive weather the accumulation of the metal could be the first event or just one of the many secondary event. Indeed, the epidemiological results which addressed the problem of Al in drinking water in connection with the incidence of AD were controversial (Reusche, 2003). In addition, many nephrologists currently use Al salts to decrease the hyperphosphatemia in uremic subjects with no major incidence of AD among these patients with respect to general population. Thus, Al itself cannot be a sufficient trigger of AD and there must be another reason for the potential AD-Al connection.

Two aspects need to be considered before approaching this issue. Firstly, Al has a complex hydrolysis pH-dependent chemistry in biological systems which can account for many inconsistencies reported in the literature on the effects of Al on animal or cellular models. As an example, when Al inorganic salts such as chloride, sulphate, hydroxide or perchlorate are dissolved in water at a calculated concentration of 1μm, the analytical Al concentration in solution is about 50 μM. The use of Al-lactate or Al-aspartate, however, increases the soluble Al concentration to 50–330 μM (Zatta, 2002). Hence, the examination of the metal bioavailability under physiological conditions has to be taken into account while designing Al studies. Secondly, a distinction has to be made between the concepts of neurotoxicity and neurodegeneration. Al has been aptly described as a neurotoxic element (Zatta, 2002) if it can not be physiologically excreted or it is in direct contact with the brain. Some studies have indeed summarized the effects of occupational exposure to Al suggesting that it induces relevant neurotoxic effects following acute or subacute exposure (see Krewski *et al.*, 2007). Nevertheless, besides the well-known neurotoxicity of Al at high concentration, the role of the metal in affecting pathways related to neurodegenerative mechanisms should be further investigated.

7.4 The Role of Metal Ions in the Aggregation and Toxicity of Aβ

Much remains to be learned about the real role of metal dysmetabolism in AD. Nevertheless, the above mentioned metals have all been advocated as modulators of Aβ aggregation with a resulting increase of neurotoxicity on neuronal cells as a consequence of the marked biophysical alteration of the Aβ properties (House *et al.*, 2004; Ricchelli *et al.*, 2005). The interest in clarifying the role of metal ions has become a crucial issue as it has been realized that in human brains and amyloid transgenic mice metal chelation could reverse the Aβ peptide aggregation dissolving the amyloid aggregates (see Bolognin *et al.,* 2009). Thus, the influence of metal ions such as Fe, Cu, and Zn in stimulating Aβ aggregation has been widely studied *in vitro* (Huang *et al.,* 2004). Zn appeared to be very efficient in inducing fast precipitation of Aβ and the formation of protease-resistant aggregates both at acidic and alkaline pH (Bush, 2003). In contrast, Cu and Fe showed a limited propensity to aggregate at physiological pH, whereas aggregation was drastically increased at slightly acidic pH (Atwood *et al.,* 2000). However, it should be emphasized that, although there is co-localization of these metal ions in the SP, this does not indicate a causative role for these elements in the pathogenesis of the disease.

The mentioned arguments led us to investigate in more detail and comparatively the adducts that are formed when reacting Aβ with Cu, Fe, Zn, and with Al; analysing the consequences of such metal binding on the Aβ aggregation profiles and the resulting effect in biological environment (Bolognin *et al.,* manuscript in preparation).

Metal complexes were prepared according to an established procedure (Drago *et al.,* 2008a) removing the metal excess by repeated washings. Immunological analysis of the various metal species by using conformation sensitive antibodies showed that Al and Fe ions were capable of inducing the formation of both fibrillary oligomers and annular protofibrils while such kinds of aggregates were not observed at all in Aβ-Zn and Aβ-Cu samples. As fibrillar oligomers and annular protofibrils have been proposed to constitute neurotoxic species, our findings strongly support the view that Aβ-Al and Aβ-Fe complexes may be far more effective that Aβ-Zn and Aβ-Cu complexes in causing neuronal cell death. Prefibrillar oligomers and annular protofibrils have been found not to correlate with cognitive dysfunction in AD brain and their levels were also elevated in some age-matched non-demented control brains (Tomic *et al.,* 2009).

Concerning the other tested metals, Zn-mediated aggregation of Aβ was rapid, giving rise to amorphous aggregates. Also Cu disfavoured the formation of fibrillary species while giving rise to a prevalent population of amorphous aggregates which, differently from Aβ-Zn, did not appear hydrophobic in character.

The mechanism by which these different Aβ aggregates interact with neurons and particularly with plasma membranes, eventually producing the activation of numerous downstream biochemical effects is under study. Small oligomeric species of Aβ could indeed readily diffuse through cell membrane and affect neuronal survival. At this regard, we have previously demonstrated that Aβ-Al increased membrane fluidity mostly in the lipid tail/polar heads border areas of cell membrane with respect to the other Aβ-metal complexes as well as metals alone (Drago *et al.,* 2008a). We can thus argue that Aβ-Al preparation, thanks to the high number of small oligomeric species, was the most effective among the tested complexes in affecting cellular membrane architecture and thus it is likely to penetrate effectively into the membrane bilayer core, potentially interacting with transmembrane proteins such as APP. Notably, in a set of microarray and real-time PCR experiments, in which we investigated the gene expression profile of SHSY-5Y cells (about 35.000 genes) treated with various Aβ-metal complexes, Aβ alone and the metal ions, we observed that the Aβ-Al complex was able to produce a selective up-regulation of APP gene (Drago *et al.,* manuscript in preparation) fitting with the up-regulated expression of the APP gene which occurs in brains of AD patients (Lukiw, 2004). Walton and Wang (2009) suggested that APP up-regulation, neuropathology, and cognitive deterioration could result from the Al accumulation in AD-vulnerable brain regions and thus, susceptible individuals with efficient Al absorption, can accumulate in neural cells sufficient amount of the metal to up-regulate APP expression. We can add that the deleterious binding of Al to Aβ may enhance this pathological alteration, favoring the focal accumulation of the metal. In this regard, Al and Aβ peptides have been showed to colocalized in the cores of Aβ fibers of SP (Yumoto *et al.,* 2009).

7.5 Aβ and Cell Membranes

Several studies indicated that Aβ neurotoxicity might be mediated through direct interaction between the peptide and cellular membranes (Kayed *et al.,* 2004; Demuro *et al.,* 2005). Notably, the amphipathic character of Aβ might explain its detrimental association with the membrane. At this regard, it has been reported that neuronal membranes can enhance Aβ-conversion into toxic oligomers (Curtain *et al.,* 2003) and that part of the critical balance between toxic and inert Aβ pools is determined by the relative amounts of lipids in the direct environment of the SP (Martins *et al.*, 2008).

Overall, these studies suggest that the relationship between Aβ and cellular membrane could be crucial in the process leading to the pathology. In accordance, perturbation in the lipid distribution was reported in many AD patients (Ji *et al.*, 2002; Pettegrew *et al.*, 2001) and hypercholesterolemia is considered an early risk factor for the development of AD (Kivipelto *et al.*, 2001; Martins *et al.*, 2009). *In vitro* studies indicated that increased cellular cholesterol levels result in the increased production of Aβ peptides (Fassbender *et al.*, 2001). Moreover, it has been reported that proteins relevant to Aβ generation localize in the membrane rafts (Reid *et al.*, 2007). Nevertheless, the possible mechanism underlying this interaction remains elusive. Many hypotheses have been made ranging from the alteration of the physiological characteristics of the membrane (Ji *et al.*, 2002), lipid peroxidation (Koppaka & Axelsen, 2000) to the formation of calcium-permeable ion channels which allow excessive Ca^{2+} influx which disrupts physiological homeostasis (Lin *et al.*, 2001). This latter event could occur either through the modulation of an existing Ca^{2+} channel or through the formation of a new cation-selective channel.

Curtain *et al.* (2003) showed that penetration of the lipid bilayer is closely related to conditions favourable to Aβ oligomerization. Moreover, oligomeric Aβ has been found to cause membrane fluctuation and transformations (Morita *et al.*, 2009). This highlights the possible role played by metal ions in promoting pathological events finally leading to AD.

We thus studied the effect of Aβ and various Aβ-metal complexes (Aβ-Al, Aβ-Cu, Aβ-Fe, Aβ-Zn) on a lipid model of cellular membrane using X-ray diffraction technique (Suwalsky *et al.*, 2009). Synthetic dimyristoylphosphatidylcholine (DMPC) and dimyristoylphosphatidylethanolamine (DMPE) bilayers were used because they represent phospholipids located in the outer and in the inner monolayer of the membrane, respectively (Boon & Smit, 2000). Aβ-Al was the most effective complex in perturbing DMPC bilayer compared to the other metal complexes. It is worth noticing that this effect was peculiar only to the Aβ-Al complex since neither the peptide alone nor the Al salt (Fig. 7.2) affected DMPC when

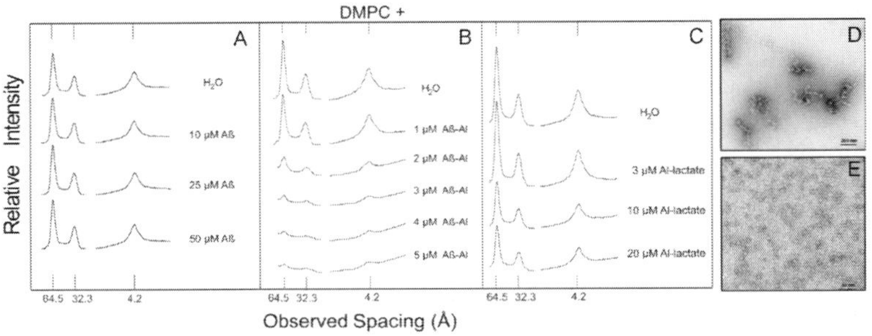

Figure 7.2 Microdensitograms from X-ray diffraction diagrams of DMPC in water and aqueous solutions of Aβ (A), Aβ-Al (B) and Al-lactate. TEM imagines of Aβ (C) and Aβ-Al (D) after the dialysis process.

incubated with concentrations similar to those of the Aβ-Al complex that produced relevant alterations. Considerably less pronounced was the effect of Aβ-Al complex on DMPE bilayer compared to that induced to DMPC. This result can be explained on the basis of their different structures: DMPC and DMPE differ only in their terminal amino groups, these being $^+N(CH_3)_3$ in DMPC and $^+NH_3$ in DMPE. Moreover, both molecular conformations are very similar in their dry crystalline phases with the hydrocarbon chains mostly parallel and extended, and the polar head groups lying perpendicularly to them. However, the gradual hydration of DMPC results in water filling the highly polar interbilayer spaces with the resulting increase of their width. This phenomenon allows for the incorporation of the Aβ-Al complex into DMPC bilayers disrupting their arrangement and consequently the whole of the bilayer structure. DMPE molecules pack tighter than those of DMPC due to their smaller polar groups and higher effective charge, resulting in a very stable bilayer system that is not significantly affected by water (Suwalsky *et al.*, 2005). On the other hand, the Al complexed with Aβ may induce a change in the net charge of the peptide which can promote abnormal lipid-peptide interactions, thus promoting pathological oligomerization of Aβ. Recently, our group has demonstrated that when Al was bound to Aβ, forming a stable metallorganic complex, the surface hydrophobicity of the peptide dramatically increased as a consequence of metal-induced conformational changes, favouring misfolding/aggregation phenomena (Ricchelli *et al.*, 2005). Aβ-Al, thanks to its higher lypophilicity compared to the other Aβ-metal complexes, could intercalate with the acyl chain region altering the bilayer arrangement.

An interesting hypothesis would be to relate the different ability of the Aβ-metal complexes in perturbing membranes to the different toxic species produced. In support of our hypothesis, Demuro *et al.* (2005) proposed that Aβs are responsible for a generalized increase in membrane permeability induced specifically by spherical amyloid oligomers.

In our experimental conditions, negligible was the effect of the sole peptide in disturbing lipid bilayer structures. On the contrary, Ambroggio and colleagues (Ambroggio *et al.*, 2005) demonstrated that $A\beta_{1-42}$ was incorporated into the membrane and remained in the lipid environment, finally altering the cohesive forces and membrane permeability. We suggest that the dependence of the behaviour of peptide-lipid monolayers on the lipid composition could account for the discrepancy between different studies.

Moreover, we determined whether complexing with Al affected the ability of radioactively iodinated Aβ to cross the *in vivo* blood brain barrier (BBB). We found that the rates of Aβ and Aβ-Al uptake were similar, but Aβ-Al entered the parenchymal space of the brain more readily (Banks *et al.*, 2006). This complex had also a longer half-life in blood and increased permeation at the striatum and thalamus suggesting that it would have more access to brain cells than the peptide alone.

7.6 Amyloid Metal Complexes and Ca

A large body of evidence reported that Ca^{2+} dyshomeostasis exerts an important role in promoting AD-related neuronal injury (Mattson, 2007; Kawahara *et al.*, 2009). Khachaturian (1994) firstly proposed a correlation between Ca^{2+} impairment and AD and recent findings confirmed this hypothesis. AD-related perturbation of Ca^{2+} signaling appears earlier than the macroscopic pathological changes which characterized the disease (LaFerla *et al.*, 2007). Increase of intracellular Ca^{2+} ($[Ca^{2+}]_i$) have been demonstrated to affect APP processing and Aβ intraneuronal accumulation (Pierrot *et al.*, 2004). In addition, it could interfere with neurofibrillary tangle formation, and cause alterations in mitochondrial functioning (Mattson, 2007). Mitochondria Ca^{2+} overload has been also linked to the increased production of reactive oxygen species (ROS) and release of pro-apoptotic factors (Bernardi *et al.*, 2007). It has been reported that mitochondria isolated from AD brains showed Aβ β-accumulation and morphological alterations (Fernandez-Vizarra *et al.*, 2004) and that Aβ may alter the structural properties of the mitochondrial membrane (Aleardi *et al.*, 2005). We thus investigated the effects of Aβ, alone and complexed with Al, Zn, Cu or Fe, on cortical neuronal cell culture $[Ca^{2+}]_i$ homeostasis (Drago *et al.*, 2008b).

We reported that, among the various Aβ-metal complexes, only the Aβ-Al complex altered glutamate-driven $[Ca]_i$ rises and was able to enhance NMDA receptor-mediated $[Ca^{2+}]_i$ dyshomeostasis. We speculate that some of the enhanced NMDA-triggered $[Ca^{2+}]_i$ rises observed in cortical neurons pre-incubated with Aβ-Al could be due to a partial impairment of mitochondrial Ca^{2+} buffering.

A large body of evidence has indicated that the accumulation of Aβ in mitochondria is associated with decreased enzymatic activity of respiratory chain complexes III and IV and reduced rate of oxygen consumption (Aleardi *et al.*, 2005; Caspersen *et al.*, 2005). Thus, we evaluated the effects of Aβ and its metal complexes on the functioning of isolated rat brain mitochondria. Aβ-Al inhibited the oxidative respiration in isolated rat brain mitochondria and it induced a decrease in state 3 respiration (Drago *et al.*, 2008b). These results appear to be in agreement with a previous study indicating that extracellular treatment with small spherical Aβ oligomers, unlike monomers and fibrils, can cause disruption of $[Ca^{2+}]_i$ (Demuro *et al.*, 2005). Our studies indicate that also Aβ induced a decrease in state 3 respiration, but the phenomenon was strongly exacerbated when the peptide was conjugated with Al.

Changes of membrane fluidity triggered by exposure to Aβ metal complexes were also evaluated by steady-state fluorescence anisotropy of mitochondria-bound 6-diphenyl-1,3,5-hexatriene (DPH) and N,N,N-trimethyl-4-(6-phenyl-1,3,5-hexatrien-1-yl) phenylammonium p-toluene sulfonate (TMA-DPH) but no perturbations of the internal lipid domains (as monitored by DPH) and polar heads group/hydrophobic tail border areas (as monitored by TMA-DPH) in the presence of Aβ-Al complex at 4 μM (the concentration that strongly inhibits state 3 respiration) was appreciated. It is likely that Aβ-Al exerted its inhibition without affecting the membrane fluidity or the low concentration of the peptide used in this study did not produce change detectable by the anisotropy assay.

Although future studies are clearly necessary to provide a more detailed insight of the alterations promoted by the Aβ-Al complex in neurons and mitochondria, these data support the idea that such complex can exert an important role in the impairment of neuronal metabolism and Ca^{2+} homeostasis.

7.7 Conclusion

Collectively, these findings suggest that the interaction between Aβ and metal ions needs to be investigated in more detail, especially in the light of the wide number of papers recently published describing the importance of the Aβ aggregational pathway and the toxicity of oligomers. Clarification of this issue will probably lead to new insights into the potential effectiveness of chelation therapy in the treatment AD. Notably, our results provide a comparative biophysical and biological characterization of Aβ and its metal complexes highlighting how the four tested metals differentially influenced the conformational properties of the peptide. TEM and immunological studies turned out to be particularly useful to qualitative discriminate the species formed during the 48h dialysis process. Cu and Zn mainly kept Aβ in an amorphous state, while Al and Fe promoted the formation of oligomers. Nevertheless, only the Aβ-Al complex formed a highly homogeneous population of smaller oligomeric species compared to the other complexes. This peculiar oligomers were responsible of a significant decrease of cell viability and activation of apoptosis on neuroblastoma cell culture (Drago *et al.,* 2008a). Moreover, Aβ-Al promoted a perturbation of $[Ca]_i$ homeostasis in cortical murine neurons which was not observed with the other metal-complexes.

Thus, it is likely that, besides the well-known neurotoxicity of the ion, these new findings could suggest a revaluation of the role of this metal as a modulator factor of Aβ aggregation. AD is most certainly a complex and multi-factorial disorder, for which aging is the main identified risk factor, while Ca^{2+} dyshomeostasis, oxidative stress, energy deficits and/or mitochondrial dysfunction are all emerging as critical co-factors. Even if Al itself cannot be the only triggering factor for AD development (see Bolognin *et al.,* 2009;

Perl & Moalem, 2006), the direct binding of Al to Aβ can give rise to stable toxic species which could detrimentally impair neuronal cell activity. Future investigations will address this still elusive issue in the attempt of better clarifying the correlation between Aβ-metal complex structure and the potential interaction with biological structures.

References

Adlard, P. A., Cherny, R. A., Finkelstein, D. I., Gautier, E., Robb, E., Cortes, M., Volitakis, I., Liu, X,. Smith, J. P., Perez, K., Laughton, K., Li, Q. X., Charman, S. A., Nicolazzo, J. A., Wilkins, S., Deleva, K., Lynch, T., Kok, G., Ritchie, C. W., Tanzi, R. E., Cappai, R., Masters, C. L., Barnham, K. J. and Bush, A. I. (2008). Rapid restoration of cognition in Alzheimer's transgenic mice with 8-hydroxy quinoline analogs is associated with decreased interstitial Abeta, *Neuron,* 59, pp. 43–55.

Aleardi, A. M., Benard, G., Augereau, O., Malgat, M., Talbot, J. C., Mazat, J. P., Letellier, T., Dachary-Prigent, J., Solaini, G. C. and Rossignol, R. (2005). Gradual alteration of mitochondrial structure and function by beta-amyloids: importance of membrane viscosity changes, energy deprivation, reactive oxygen species production, and cytochrome *c* release, *J. Bioenerg. Biomembr.*, 37, pp. 207–225.

Alfrey, A. C., LeGrendre, G. R. and Kaehny, W. D. (1976). The dialysis encephalophaty syndrome. Possible aluminum intoxication, *N. Engl. J. Med.,* 294, pp. 184–188.

Altamura, S. and Muckenthaler, M. U. (2009). Iron toxicity in diseases of aging: Alzheimer's disease, Parkinson's disease and atherosclerosis. *J. Alzheimers Dis.,* 16, pp. 879–895.

Ambroggio, E. E., Kim, D. H., Separovic, F., Barrow, C. J., Barnham, K. J., Bagatolli, L. A. and Fidelio, G. D. (2005). Surface behavior and lipid interaction of Alzheimer beta-amyloid peptide 1–42: a membrane-disrupting peptide, *Biophys. J.,* 88, pp. 2706–2713.

Atwood, C. S., Scarpa, R. C., Huang, X., Moir, R. D., Jones, W. D., Fairlie, D. P., Tanzi, R. and Bush, A. I. (2000). Characterization of copper interactions with Alzheimer amyloid beta peptides identification of an attomolar-affinity copper binding site on amyloid beta 1–42, *J. Neurochem.*, 75, pp. 1219–1233.

Banks, W. A., Niehoff, M. L., Drago, D. and Zatta, P. (2006). Aluminum complexing enhances amyloid beta protein penetration of blood-brain barrier, *Brain Res.,* 1116, pp. 215–221.

Bayer, T. A., Schäfer, S., Breyhan, H., Wirths, O., Treiber, C. and Multhaup, G. (2006). A vicious circle: role of oxidative stress, intreneuronal Abeta and Cu in Alzheimer's disease, *Clin. Neuropathol.,* 25, pp. 163–171.

Bayer, T. A., Schäfer, S., Simons, A., Kemmling, A., Kamer, T., Tepest, R., Eckert, A., Schüssel, K., Eikenberg, O., Sturchler-Pierrat, C., Abramowski, D., Staufenbiel, M. and Multhaup, G. (2003). Dietary Cu stabilizes brain superoxide dismutase 1 activity and reduces amyloid Abeta production in APP23 transgenic mice, *Proc. Natl. Acad. Sci. U.S.A.,* 100, pp. 14187–14192.

Bernardi, P., Krauskopf, A., Basso, E., Petronilli, V., Blachly-Dyson, E., Di Lisa F. and Forte, M. A. (2006). The mitochondrial permeability transition from in vitro artifact to disease target, *FEBS J.,* 273, pp. 2077–2099.

Bolognin, S., Messori, L. and Zatta, P. (2009). Metal ions physiopathology in neurodegenerative disorders, *Neuromolecular Med.* 11, pp. 223–238.

Boon, J. M. and Smith B. D. (2000). Chemical control of phospholipid distribution across bilayer membranes, *Med. Res. Rev.,* 22, pp. 251–281.

Bush, A. I. (2003). The metallobiology of Alzheimer's disease, *Trends Neurosci.* 26, pp. 207–214.

Cahill, C. M., Lahiri, D. K., Huang, X. and Rogers, J. T. (2009). Amyloid precursor protein and alpha synuclein translation, implications for iron and inflammation in neurodegenerative diseases. *Biochim. Biophys. Acta,* 1790, pp. 615–628.

Caspersen, C., Wang, N., Yao, J., Sosunov, A., Chen, X., Lustbader, J. W., Xu, H. W., Stern, D., McKhann, G. and Yan, S. D. (2005). Mitochondrial Abeta: a potential focal point for neuronal metabolic dysfunction in Alzheimer's disease, *FASEB J.,* 19, pp. 2040–2041.

Caughey, B. and Lansbury, P. T. (2003). Protofibrils, pores, fibrils, and neurodegeneration: separating the responsible protein aggregates from the innocent bystanders, *Ann. Rev. Neurosci.* 26, pp. 267–298.

Cheng, I. H., Scearce-Levie, K., Legleiter, J., Palop, J. J., Gerstein, H., Bien-Ly, N., Puolivali, J., Lesne, S., Ashe, K. M., Muchowski, P. J. and Mucke, L. (2007). Accelerating amyloid-beta fibrillization reduces oligomer levels and functional deficits in Alzheimer disease mouse models, *J. Biol. Chem.,* 282, pp. 23818–23828.

Crichton, R. R., Dexter, D. T. and Roberta, J. W. (2008). Metal based neurodegenerative diseases—from molecular mechanisms to therapeutic strategies, *Coord. Chem. Rev.,* 251, pp. 1189–1199.

Crouch, P. J., Hung, L. W., Adlard, P. A., Cortes, M., Lal, V., Filiz, G., Perez, K. A., Nurjono, M., Caragounis, A., Du, T., Laughton, K., Volitakis, I., Bush, A. I., Li, Q. X., Masters, C. L., Cappai, R., Cherny, R. A., Donnelly, P. S., White, A. R. and Barnham, K. J. (2009). Increasing Cu bioavailability inhibits Abeta oligomers and tau phosphorylation, *Proc. Natl. Acad. Sci.,* 106, pp. 381–386.

Cuajungco, M. P. and Faget, K. Y. (2003). Zinc takes the center stage: its paradoxical role in Alzheimer's disease, *Brain Res. Rev.,* 41, pp. 44–56.

Curtain, C. C., Ali, F. E., Smith, D. G., Bush, A. I., Masters, C. L. and Barnham, K. J. (2003). Metal ions, pH, and cholesterol regulate the interactions of Alzheimer's disease amyloid-beta peptide with membrane lipid, *J. Biol. Chem.*, 278, pp. 2977–2982.

Deibel, M. A., Ehmann, W. D. and Markesbery, W. R. (1996). Copper, iron, and zinc imbalances in severely degenerated brain regions in Alzheimer's disease: possible relation to oxidative stress, *J. Neurol. Sci.*, 143, pp. 137–142.

Demuro, A., Mina, E., Kayed, R., Milton, S. C., Parker, I. and Glabe, C. G. (2005). Calcium dysregulation and membrane disruption as a ubiquitous neurotoxic mechanism of soluble amyloid oligomers, *J. Biol. Chem.,* 280, pp. 17294–17300.

Deshpande, A., Kawai, H., Metherate, R., Glabe, C. G. and Busciglio, J. (2009). A role for synaptic zinc in activity-dependent Abeta oligomer formation and accumulation at excitatory synapses, *J. Neurosci.,* 29, pp. 4004–4015.

Dong, J., Robertson, J. D., Merkesbery, W. R. and Lovell, M. A. (2008). Serum zinc in the progression of Alzheimer's Disease, *J. Alzheimers Dis.*, 15, pp. 443–450.

Drago, D., Bettella, M., Bolognin, S., Cendron, L., Scancar, J., Milacic, R., Ricchelli, F., Casini, A., Messori, L., Tognon, G. and Zatta, P. (2008). Potential pathogenic role of β-amyloid$_{1-42}$-aluminum complex in Alzheimer's disease, *Int. J. Biochem. Cell Biol.,* 40, pp. 731–746, a.

Drago, D., Cavaliere, A., Mascetra, N., Ciavardelli, D., Ilio, C., Zatta, P. and Sensi, S. L. (2008). Aluminum modulates effects of beta amyloid(1–42) on neuronal calcium homeostasis and mitochondria functioning and is altered in a triple transgenic mouse model of Alzheimer's disease, *Rejuvenation. Res.*, 11, pp. 861–871, b.

Fassbender, K., Simons, M., Bergmann, C., Stroick, M., Lutjohann, D., Keller, P., Runz, H., Kuhl, S., Bertsch, T., von Bergmann, K., Hennerici, M., Beyreuther, K. and Hartmann, T. (2001). Simvastatin strongly reduces levels of Alzheimer's disease beta-amyloid peptides Abeta 42 and Abeta 40 in vitro and in vivo, *Proc. Natl. Acad. Sci. USA*, 98, pp. 5856–5861.

Fernandez-Vizarra, P., Fernandez, A. P., Castro-Blanco, S., Serrano, J., Bentura, M. L., Martinez-Murillo, R., Martinez, A. and Rodrigo, J. (2004). Intra- and extracellular Abeta and PHF in clinically evaluated cases of Alzheimer's disease, *Histol. Histopathol.*, 19, pp. 823–844.

Frederickson, C. J., Koh, J. Y. and Bush, A. I. (2005). The neurobiology of zinc in health and disease, *Nat. Rev. Neurosci.*, 6, pp. 449–462.

Gerhardsson, L., Lundh, T., Minthon, L. and Londos, E. (2008). Metal concentrations in plasma and cerebrospinal fluid in patients with Alzheimer's disease, *Dement. Geriatr. Cogn. Disord.*, 25, pp. 508–515.

Haass, C. and Selkoe, D. J. (2007). Soluble protein oligomers in neurodegeneration: lessons from the Alzheimer's amyloid beta-peptide, *Nat. Rev. Mol. Cell Biol.*, 8, pp. 101–112.

Hardy, J. and Higgins, G. A. (1992). Alzheimer's disease: the amyloid cascade hypothesis, *Science (New York, NY)*, 256, pp. 184–185.

House, E., Collingwod, J., Khan, A., Korchazkina, O., Berthon, G. and Exley, C. (2004). Aluminium, iron, zinc and copper influence the *in vitro* formation of amyloid fibrils of Abeta42 in a manner which may have consequences for metal chelation therapy in Alzheimer's disease, *J. Alzheimer. Dis.*, 6, pp. 291–301.

Hu, W. P., Chang, G. L., Chen, S. and Kuo, Y. M. (2006). Kinetic analysis of beta-amyloid peptide aggregation induced by metal ions based on surface plasmon resonance biosensing, *J. Neurosci. Methods*, 154, pp. 190–197.

Huang, X., Atwood, C. S., Moir, R. D., Hartshorn, M. A., Tanzi, R. E. and Bush, A. I. (2004). Trace elements contamination initiates the apparent auto-aggregation, amyloidosis, and oligomerization of Alzheimer's Abeta peptides, *J. Biol. Inorg. Chem.* 9, pp. 954–960.

Hureau, C. and Faller, P. (2009). Abeta-mediated ROS production by Cu ions: structural insights, mechanisms and relevance to Alzheimer's disease, *Biochim.,* 9, pp. 1212–1217.

Hwang, E. M., Kim, S. K., Sohn, J. H., Lee, J. Y., Kim, Y., Kim, Y. S. and Mook-Jung I. (2006)., *Biochem. Biophys. Res . Commun.,* 349, pp. 654–659.

Iqbal, K., Liu, F., Gong, C. X., Alonso Adel, C. and Grundke-Iqbal, I. (2009). Mechanism of tau-induced neurodegeneration, *Acta Neuropathol,* 118, pp. 53–69.

Ji, S. R., Wu, Y. and Sui, S. F. (2002). Cholesterol is an important factor affecting the membrane insertion of beta-amyloid peptide (A beta 1–40), which may potentially inhibit the fibril formation, *J. Biol. Chem.*, 277, pp. 6273–6279.

Kawahara, M., Negishi-Kato, M. and Sadakane, Y. (2009). Calcium dyshomeostasis and neurotoxicity of Alzheimer's beta-amyloid protein, *Expert. Rev. Neurother.*, 9, pp. 681–693.

Kayed, R., Sokolov, Y., Edmonds, B., McIntire, T. M., Milton, S. C., Hall, J. E. and Glabe, C. G. (2004). Permeabilization of lipid bilayers is a common conformation-dependent activity of soluble amyloid oligomers in protein misfolding diseases, *J. Biol. Chem.,* 279, pp. 46363–46366.

Kessler, H., Bayer, T. A., Bach, D., Schneider-Axmann, T., Supprian, T., Herrmann, W., Haber, M., Multhaup, G., Falkai, P. and Pajonk, F. G. (2008). Intake of copper has no effect on cognition in patients with mild Alzheimer's disease: a pilot phase 2 clinical trial, *J. Neural Transm.,* 115, pp. 1181–1187.

Khachaturian, Z. S. (1994). Calcium hypothesis of Alzheimer's disease and brain aging, *Ann. N. Y. Acad. Sci.*, 747, pp. 1–11.

Kivipelto, M., Helkala, E. L., Hänninen, T., Laakso, M. P., Hallikainen, M., Alhainen, K., Soininen, H., Tuomilehto, J. and Nissinen, A. (2001). Midlife vascular risk factors and Alzheimer's disease in later life: longitudinal, population based study, *BMJ*, 322, pp. 1447–1451.

Koppaka, V. and Axelsen, P. H. (2000). Accelerated accumulation of amyloid beta proteins on oxidatively damaged lipid membranes, *Biochem.*, 39, pp. 10011–10016.

Krewski, D., Yokel, R. A., Nieboer, E., Borchelt, D., Cohen, J., Harry, J., Kacew, S., Lindsay, J., Mahfouz, A. M. and Rondeau, V. (2007). Human health risk assessment for aluminium, aluminium oxide, and aluminium hydroxide, *J. Toxicol. Environ. Health. B. Crit. Rev.*, 10, pp. 1–269.

LaFerla, F. M., Green, K. N. and Oddo, S. (2007). Intracellular amyloid-beta in Alzheimer's disease, *Nat. Rev. Neurosci.*, 8, pp. 499–509.

Lavados, M., Guillon, M., Mujica, M. C., Rojo, L. E., Fuentes, P. and Maccioni R. B. (2008). Mild cognitive impairment and Alzheimers patients display different levels of redox-active CSF iron, *J. Alzheimer's Dis.*, 13, pp. 225–232.

Lee, J. Y., Cole, T. B., Palmiter, R. D., Sush, S. W. and Koh, J. Y. (2002). Contribution by synaptic zinc to the gender-disparate plaque formation in human Swedish mutant APP transgenic mice, *Proc. Nat. Acad. Sci. USA*, 99, pp. 7705–7710.

Lee, J. Y., Friedman, J. E., Angel, I., Kozak, A. and Koh, J. Y. (2004). The lipophilic metal chelator DP-109 reduces amyloid pathology in brains of human beta-amyloid precursor protein transgenic mice, *Neurobiol. Aging*, 25, pp. 1315–1321.

Lesne, S., Koh, M. T., Kotilinek, L., Kayed, R., Glabe, C. G., Yang, A., Gallagher, M. and Ashe, K. H. (2006). A specific amyloid-beta protein assembly in the brain impairs memory, *Nature*, 440, pp. 352–357.

Lin, H., Bhatia, R. and Lal, R. (2001), Amyloid beta protein forms ion channels: implications for Alzheimer's Disease pathophysiology, *FASEB J.*, 15, pp. 2433–2444.

Linkous, D. H., Adlard, P. A., Wanschura, P. B., Conko, K. M. and Flinn, J. M. The effects of enhanced Zinc on spatial memory and plaque formation in transgenic mice. J. Alzheimer's Dis. In press.

Loeffler, D. A., LeWitt, P. A., Juneau, P. L., Sima, A. A., Nguyen, H. U., DeMaggio, A. J., Brickman, C. M., Brewer, G. J., Dick, R. D., Troyer, M. D. and Kanaley L. (1996). Increased regional brain concentrations of ceruloplasmin in neurodegenerative disorders, *Brain Res.*, 738, pp. 265–274.

Lovell, M. A., Robertson, J. D., Teesdale, W. J., Campbell J. L. and Markesbery W. R. (1998), Copper, iron and zinc in Alzheimer's disease senile plaques., *J. Neurol. Sci.*, 158, pp. 47–52.

Lukiw, W. J. (2004). Gene expression profiling in fetal, aged, and Alzheimer hippocampus: A continuum of stress-related signaling, *Neurochem. Res.*, 29, pp. 1287–1297.

Martins, I. C., Kuperstein, I., Wilkinson, H., Maes, E., Vanbrabant, M., Jonckheere, W., Van Gelder, P., Hartmann, D., D'Hooge, R., De Strooper, B., Schymkowitz, J. and Rousseau, F. (2008). Lipids revert inert Aβ amyloid fibrils to neurotoxic protofibrils that affect learning in mice, *EMBO J.*, 24, pp. 224–233.

Martins, I. J., Berger, T., Sharman, M. J., Verdile, J., Fuller, S. G. and Martins, R. N. (2009). Cholesterol metabolism and transport in the pathogenesis of Alzheimer's Disease, *J. Neurochem.*, 9, pp. 1275–1308.

Mattson, M. P. (2007). Calcium and neurodegeneration, *Aging Cell*, 26, pp. 337–350.

Miller, L. M., Wang, Q., Telivala, T. P., Smith, R. J., Lanzirotti, A. and Miklossy, J. (2006). Synchroton-based infrared and x-ray imaging shows focalized accumulation of Cu and Zn co-localized with beta-amyloid deposits in Alzheimer's disease, *J. Struct. Biol.,* 155, pp. 30–37.

Mo, Z. Y., Zhu, Y., Zhu, H., Fan, J., Chen J. and Liang Y. (2009). Low micromolar zinc accelerates the fibrillization of human tau via bridging of Cys-291 and Cys-322, *J. Biol. Chem.,* 284, pp. 34648–34657.

Molina, J. A., Jiménez-Jiménez, F. J., Aguilar, M. V., Meseguer, I., Mateos-Vega, C. J., González-Muñoz, M. J., de Bustos, F., Porta, J., Ortí-Pareja, M., Zurdo, M., Barrios, E. and Martínez-Para, M. C. (1998). Cerebrospinal fluid levels of transition metals in patients with Alzheimer's disease, *J. Neural. Transm.*, 105, pp. 479–488.

Morita, M., Vestergaard, M., Hamada, T. and Takagi, M. (2009). Real-time observation of model membrane dynamics induced by Alzheimer's amyloid beta, *Biophys. Chem.*, in press.

Näslund, J., Haroutunian, V., Mohs, R., Davis, K. L., Davies, P., Greengard, P. and Buxbaum, J. D. (2000). Correlation between elevated levels of amyloid beta-peptide in the brain and cognitive decline. JAMA, 283, pp. 1571–1577.

Nordberg, A. (2008). Amyloid plaque imaging in vivo: current achievement and future prospects, *Eur. J. Nucl. Med. Mol. Imag.,* 35, pp. 46–50.

Oddo, S., Caccamo, A., Kitazawa, M., Tseng, B. P. and LaFerla, F. M. (2003), Triple-transgenic model of Alzheimer's disease with plaques and tangles: intracellular Abeta and synaptic dysfunction, *Neuron,* 39, pp. 409–421.

Opazo, C., Huang, X., Cherny, R. A., Moir, R. D., Roher, A. E., White, A. R., Cappai, R., Masters, C. L., Tanzi, R. E., Inestrosa, N. C. and Bush, A. I. (2002). Metalloenzyme-like activity of Alzheimer's disease beta-amyloid. Cu-dependent catalytic conversion of dopamine, cholesterol, and biological reducing agents to neurotoxic H(2)O(2), *J. Biol. Chem.,* 277, pp. 40302–40308.

Pantopoulos, K. and Hentze, M. W. (1998). Activation of iron regulatory protein-1 by oxidative stress in vitro. Proc. *Natl. Acad. Sci. U.S.A.,* 95, pp. 10559–10563.

Perl, D. P. and Moalem, S. (2006). Aluminum Alzheimer's disease and the geospatial occurence of similar disorders, *Med. Miner. and Geochemistry,* 64, pp. 115–134.

Pettegrew, J. W., Panchalingam, K., Hamilton, R. L. and McClure, R. J. (2001). Brain membrane phospholipid alterations in Alzheimer's disease, *Neurochem. Res.,* 26, pp. 771–782.

Pierrot, N., Ghisdal, P., Caumont, A. S. and Octave, J. N. (2004). Intraneuronal amyloid-beta1-42 production triggered by sustained increase of cytosolic calcium concentration induces neuronal death, *J. Neurochem.,* 88, pp. 1140–1150.

Quinn, J. F., Crane, S., Harris, C. and Wadsworth, T. L. (2009). Copper in Alzheimer's disease: too much or too little? *Expert. Rev. Neurother.,* 9, pp. 631–637.

Rajendran, R., Minqin, R., Ynsa, M. D., Casadesus, G., Smith, M. A., Perry, G., Halliwell, B., and Watt, F. (2009), *Biochem. Biophys. Res. Commun.* 382, pp. 91–95.

Reid, P. C., Urano, Y., Kodama, T. and Hamakubo, T. (2007). Alzheimer's disease: cholesterol, membrane rafts, isoprenoids and statins, *J. Cell Mol. Med.,* 11, pp. 383–392.

Reusche, E. (2003). Aluminium and central nervous system morphology in hemodialysis, in: Metal ions and neurodegenerative disorders. Singapore, London, World Scientific pp. 117–138.

Ricchelli, F., Drago, D., Filippi, B., Tognon, G. and Zatta, P. (2005). Aluminum-triggered structural modifications and aggregation of beta-amyloids, *Cell Mol. Life Sci.,* 62, pp. 1724–1733.

Roychaudhuri, R., Yang, M., Hoshi, M. M. and Teplow, D. B. (2009). Amyloid beta- protein assembly and Alzheimer's Disease, *J. Biol. Chem.,* 284, pp. 4749–4753.

Rulon, L. L., Robertson, J. D., Lovell, M. A., Deibel, M. A., Ehmann, W. D. and Markesber, W. R. (2000). Serum zinc levels and Alzheimer's disease, *Biol. Trace Elem. Res.*, 75, pp. 79–85.

Schipper, H. M. (2004). Heme oxygenase expression in human central nervous system disorders, *Free Radic. Biol. Med.,* 37, pp. 1995–2011.

Shankar, G. M., Li, S., Mehta, T. H., Garcia-Munoz, A., Shepardson, N. E., Smith, I., Brett, F. M., Farrell, M. A., Rowan, M. J., Lemere, C. A., Regan, C. M., Walsh, D. M., Sabatini, B. L. and Selkoe, D. J. (2008). Amyloid-beta protein dimers isolated directly from Alzheimer's brains impair synaptic plasticity and memory, *Nat. Med.,* 14, pp. 837–842.

Silvestri, L. and Camaschella, C. (2008). A potential pathogenetic role of iron in Alzheimer's disease. *J. Cell Mol. Med.,* 12, pp. 1548–1550.

Smith, D.G., Cappai, R. and Barnham, K. J. (2007). The redox chemistry of the Alzheimer's disease amyloid beta peptide, *Biochim. Biophys. Acta,* 1768, pp. 1976–1990.

Smith, M. A., Zhu, X., Tabaton, M., Liu, G., McKeel, D. W., Cohen, M. L., Wang, X., Siedlak, S. L., Dwyer, B. E., Hayashi, T., Nakamura, M., Nunomura, A. and Perry, G. (2010). Increased iron and free radical generation in preclinical Alzheimer Disease and mild cognitive impairment, *J. Alzheimers Dis.,* 19, pp. 363–372.

Speziali, M. and Orvini, E. (2003). Metals distribution and regionalization in the brain, in: Metal ions and neurodegenerative disorders. Singapore, London, World Scientific, pp. 15–65.

Stoltenberg, M., Bruhn, M., Sondergaard, C., Doering, P., West, M. J., Larsen, A., Troncoso, J. C. and Danscher, G. (2005), Immersion autometallographic tracing of zinc ions in Alzheimer beta-amyloid plaques, *Histochem. Cell Biol.,* 123, pp. 605–611.

Suwalsky, M., Bolognin, S. and Zatta, P. (2009). Interaction between Alzheimer β-amyloid and β-amyloid-metal complexes with cell membranes, *J. Alzheimer Dis.*, 17, pp. 81–90.

Suwalsky, M., Martínez, F., Cárdenas, H., Grzyb, J. and Strzalka, K. (2005). Iron affects the structure of cell membrane molecular models, *Chem. Phys. Lipids*, 134, pp. 69–77.

Tomic, J. L., Pensalfini, A., Head, E. and Glabe, C. G. (2009). Soluble fibrillar oligomer levels are elevated on Alzheimer's disease brain and correlate with cognitive dysfunction, *Neurobiol. Dis.,* 35, pp. 352–358.

Tsai, J., Grutzendler, J., Duff, K. and Gan, W. B. (2004) Fibrillar amyloid deposition leads to local synaptic abnormalities and breakage of neuronal branches. *Nat. Neurosci.,* 7, pp. 1181–1183.

Walton, J. R. and Wang, M. X. (2009). APP expression, distribution and accumulation are altered by aluminum in a rodent model for Alzheimer's disease, *J. Inorg. Biochem.,* 103, pp. 1548–1554.

Yumoto, S., Kakimi, S., Ohsaki, A. and Ishikawa, A. (2009). Demonstration of aluminum in amyloid fibers in the cores of senile plaques in the brains of patients with Alzheimer's disease. *J. Inorg. Biochem.,* In press.

Zambenedetti, P., Giordano, R. and Zatta P. (1998). Metallothioneins are highly expressed in astrocytes and microcapillaries in Alzheimer's disease, *J. Chem. Neuroanat.,* 15, pp. 21–26.

Zatta, P. (2006). Aluminum and Alzheimer's disease: a vexata question between uncertain data and a lot of imagination, *J. Alzheimers Dis.,* 10, pp. 33–37.

Zatta, P. (Ed.) (2002). Recent topics in aluminium chemistry, *Coord. Chem. Rev.* 228, pp. 391–396.

Zatta, P., Drago, D., Bolognin, S. and Sensi, S. L. (2009). Alzheimer's disease, metal ions and metabolic homeostatic therapy, *Trends Pharmacol. Sci.,* 30, pp. 346–355.

Zatta, P. and Frank, A. (2006). Copper deficiency and neurological disorders in man and animals, *Brain Res. Rev.*, 54, pp. 19–33.

Zatta, P., Zambenedetti, P., Reusche, E., Stellmacher, F., Cester, A., Albanese, P., Meneghel, G. and Nordio, M. (2004). A fatal case of aluminium encephalopathy in a patient with severe chronic renal failure not on dialysis, *Nephrol. Dial. Transplant.,* 19, pp. 2929–2931.

Chapter 8

Prion Diseases, Metals and Antioxidants

Paul Davies and David R. Brown
Department of Biology and Biochemistry, University of Bath, Bath, BA2 7AY, United Kingdom

Prion diseases are rare neurodegenerative diseases but have gained considerable notoriety because of BSE and its links to the human disease, variant CJD. However, prion diseases, like other amyloidogenic diseases, are linked to changes to a key protein. The prion protein's change in conformation is considered the pivotal event that leads to prion disease. However, this mysterious protein has been hard to understand in terms of its role in the cell and the events that trigger this conformational change. One of the key findings that have changed the way we view these diseases is discovery of the ability of the prion protein to bind metals. In particular the normal isoform of the prion protein binds copper. Expression of the prion protein by cells has also been shown to protect cells, particularly from oxidative damage. Binding of the "wrong" metal, manganese plays a role in protein conversion. Therefore the study of metals in prion disease has proven a very rich ground for understanding these complex diseases and the enigmatic protein associated with them, the prion protein. This chapter provides an up-do-date review of the current knowledge of the prion protein as a metalloprotein.

Brain Diseases and Metalloproteins
Edited by David R. Brown

ISBN 978-981-4316-01-9 (Hardcover), 978-981-4364-07-2 (eBook)
www.panstanford.com

8.1 The Transmissible Spongiform Encephalopathies

The transmissible spongiform encephalopathies (TSE's) or prion diseases are a range of neurodegenerative disorders in which a cell surface glycoprotein known as the prion protein (PrP) has been directly implicated in pathogenesis. These diseases include Creutzfeldt–Jacob disease (CJD), Gerstmann Straussler syndrome (GSS), Familial Insomnia (FFI) and kuru in humans as well as bovine spongiform encephalopathy (BSE) in cattle, chronic wasting disease (CWD) in deer and elk and scrapie in sheep (Gajdusek, 1991; Prusiner, 1996, 1998a, b). The diseases have a distinctive neuropathology where spongiform degeneration occurs within the brain accompanied by deposition of abnormal aggregated PrP. Symptomatically, there is a rapid onset of dementia, loss of coordination, paralysis and invariably death usually from secondary complications.

The earliest reported cases of TSE in animals date back to the mid eighteenth century (Brown and Bradley, 1998), where the first incidents of scrapie in sheep were noted. In contrast it was not until the 1980's that the now familiar BSE emerged (Hope *et al.,* 1988) where cattle developed scrapie-like symptoms, probably as a result of eating feed containing infected offal. The British epidemic of BSE, labelled by the media as 'mad cow disease', caused the deaths of nearly 200,000 cattle in the late 1980's and 1990's (Anderson *et al.,* 1996). It is generally now accepted that the epidemic was as a result of the introduction of new rendering procedures in the preparation of cattle feed (Harris, 1999). Since the epidemic arose, government restrictions on how cattle feed is produced has led to a significant reduction in the number of cases on a year on year basis. The disease, however, has yet to be fully eradicated. In addition, since 1996, a human form of disease known as new variant CJD, has been linked with consumption of infected material from BSE cattle (Will *et al.,* 1996). Inter human transmission of CJD has also been linked to tissue grafts and other forms of surgery as well as blood transfusion (Brown *et al.,* 1992; Brown, 1998).

Human forms of the TSE's can be put into three epidemiological categories, infectious (5%), sporadic (80%), and inherited (15%) (McKintosh *et al.,* 2003). The sporadic form of the human disease, sCJD, occurs with an incidence of around 1 case per 2 million people.

This possibly results from a single PrP misfolding event. Alternatively, the condition may occur as a result of somatic mutations in the prion gene, Prnp (Will *et al.*, 1996). GSS, familiar CJD and FFI are inherited forms of disease with a pattern of autosomal dominance (Collins *et al.*, 2001). The most common inherited form, GSS, usually occurs in middle age as a chronic cerebral ataxia (McKintosh *et al.*, 2003). Dementia then follows with death occurring around five years following onset of symptoms. The main mutation causing GSS was first identified in 1989 as being P102L, although several other mutations in the prion protein gene have since been shown to cause the disease (Hsiao *et al.*, 1992). FFI is typified by an increasing inability to sleep along with a steady decline in autonomic function (Lugaresi *et al.*, 1986a; Lugaresi *et al.*, 1986b). FFI patients exhibit atrophy of the anterior-ventral and medial-dorsal thalamic nuclei and carry mutations within codon 178 in the prion gene (Medori *et al.*, 1992). Of note with these inherited forms is that it has been shown that the disease prions formed in the brains of these patients are still infectious and can be used to infect other individuals (Medori *et al.*, 1992; Kretzschmar *et al.*, 1995).

The infectious forms of TSE are by far the rarest and depend upon acquisition of an infectious form of the protein and consequent transfer across the blood brain barrier. This acquisition can be through ingestion or direct blood or tissue transfer from an infected individual (iatrogenic CJD). For direct modes of transmission, there are recorded cases of infection from contaminated dura mater grafts, corneal transplants, neurosurgical instruments, and cadaveric growth hormone (Brown *et al.*, 1992). Transmission via ingestion was first recorded for animal populations and later for human populations through the discovery of Kuru (McKintosh *et al.*, 2003). These cases involved intra-species transmission and exhibited extremely high communicability with a relatively consistent incubation before symptom onset. Where different species are concerned, infection is limited considerably by a species barrier. This is exemplified when material is taken from an infected individual and used to challenge an individual from another species. The rate of infection is significantly less and also carries a much longer incubation period (McKintosh *et al.*, 2003). This species barrier has been identified as being caused by differing PrP primary structures between species (Collinge and Palmer, 1994). However, this is not always the case as BSE can be transmitted readily to species where the protein has a significantly different sequence (Bruce *et al.,* 1994). The most relevant example of this phenomenon is the transmission of prion disease to humans from cattle, leading to vCJD.

Transmission itself is thought to involve only PrP and no genetic material, hence the formation of the 'protein only' hypothesis proposed by Griffiths in 1967 (Griffith, 1967) and later adapted by Stanley Prusiner (Prusiner, 1982). This hypothesis sees an abnormally folded form of the prion protein (PrP^{Sc}) cause the normal cellular form (PrP^{c}) to become misfolded itself. This then leads to a chain reaction in which PrP^{Sc} leads to the conversion and formation of more PrP^{Sc} using endogenous PrP^{c} as substrate. With this knowledge, transmission via surgical procedure has been minimised by new procedures and techniques such as recombinant forms of donor hormone and familial screening of potential tissue donors (McKintosh *et al.,* 2003). Transmission to humans via the ingestion of infected material caused vCJD cases to peak in 2000 with 28 deaths but has been declining exponentially since (Andrews, 2009). For humans at least, the possibilities of an epidemic now seem unlikely. For the animal population, however, cases still persist.

The possible modes of prion disease pathogenesis can be divided into three categories; loss of PrP function, gain of PrP function or subversion of PrP function (Westergard *et al.,* 2007). The exact cause of pathogenesis is hotly debated, but may involve one or more of these categories. Loss of function centres on PrP^{c} being converted to PrP^{Sc} and losing an essential physiological role. The most commonly suggested role for PrP^{c} in this regard is one of anti-apoptosis leading to a loss of neuronal cells. Subversion of function involves the conversion of PrP from a transducer of neuro-protective signals to a transducer of neuro-toxic signals. Gain of function would see PrP^{Sc} act directly as a toxic entity, possibly through the aggregation of the protein leading to disruption of axonal transport and synapse transmission or the triggering of apoptotic pathways (Westergard *et al.,* 2007).

The different classifications or strains of the TSE's are identified by several means. Firstly, each type of TSE produces symptoms specific to the strain type. Post mortem analysis can also reveal different patterns of damage to the brain. Perhaps the most distinguishing feature however, is the ability to identify strain types on a polyacrylamide gel. Following digestion using the protease proteinase K, PrP^{Sc} has different migration characteristics specific to the strain type (Wadsworth *et al.,* 1999). This allows for an accurate determination of the source of infection.

8.2 The Prion Protein

The prion protein (PrP) is one of an emerging family of proteins. Two homologous proteins are doppel and shadoo (Moore *et al.,* 1999; Premzl *et al.,* 2003). While the prion diseases are restricted to mammals, there are similar proteins found in all lower vertebrates (Calzolai *et al.,* 2005). These proteins are all cell surface associated glycoproteins (Meyer *et al.,* 1986). An altered isoform of mammalian PrP is considered to be the agent of infection in prion diseases (Prusiner, 1998b). As such understanding the nature of the molecular switch that converts a normal brain glycoprotein to aggregating, neurotoxic isoform is the key to any possible treatment for the diseases. Additionally, understanding the molecular biology of the normal isoform of PrP could provide insights into why, under certain circumstances, misfolding and aggregation of this protein leads to disease. As these diseases a predominantly seen in patients over 60, then it is quite possible the changes occur as a result of cellular aging.

The cellular isoform of PrP (PrP^c) is a globular protein with a long unstructured N-terminal tail (Riek *et al.,* 1996; Riek *et al.,* 1997). The protein is anchored to the outer surface of the cell membrane of glycosyl-phosphotidyl-inositol (GPI) anchor (Stahl *et al.,* 1992) and has a half-life at the surface of the cell of approximately one hour (Pauly and Harris 1998). The globular domain is formed principally of three alpha helices, two of which are linked by a disulphide bridge formed between two cysteins. The cellular protein is also glycosylated at two asparagine residues and can exist as di- mono- and non-glycosylated forms. The primary sequence is highly conserved in all mammals. This is important as only mammals develop prion diseases (Wopfner *et al.,* 1999). The protein is around 25 kD in the unglycosylated form and has two highly conserved domains. The first of these is the octameric repeat region, within the N-terminal domain which has 4-6 repeats of 8 amino acid residues which includes one histidine in each. The second conserved region is a hydrophobic region towards the end of the N-terminal domain which contains a palendromic sequence (Arg-Gly-Arg-Arg-Arg-Arg-Gly-Arg). The protein is cleaved during break down principally within this domain but also toward the end of the octameric repeat region (Vincent *et al.*, 2001; Abdelraheim *et al.*, 2006).

The PrP was first identified as an aggregated protein in brains of rodents with scrapie (Bolton *et al.*, 1982). Analyses of the protein eventually lead to the identification of the Prnp gene (Basler *et al.*, 1986), and that a cellular protein existed, with the same protein sequence. The abnormal isoform of PrP is usually termed PrP^{Sc}, although variations exist such as PrP* and PrP^{d} (Jeffrey and Gonzalez, 2007). The abnormal isoform has higher beta-sheet content and is protease resistant (Pan *et al.*, 1993). It is also often cleaved at the N-terminus around residue 90. In sporadic cases of prion diseases the sequence of the protein is no different suggesting that the disease is principally a protein conformation disorder. The abnormal isoform of PrP is considered the agent of disease as transference of the protein between individuals appears to be sufficient for disease transmission (Prusiner, 1982). Additionally, animals knocked out for the prion protein are resistant to prion disease, suggesting that host expression of PrP is essential for disease (Bueler *et al.*, 1993). This highlights the critical nature of conversion of host cellular PrP to the abnormal isoform to prion disease. Therefore, understanding this process and factors that influence this process are at the heart of research strategies to understand the cause and possible treatment of prion diseases.

8.3 PrP and Copper Binding

It is now well established that PrP is a metalloprotein (Brown *et al.*, 1997; Jackson *et al.*, 2001; Kramer *et al.*, 2001; Burns *et al.*, 2003; Thompsett *et al.*, 2005; Davies *et al.*, 2009; Nadal *et al.*, 2009). Initial studies identified PrP as a copper binding protein based on the study of the N-terminus (Brown *et al.*, 1997). The interaction between copper and PrP was principally through the N-terminal octameric repeats but another site was also proposed even at this early stage. The octameric repeat region can bind up to four atoms of copper, the co-ordination of each involving a key histidine as deletion of these histidines results in loss of copper binding. Subsequently, a site for binding of a fifth copper atom has been identified in association with histidines near the octameric repeat region (hisitidines at amino acid residues 96 and 111 in the human sequence) (Jones *et al.*, 2004).

8.3.1 *The Octameric Repeat Region – Coordination*

Figure 8.1 shows a schematic of the primary structure of mouse PrP. The mature prion protein, when cleaved from its signal sequence and GPI motif, is some 209 amino acids long with a structured C terminus and, in the absence of copper at least, an unstructured N terminus. The first real hard evidence for this association of copper with PrP came from the work of Brown *et al.* in 1997. By carrying out equilibrium dialysis on the recombinant N terminal region (residues 23–98), they were able to show that between five and six atoms of copper bound per recombinant fragment. Further detailed studies involving electron paramagnetic resonance imaging (EPR)

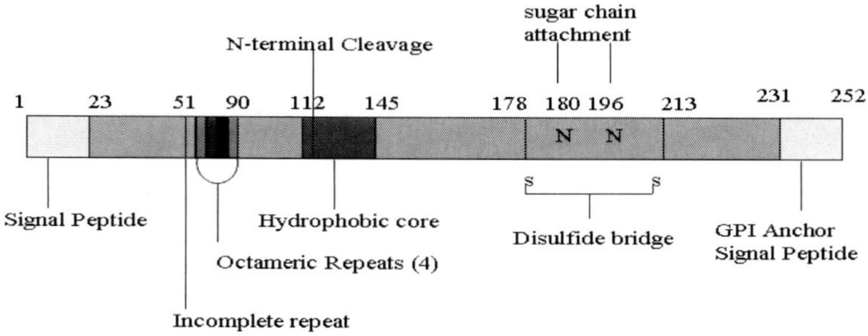

Figure 8.1 The primary structure of the mouse prion protein. This protein is anchored to the cell membrane by a GPI anchor. The signal peptide for entry into the endoplasmic reticulum and the GPI signal peptide are cleaved off before the protein reaches the cell surface. Glycosylation can occur on one, two or none of the asparagine residues indicated. A hydrophobic region envelopes a cleavage point where the protein is cleaved during normal metabolic breakdown. A disulphide bond links two regions of the protein which form separate alpha-helices in the three dimensional structure of the protein. The complete octarepeats can bind up to four copper atoms. Most mammals also have an incomplete repeat prior to this. See also Colour Insert.

and X-Ray crystallography on recombinant peptide fragments have demonstrated that a single copper is coordinated by each octarepeat segment in a pentacoordinate complex involving residues HGGGW only (Aronoff-Spencer *et al.,* 2000; Burns *et al.,* 2002). This study by Arnoff-Spencer *et al.,* 2000 revealed an equatorial coordination involving the Histidine imidazole, deprotonated amides from the adjacent two glycines and a deprotonated carbonyl from the last glycine. A water molecule was also identified as being involved by allowing oxygen to coordinate axially forming a bridge to the NH of the indole on the last tryptophan. Further work by this group, however, has revealed a more complex binding system. In 2005, the same group demonstrated that the coordination of copper was dependent on the degree of copper occupancy on the protein (Chattopadhyay *et al.,* 2005). They revealed three distinct coordination modes, clearly distinguishable at different relative concentrations of copper. Using X-band and S-band EPR and electron spin echo envelope modulation (ESEEM) to analyse a library of modified peptides, they showed a multiple His coordination mode at low copper occupancy, moving through a transitional coordination to the maximal occupancy at a physiological pH 7.4.

Component 1 coordination dominates at high copper occupancy of 2 copper equivalents and above. As the group previously revealed, a 3N1O arrangement coordinated a single copper per HGGGW motif. Figure 8.2 shows a three dimensional model of this mode for clarity.

EPR also revealed evidence of dipolar copper-copper centres in approximately 20% of the spectra suggesting a close copper-copper proximity of between 3.5 Å and 6 Å, close enough for van der Waals interactions. These interactions may be responsible for driving a hydrophobic collapse and consequent N terminal structural organisation at full copper occupancy. Component two coordination is present only as a transition between low copper occupancy of one atom or less and full copper occupancy. The precise coordination mode of this component was extremely difficult to accurately characterise as the authors found it to be mixed with other modes in all the conditions tested. However, by methylating the second glycine residue in each octarepeat, they were able to successfully block component 1 formation and resolve the coordination mode for component 2. The results suggest a 2N2O arrangement at the

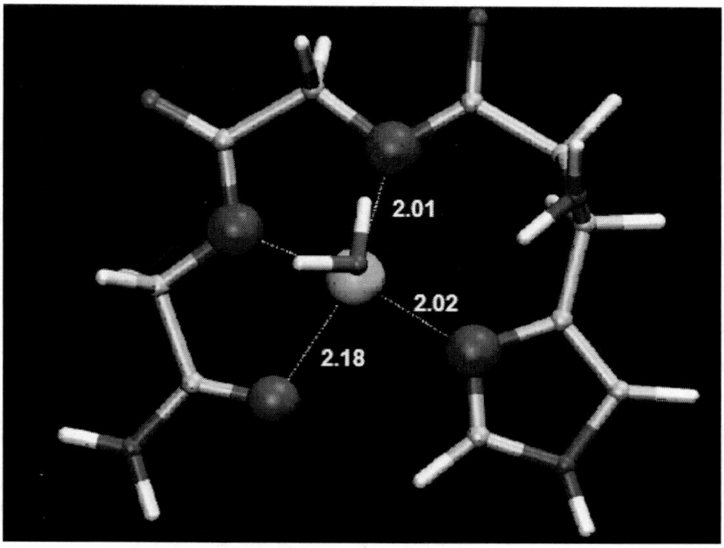

Figure 8.2 Three-dimensional model representing component 1 of the equatorial coordination mode of copper binding to the octarepeat region Bond lengths shown are in angstroms (1 = 0.1 nm). See also Colour Insert.

expected neutral charge state. These findings suggest an intermediate coordination involving the His imidazole and its exocyclic nitrogen. Further contributions appear to stem from the oxygens of two water molecules within the equatorial arrangement. Component 3, present at low copper occupancy only, is likely to provide the highest affinity copper binding within the octarepeat due to its multi-His coordination. This coordination mode involves either a 3N1O or 4N arrangement and is only available at above pH 6.5 and when multiple histidine residues are present. This groups combined data points towards a mechanism for dramatic structural change within the N terminus of PrP that is entirely dependent on the amount of copper bound to the protein.

Recent work by Weiss *et al.,* 2007 employed extended X-ray absorbance fine structure spectroscopy (EXAFS), EPR, electron nuclear double resonance (ENDOR) and molecular modelling to resolve the copper coordination to PrP. In contrast to many of the previous studies (Viles *et al.,* 1999; Aronoff-Spencer *et al.,* 2000; Burns *et al.,* 2002; Morante *et al.,* 2004; Chattopadhyay *et al.,* 2005), they concentrated on spectra from full length recombinant human PrP.

Although the authors report multiple configurations dependent on copper occupancy, they only identified two distinct modes of binding, models. As found previously (Mentler *et al.,* 2005), a coordination identified as species 1 was evident that utilised a 3N1O configuration. Contributions are from the imidazole nitrogen, two glycine amides and a carbonyl oxygen from the last glycine.

It is clear that this is exactly the same model as that was proposed previously (Aronoff-Spencer *et al.,* 2000; Burns *et al.,* 2002). The second model, or species 2, was obtained through analysis of the super hyperfine element of the EPR spectra, based on the principle that multiplicity relates to the number of nitrogen ligands. They revealed a 4N arrangement when looking at low to intermediate copper occupancy on full length protein and the entire octarepeat segment. This in itself is not strikingly different from previous findings, however the author's conclusions, in light of their ENDOR data are dramatically at odds with anything previously proposed. An intermediate structure involving two imidazole nitrogens and two glycine nitrogens was proposed.

A very interesting article utilising *ab initio* simulations (Furlan *et al.,* 2007) produced some coordination models that were very similar to those proposed experimentally. They supported the idea of multi-His coordination under certain conditions. Interestingly, however, they suggested that the axial water thought to be involved in a pentacoordination was actually only bound to the indole of tryptophan and not directly involved. They went on to verify this and confirmed that the Cu-water interaction was extremely weak.

The relative difficulty of resolving structures accurately by nuclear magnetic resonance imaging (NMR) when paramagnetic elements, such as copper, are present has led to an interest in finding copper analogues to help elucidate the Copper-octarepeat environment. One group investigated Ni^{2+}, Pd^{2+}, Pt^{2+} and Au^{3+} ions for their suitability as a diamagnetic probe of Cu^{2+} binding (Garnett *et al.,* 2006). The authors found that Pd^{2+} would form a square planar complex (the other ions would not) but in a different coordination than that seen for copper. Hence, they conclude that for the octameric repeat at least, there are no suitable analogues. Another study (Shearer and Soh, 2007b) also showed Ni^{2+} was an unsuitable analogue.

8.3.2 *The 5th Site – Coordination*

A large body of literature now exists that demonstrates that copper is able to bind outside of the octarepeat region of PrP (Hasnain *et al.*, 2001; Jackson *et al.*, 2001; Kramer *et al.*, 2001; Burns *et al.*, 2003; Jones *et al.*, 2004; Jones *et al.*, 2005a). Work by Jones *et al.*, in 2004 and 2005 highlighted these copper binding regions as His96 and His111 in the human protein. They also identified the minimum sequence necessary for Copper binding to this region of PrP are amino acids 92–96 and 107–111. Recently, the same group utilised NMR and visible circular dichroism (Vis-CD) to fully elucidate the coordination of copper to this so called 5th site (Klewpatinond and Viles, 2007). Interestingly, they found that the coordination of copper changed dramatically dependent on chain length and pH. The Vis-CD spectra for 90–126 was strikingly different from 91–115, despite only two sites being present on both fragments. By studying each individual site on each fragment by replacing the histidine at either 96 or 111 with alanine and comparing the spectra of these mutant fragments with that from the original fragments, they determined that this change in spectra was not caused by differences in coordination. They concluded that the spectra differences were caused by a change in relative affinity of His96 and His111 for copper. Although His111 seems to display the highest affinity for copper, His96 affinity increases dramatically on the addition of the eleven amino acid hydrophobic segment. Interestingly, the relative affinity of these two sites is reversed for nickel, with His96 demonstrating the tightest binding. They also discovered a multi coordination mode that was strongly influenced by pH. Although all display a square planar geometry, the differences are clear. At pH 7.5 and above, a 4N complex dominates, while at pH6, a ligand rearrangement shifts the coordination to a 3N1O configuration. At low pH, a multi His 2N2O coordination dominates. By combining this EPR data with Vis-CD, the authors conclude that the histidines at 96 and 111 bind copper independently except at low pH, where it appears that both histidines are involved in the coordination of a single copper atom. Analysis by ^{1}H-NMR using nickel as a probe further confirms these findings. Another study also reported the key binding site as being present between residues 106 and 114 (Shearer and Soh, 2007a).

In stark contrast with the finding that His111 is most important for copper binding, another group reported that His96 was the key site involved (Treiber *et al.*, 2007). This was identified using real time surface plasmon resonance (SPR) analysis on synthetic peptides and recombinant protein.

8.3.3 *The Affinity of Copper for PrP*

Clearly, in order to understand the physiological importance of copper binding to PrP, it is necessary to know the affinity of the metal for the protein. With copper so tightly regulated in the body (Linder and Hazegh-Azam, 1996), any functional copper protein must be able to chelate the copper and keep hold of it through its functional life. There have been a great number of attempts made to calculate the disassociation constants of copper and other metals from PrP but there is, unfortunately, much difference of opinion.

The first real attempt to assess the affinity of copper for PrP was in the mid 1990s, first on synthetic peptides (Hornshaw *et al.*, 1995) by X-ray fluorescence, revealing a K_d of 6.7 μM and then on full length protein (Stockel *et al.*, 1998) yielding a K_d of 14 μM. Following on from this, another group used various spectroscopic techniques and found the affinity for copper within the octarepeat to be k_d~6 μM hen two octarepeat segments present (Viles *et al.*, 1999). When three or four coppers were present, cooperative binding was observed. These early results were all in close agreement. In 2001, however, Jackson *et al.*, utilised fluorescence quenching and discovered a very different story. They discovered that a very tight binding event was followed by a significantly weaker event. Using glycine competition, they calculated the initial binding event occurred with a k_d~8 fM and subsequent binding with a K_d~15 μM. Another group then published further data in support of positive cooperation (Garnett and Viles, 2003). Using CD spectroscopy and competitive chelators, they ruled out the previously reported femtomolar affinity by showing that glycine was successfully able to compete with PrP for copper. As glycine has a nanomolar affinity for copper, this meant that the affinity range had to be between the micromolar to nanomolar range. Further work (Walter *et al.*, 2006) sought to assess the affinities for copper within each of the coordination modes they discovered.

They found evidence for negative cooperativity, reporting a high affinity initial binding event in the nanomolar range followed by 3 subsequent events in the micromolar range. Data obtained by a two different methods (Thompsett *et al.*, 2005) also demonstrated a negative cooperation between binding events in the octarepeat region. By utilising isothermal titration calorimetry (ITC) and competitive metal capture analysis (CMCA) they reported an initial binding event in the low femtomolar range followed by three subsequent events in the picomolar range. Recently, Treiber *et al.*, 2007 used RT-SPR (real time surface plasmon resonance) to assess the affinity of the octarepeat for copper and found the K_d to be in the nanomolar range.

The affinity of the 5th site for copper is also a subject of much debate. Jones *et al*, 2004, spectroscopically assessed the 5th site to bind copper with nanomolar affinity. The same group then showed that the affinity of the 5th site for copper was higher than that of the octarepeat for copper (Jones *et al.*, 2005a). This was considerably more conservative than the femtomolar affinity proposed by Jackson *et al.*, 2001. Then, in 2007, two independent groups both produced data from very different methods suggesting the affinity of the 5th site was in the mid-micromolar range (Shearer and Soh, 2007a; Treiber *et al.*, 2007). The later group also compared the 5th site with the octarepeat and found that the relative affinity was higher in the octarepeat region.

More recent studies using either full length recombinant protein on its own or in conjunction with peptides have largely reached a consensus that the dissociation constant lies within the low nanomolar range (Nadal *et al.*, 2009). This fits quite well with early studies that examined copper uptake into cells that was associated with the internalisation of PrP. These studies suggested a Km in the nanomolar range (Brown, 1999; 2004). Studies with purified native PrP have shown that the affinity of copper for PrP is not affected by glycosylation (Davies *et al.*, 2009).

8.4 The Implications of Copper Binding

A key factor in the physiological relevance of copper binding to PrP is the effect that copper has when bound. There is an enormous body of evidence reporting on the ability of PrP to aggregate, form amyloid, alter its resistance to proteases, carry out electrochemical activities and transport metals through membranes. The next section looks at the evidence for the effects of copper on PrP.

8.4.1 *Protein Electrochemistry*

There has been some recent work on the electrochemistry of the protein when bound to copper. It has been shown that copper is reduced form Cu^{2+} to Cu^{+} on binding to PrP (Miura *et al.,* 2005). Brown *et al.,* 1999 first suggested that PrP displayed superoxide dismutase like activity by carrying out assays with both native and recombinant mouse protein. They demonstrated, by modifying sections of the protein, that it was the copper bound octarepeat that was responsible for the enzymatic activity of PrP. They further demonstrated this in 2001 with native protein and showed that it was able to protect against oxidative stress (Brown *et al.,* 2001). A study in the same year showed that PrP from scrapie infected mice demonstrated a dramatic reduction in this antioxidant activity (Wong *et al.,* 2001a). Also in this year, a study demonstrated that PrP knock out mice have an increase in markers of oxidative stress such as lipid and protein oxidation (Klamt *et al.,* 2001). Another study produced evidence that the 5th site was redox active by employing cyclic voltammetry on copper bound PrP fragments encompassing the 5th site region (Shearer and Soh, 2007a). Their data clearly demonstrate a quasi-reversible reaction where the 5th site is able to cycle electrons without becoming permanently oxidised or reduced. A separate study also demonstrated a quasi-reversible reaction for copper bound peptides corresponding to the octarepeat region (Hureau *et al.,* 2006). These studies combined suggest that all copper centres on the protein are able to undergo cyclic redox chemistry. In line with this evidence, an independent study showed that PrP does not redox silence copper (Nadal *et al.,* 2007). They found that the protein was able to dramatically reduce the amount of hydroxyl radicals present in a Cu^{2+}/ascorbate/oxygen system without affecting hydrogen peroxide levels. The conclusion is clearly that the protein is quenching these radicals by a method other than Fenton chemistry and is doing so sacrificially. Interestingly, they also showed that the octarepeat region was protective against residue oxidation within the 5th site, suggesting that the 90–231 form of the protein may be more susceptible to oxidative damage. This evidence for electrochemical activity with PrP is supported by numerous others articles including (Walz *et al.,* 1999; White *et al.,* 1999; Frederikse *et al.,* 2000; Wong *et al.,* 2000a; Dupuis *et al.,* 2002; Huber *et al.,* 2002; Miele *et al.,* 2002; Curtis *et al.,* 2003; Rachidi *et al.,* 2003; Zeng *et al.,* 2003)

More recently, studies involving our research group have focused more on the quantifying and qualifying the precise mode and action of the transfer of electrons involving the copper centres of PrP. The first highlighted the effects of disrupting the proteins electrochemistry by exposing PrP to manganese (Brazier *et al.,* 2008). The study clearly demonstrated that it was the copper within the octarepeat region that was necessary for redox cycling. A more detailed study (Davies *et al.,* 2009) then defined the electron transfer kinetics and pH dependence of PrP redox activity when copper is bound. The results demonstrated that PrP was able to undergo stable redox cycling and provided evidence for a redox related function for PrP. In contrast to this rather compelling evidence for the electrochemical properties of PrP, two other studies have cast serious doubt on the ability of PrP to act as a superoxide dismutase, demonstrating that neither the *in vivo* protein (Hutter *et al.,* 2003) or recombinant protein (Jones *et al.,* 2005b) posses any superoxide dismutase like activity. Unfortunately, the studies disputing these finding are based on analyses with commercial kits employing high concentrations of copper chelators that would easily strip PrP of any copper bound (Hutter *et al.,* 2003; Jones *et al.,* 2005b). The activity of PrP is sensitive to refolding conditions and loses activity if a disulfide bridge is not formed (Wong *et al.,* 2000b). Furthermore, a definitive study of copper bound PrP using cyclic voltammetry clear shows highly stable redox cycling with uniform reductive and oxidative components and a reversal potential similar to other superoxide dismutases (Davies *et al.,* 2009). Such an observation confirms the potential of PrP to act as a superoxide dismutase. Given these findings and the overwhelming evidence for PrP antioxidant activity *in vivo* it seems beyond reasonable doubt that PrP has this function in the brain. This activity is lost when PrP^{c} is converted to PrP^{Sc} during disease (Wong *et al.,* 2001a; Thackray *et al.,* 2002).

8.4.2 *Protein Behaviour and Turnover*

There have been several publications that have looked into the physical behaviour of PrP when exposed to copper. Pauly and Harris, 1998, showed that exposure to copper at the cell surface increased the rate of PrP internalisation. They also proved that the N-terminal region of the protein was important for the rate increase of internalisation. This data was later supported by other work (Lee *et al.,* 2001; Marella *et al.,* 2002). Additionally, another area of the protein located at the far N-terminal region of basic residues was also identified as key to the internalisation process (Lee *et al.,* 2001; Walmsley *et al.,* 2003). These findings were supported by a study that showed that copper induced PrP internalisation was abolished in the disease state (Perera and Hooper, 2001). Another group showed that copper accelerated internalisation to transferrin containing early endosomes and golgi departments (Brown and Harris, 2003). Work by Brown in 1999 showed that PrP expression aided cellular copper uptake and in 2004 further supported this by showing that PrP expression influenced cellular uptake in astrocytes (Brown, 2004). Further work by our group discovered that physiological levels of copper were sufficient to drive the internalisation process and that no other metal could stimulate it (Haigh *et al.,* 2005).

8.4.3 *Copper Sequestration / Buffering / Sensing*

The extensive work on the metal binding characteristics of both recombinant and native PrP (Aronoff-Spencer *et al.,* 2000; Burns *et al.,* 2002; Burns *et al.,* 2003; Morante *et al.,* 2004; Chattopadhyay *et al.,* 2005; Redecke *et al.,* 2005) all suggest PrP has multiple metal binding sites with a high specifity for copper. The question remains, however, is the affinity of PrP for copper high enough to directly chelate copper from copper serum transporters with disassociation constants in the region of 10^{-10} to 10^{-11} (Linder and Hazegh-Azam, 1996; Linder, 2001). The evidence for the affinity of PrP for copper is simply too variable to make a reasoned assessment. The highest values given of femtomolar (Jackson *et al.,* 2001; Thompsett *et al.,* 2005) would allow for a sequestration role whereas values in the micromolar range (Viles *et al.,* 1999) or nanomolar range (Garnett and Viles, 2003; Walter *et al.,* 2006) would not. Additionally, does the protein become loaded with copper inside the cells, transporting to the outside for a protective or catalytic purpose (Brown *et al.*, 1998) or is the protein presented to the extracellular environment without copper loaded or partially loaded so as to chelate copper from the outside? Certainly PrP is involved in copper (II) uptake in cells (Brown 1999, 2004). It would seem plausible given the multiple reports of a negative cooperation and multiple disassociation constants within the octarepeat binding region (Jackson *et al.*, 2001; Garnett and Viles 2003; Thompsett *et al.*, 2005; Walter *et al.*, 2006) that there would be at least some available copper binding sites free on presentation to the extracellular environment. This would leave open the interestingly possibility that the dramatic restructuring seen in the N-terminus of the protein during maximum copper occupancy (Chattopadhyay *et al.*, 2005; Weiss *et al.*, 2007) could be a trigger for internalisation of the protein. Clearly there is much direct evidence showing that copper increases the rate of internalisation (Pauly and Harris, 1998) and that the N terminus was key to this process (Lee *et al.*, 2001; Marella *et al.*, 2002). Additionally, it is now known that physiologically relevant amounts of copper are sufficient to drive the internalisation process (Haigh *et al.*, 2005). This may suggest that PrP is able to function as a concentration sensitive sequester of copper, activated when extracellular copper reaches peak levels during synaptic transmission and depolarisation of between 15 μM and 300 μM (Kardos *et al.*, 1989). This copper may then be transferred back into the cells for recycling.

8.5 Other Metals

While copper is acknowledge as the most important metal of interest in metal binding to PrP, there have been reports of other metals also binding to the protein. Of these, the three most important are zinc, iron and manganese. Most of the data in regards to zinc indicate that increased levels of zinc alter the behaviour of the protein, such as internalisation. However, analysis of the binding of zinc to PrP did not indicate significant affinity between the metal and the protein (Davies *et al.*, 2009). This suggests that any interaction between zinc and PrP in regards to cellular metabolism is probably indirect or that PrP is significantly downstream of changes directly induced by zinc. Similarly, research relating PrP and iron has mostly been indirect association of altered iron metabolism to changes in PrP expression (Singh *et al.*, 2009). Analysis of iron binding to PrP shows low affinity and suggests that most other iron binding proteins would out-compete PrP for metal occupancy.

Manganese binding to PrP was first suggested in 2000 (Brown *et al.*, 2000). Manganese is a much rarer trace element than copper or iron and its levels in cells are about one tenth of those of copper. Therefore, the suggestion that it could bind manganese seemed rather unlikely when it was first suggested. However, analysis of the brains of mice experimental infected with prion diseases suggested that PrP could be isolated in which manganese had replaced copper on the protein (Thackray *et al.*, 2002). It was also possible that this replacement of copper by manganese could be somehow related to the conversion process. Furthermore, cells grown in the presence of manganese occasionally showed expression of protease resistant PrP (Brown *et al.*, 2000). This has also been observed in a yeast based system (Treiber *et al.*, 2006). Protease resistance is one of the characteristics of PrP^{Sc}. This implies that under certain conditions, PrP and manganese could interact.

A thorough analysis of metal binding by PrP was carried out using isothermal titration calorimetry (Brazier *et al.*, 2008). This study showed that the affinity of manganese for PrP was in the same range as that of other manganese binding proteins. It also identified two binding sites on the protein, one associated with the octameric repeat region and the second associated with the histidine at amino acid residue 95 (Fig. 8.3). This is the high affinity site of the protein with a dissociation constant of 63 mM. This affinity places PrP in the range of other known manganese binding proteins where manganese is a known co-factor. Few other studies have looked at the affinity of manganese for the protein. One of these studies confirmed binding but indicated a slightly lower affinity (Jackson *et al.*, 2001). Another study was based on a peptide of the octameric repeat region but showed no binding (Garnett and Viles, 2003). A third study suggested that binding occurs within the 106–126 region of PrP, centred on histidine 111 (Gaggelli *et al.*, 2005). A study using surface plasmon resonance indicated that while peptides from the octameric repeat did not bind manganese, that the full length protein may possess a conformational binding site with a nanomolar dissociation constant (Treiber *et al.*, 2007).

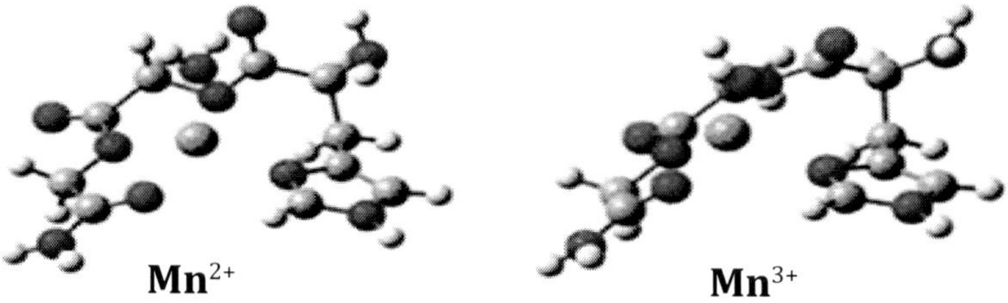

Figure 8.3 Mn binding to PrP. The main binding site of PrP is in association with histidine at position 95 in the mouse sequence. The above model shows the theoretical co-ordination of the metal to the imidazole and the two proceeding amino acid side chains. PrP would bind manganese in the reduced form but this would be rapidly oxidised (Mn^{3+}). Computer modelling shows how this oxidation would be sufficient to cause conformational change in the protein. See also Colour Insert.

These differences are likely due to the different techniques used or the use of non-physiological peptides.

While there is no evidence that manganese binding to PrP plays any known functional role, the possibility remains that under normal cellular conditions manganese could bind in place of or as well as copper. This was demonstrated in the same study (Brazier *et al.,* 2008). Copper saturated PrP was analysed with isothermal titration calorimetry and manganese was found to bind to histidine 95 with the same affinity as the apo-protein. As well as binding to PrP on the surface of the cell, manganese might interact with PrP during the refolding process during synthesis. Using a refolding method during the preparation of recombinant protein, it has been shown that manganese can bind to the octameric repeat region (Brown *et al.,* 2000). The results of this study differed from those of the colorimetry study, however the difference is likely due to the methods used to allow binding to occur. Verification of manganese binding to PrP comes from studies that identify conformational changes in the protein as a result of binding (Thompsett and Brown, 2007; Zhu *et al.,* 2008). Unlike, copper binding, manganese binding causes conformational transitions that occur over an extended time period (Tsenkova *et al.,* 2004). Over a 90 minute period PrP with manganese bound changed from a conformation similar to copper bound PrP to one that was destabilised with an increased exposure of protein residues to the water environment allowing rapid oxidation of the protein. Such changes in the protein would explain the ability of manganese to cause PrP to take on increased beta-sheet content and aggregate.

8.6 Metals and Aggregation

The mechanism considered most likely to describe the aggregation of PrP is the nucleation or seeding hypothesis. In this model the limiting step is the formation of a seed (Cohen and Prusiner, 1998). This makes this model attractive because of the long incubation period associated with prion diseases. PrP aggregation occurs in a number of forms. Fibrils have been commonly studied because the formation of prion fibrils is easy to detect in both the brain and preparation of PrP^{Sc} isolated from the brain. Additionally, oligomeric species have also been identified and can also be generated using *in vitro* assays of PrP aggregation. It has been suggested that the formation of smaller oligomeric structures in misfolding pathways are the key components to prion diseases (Baskakov *et al.*, 2001; Morillas *et al.*, 2001; Baskakov *et al.*, 2002; Sokolowski *et al.*, 2003; Govaerts *et al.*, 2004) and that these intermediates are in fact the molecule responsible for infection (Weissmann, 1991). Initially, it was supposed PrP formed distinct, β-rich oligomers prior to amyloid formation (Baskakov *et al.*, 2001), but data now seems to support the idea that β-oligomers are not on the pathway to amyloid fibrils, but rather exist in a dynamic state with α-helical monomers (Baskakov *et al.*, 2002; Bocharova *et al.*, 2005a; Teplow *et al.*, 2006; Haass and Selkoe, 2007; El Moustaine *et al.*, 2008; Soto and Estrada, 2008). β-structures formed under acidic conditions have different properties than those formed under neutral conditions—that is, β-oligomers formed at acid pH do not have amyloid characteristics (Bocharova *et al.*, 2005a; Vendrely *et al.*, 2005).

Assays for aggregation can be cell culture based but the majority are based on assays of recombinant protein. Once a system is established researchers generally attempt to identify factors that alter the rate of aggregation or increase the formation of particular species such as oligomers. As metal binding is able to impart structural changes to PrP then, logically, metal binding has been studied in relation to aggregation. While copper binding under physiological conditions is associated with the normal activity of PrP, it has been shown that high concentrations can cause aggregation (Requena *et al.*, 2001; Rezaei *et al.*, 2002; Rezaei *et al.*, 2005; Shiraishi *et al.*, 2005; Tsiroulnikov *et al.*, 2006; Redecke *et al.*, 2007; Fitzmaurice *et al.*, 2008). Particularly interactions with residues in the vicinity of the hydrophobic domain of the protein are able to initiate structural changes in bring about aggregation. Additionally, copper has been shown to accelerate aggregation of PrP^{Sc} and during the disease process, chelation of copper can cause lengthening of the incubation period. In contrast, copper can inhibit aggregation (Giese *et al.*, 2004; Bocharova *et al.*, 2005b; Ricchelli *et al.*, 2006; Liu *et al.*, 2008) and cell-free conversion (Orem *et al.*, 2006). The difference in these results probably relates to the concentration used as high concentrations are likely to bind non-specifically. Additionally, once protein conversion has begun beta-sheet elements would out-compete true binding sites and this would accelerate the process of auto-catalytic aggregation resulting in more rapid conversion of the protein. Given these considerations it is possibly more reasonable to assume that physiological concentrations of copper are protective in regards conversion and that aggregation induced by copper might occur once the disease process has begun but copper is unlikely to be an initiating factor.

As discussed above, there is now reliable evidence that manganese can bind to PrP. Analysis using recombinant protein has shown that this manganese-bound PrP will become protease resistant (Brown *et al.*, 2000; Abdelraheim *et al.*, 2006). Analysis using Near Infra-Red spectroscopy has shown in real time that after initial binding of manganese to PrP that the protein undergoes a gradual structural transition that is not seen with copper-bound PrP (Tsenkova *et al.*, 2004). Studies with other techniques such as Raman spectroscopy and dual polarization inferferometry confirm that manganese loaded PrP becomes structurally distinct from either holo-PrP or copper bound PrP and has a more compact structure (Thompsett and Brown, 2007; Zhu *et al.*, 2008). With different exposure of key residues to the environment, the protein is more able to be influenced by oxidative radicals that potentially play a role in structural changes. Analyses with cyclic voltammetry clearly illustrate the difference in this regard because of the high propensity of the metal to be oxidised irreversibly (Brazier *et al.*,

2008). The response of PrP to manganese in terms of change in protease resistance and aggregation is altered by the number of octameric repeats present. This might alter the number of manganese atoms bound by the protein and increase the rate of its oxidation (Li *et al.*, 2009).

Manganese-loaded PrP has been shown to initiate aggregation of PrP (Brown *et al.*, 2000; Giese *et al.*, 2004; Lekishvili *et al.*, 2004; Kim *et al.*, 2005). This has been shown to be the result of the formation of "seeds" which are then able to catalyse further aggregation. Addition of manganese during the infection process enhances infectivity in that 100 fold (i.e., 2 logs) less infectious agent is required to infect cultured cells (Davies and Brown 2009). Therefore it is quite likely that manganese is able to stabilise the "seeds" of PrP that are able to enter cells and induce the aggregation process with the host PrP. Other studies have shown that chelation of manganese out of PrP seeds strips them of their potential to cause aggregation. While manganese is able to stabilise prion seeds it does not appear to influence the rate of aggregation in the same was as excess copper concentration does. Indeed, some findings suggest that manganese delays the aggregation of PrP82–146 (Ricchelli *et al.*, 2006) and does not affect cell-free conversion (Orem *et al.*, 2006) wherein purified PrP27–30 derived from infected brain homogenate is used to convert purified PrP^C. Manganese also does not affect aggregation of partially denatured PrP (Bocharova *et al.*, 2005b) which may be attributed to manganese being unable to bind to the denatured protein properly. However, observing a specific effect of manganese under quite defined and unique conditions (i.e., seed formation) suggests a unique mechanism of action. This makes the finding more likely to be significant unlike the more artefactual effects of copper in regards to PrP aggregation.

One of the consequences of substitution of copper by manganese and the subsequent aggregation of the protein is the conversion of the protein to a neurotoxic form. Studies have shown that manganese induced aggregates are highly toxic to both cultured primary neurons and neuronal cell lines (Uppington and Brown, 2008). As neuronal loss is the major cause of pathological changes in prion disease this observation is quite significant. The toxicity observed was not a result of the manganese bound to the protein, but was a consequence to changes in the protein. The concentration of protein required for toxicity was significantly lower than those used in other studies (such as with PrP106–126) and the mechanism of action was similar to that observed for PrP^{Sc}. While binding of manganese to PrP might not be the only way to generate toxic forms of PrP, it does provide a simple method to generate appropriately toxic protein for further study.

8.7 Changes in Brain Metals

As PrP binds metals then it is logical to assume that altering PrP might alter metal regulation and possibly basal levels of metals in the brain. As mentioned above, altered expression of PrP alters these properties in cells. Changes in brain copper in experimental rodents showed differences only with increasing age of the mice (Brown, 2003). While the copper content of the brains of mice knocked out for PrP, or expressing a form of PrP lacking Cu binding, ability showed no change with increasing age, brains from both wild-type mice and mice overexpression PrP ten fold showed elevated copper levels in the brain with increased age. The highest levels observed were those of 18 month old mice that overexpressed PrP. PrP is expressed at the highest levels in the synapses of neurons and synaptosomal preparations from transgenic mice showed altered copper levels. This again supports the notion that PrP levels alter the levels of copper observed. There have been differences in the findings of another research groups in this regard (Waggoner *et al.*, 2000). However, this is likely due to preparation techniques as synaptosomal isolates rapidly lose their copper content if not handled appropriately.

Expression of PrP by cells alters the handling of copper as suggested above. Additionally manganese transporting proteins are altered by the expression levels of PrP as well (Kralovicova *et al.*, 2009). Furthermore, prion infection also alters these proteins and increases manganese retention in cells. There is likely to be a feed-forward mechanism in this case as increased manganese levels are likely to alter other cellular processes such as glycosylation which is manganese dependent (Kaufman *et al.*, 1994) and also the activity of the proteasome which is inhibited by increased manganese levels (Zhou *et al.*, 2004; Cai *et al.*, 2007). Reduction of proteasomal degradation of PrP may then result in increased levels of PrP^{Sc} as has been observed (Choi *et al.*, 2010). Reversing proteasomal inhibition has been shown to greatly decrease PrP^{Sc} levels and possibly inhibit cellular infection altogether (Kristiansen *et al.*, 2007; Webb *et al.*, 2007).

Changes at the cellular level are likely to have consequences for the brain. Studies of the brains of experimental infected mice showed distinct elevation of manganese in the brain as well as in blood (Thackray *et al.*, 2002). The mechanistic explanation of how prion infection alters blood manganese is currently lacking. However, it is possible that one consequence of the infection is increased manganese retention in the blood due to the prion protein present there. Alternatively, prion disease could inhibit manganese clearance from the body which has the secondary effect of elevating blood manganese. It was also noted that elevation of manganese occurred preclinically (before onset of symptoms)(Thackray *et al.*, 2002). This prompted further study of the changes in manganese with the possibility that these changes could be used as a diagnostic test. This is a possibility that has never been realised despite considerable evidence of the relevance of increased manganese levels for all forms of prion disease.

Studies of large numbers of experimental BSE animals and BSE field cases showed that manganese was elevated in brain stem and spinal cord but not frontal cortex (Hesketh *et al.*, 2007). The elevation occurred before onset of clinical signs. Similarly, sheep with scrapie showed elevated manganese in the cerebellum and the brain stem (obex). In both cases the elevation was in areas associated with pathology in the diseases. Changes in blood were also observed for scrapie and BSE before the onset of clinical signs. In the cases of scrapie such changes were also observed in so called "scrapie resistant sheep" experimentally challenged with scrapie but that did not come down with the disease. In this case, the elevated manganese levels were not maintained and dropped back to control levels after some time. This suggests possible preclinical changes associated with infection rather than disease progress result in the elevated manganese levels. Studies of elk with chronic wasting disease have also suggested an association with increased manganese levels, although in this case there was also an observed decrease in magnesium (White *et al.*, 2010).

Altered manganese levels have also been observed in cases of CJD (Wong *et al.*, 2001b; Hesketh *et al.*, 2008). In this case preclinical studies have not been possible. However, as with the animal prion diseases, specific brain regions such as the frontal cortex and cerebellum showed elevation of manganese levels in both sporadic and variant CJD cases. Similarly, blood manganese levels were also elevated in CJD (Hesketh *et al.*, 2008). This elevated blood manganese distinguished CJD from numerous other neurodegenerative diseases such as Alzheimer's disease and Parkinson's disease and also many blood born diseases. The only other diseases to show elevated manganese levels were haemocromatosis and blood brain barrier diseases.

Mechanistically, it remains unclear as to why manganese levels go up in brain and blood in prion diseases. It is highly unlikely that these increases relate to increased uptake. Also given that experimental diseases and sporadic diseases show the same trend it is also relatively easy to dismiss differential exposure to elevated environmental levels of manganese. However, this does not preclude a role of the environment in the initiation or development of the disease. It is far more likely that prion disease causes retention of manganese or possibly a redistribution of the metal from other tissues. Viewed on their own, these finding could at most be seen as a phenomenon but when considered alongside the other data about the association of PrP and manganese then these findings become another part of the same puzzle.

If manganese is important to prion disease then reduction of brain manganese should have some consequence for disease progress. A recent new study has shown this to be true. Mice infected with a low dose of prion disease showed a significant increase in survival time when treated with the manganese chelator CDTA (Brazier *et al.*, 2010). This is likely to be a result of the CDTA stripping manganese from PrP^{Sc} and destabilising the molecule and allowing its degradation. Manganese is an inhibitor of the proteasome and reducing brain manganese is likely to increase the success of degradation of misfolded proteins. In cell experiments where manganese, a known proteasome inhibitor, was applied to cells increased PrP^{Sc} levels have been noted (Choi *et al.*, 2010). The consequences of these experiments are very important when considering the interaction of PrP and manganese and its relation to prion disease. While elevated manganese levels could be only an indirect consequence of the disease, these findings suggest that prognosis is improved if manganese levels are reduced. Assuming that infection of an individual with prion disease is independent of manganese, reducing manganese during these early stages could potentially make the difference between infection leading to disease progression or not.

Levels of manganese present in cell culture medium during experimental infection of cells shows that infectivity can be enhanced 100 fold (Davies and Brown, 2009). Assuming that not all infectious PrP contains manganese, it does imply that the binding of manganese would give some forms of PrP a great selective advantage during the infection process and that manganese bound PrP is more likely to cause infection. As PrP can bind manganese and clay matrices rich in manganese can trap and preserve PrP (Davies and Brown, 2009), then it is highly likely that injection of PrP with manganese bound from environmental sources occurs.

Analysis of disease progression in mice on manganese rich diets indicated an increase in neuronal loss in this model (Hortells *et al.*, 2009). This loss was less than when the animals were fed a copper depleted diet but indicated the clear influence of diet on neuronal vulnerability to death during prion disease progression. Combining all these pieces of the puzzle gradually creates a picture as to how manganese can influence prion disease and highlights points at which the interaction of the protein and the metal makes a significant contribution to what is observed. The overall picture generated indicated that the metal both induces an abnormal conformation and stabilises it to increase the likelihood of both protein survival and infection. As chelation of the metal during disease alters disease progress it clearly also plays a role in the time course of the disease providing some evidence for the relevance of elevated levels of manganese observed in the brain.

8.8 PrP Survival in the Environment

Despite concerted efforts by many affected countries to eradicate TSEs, it still remains a major problem. Clusters of outbreaks suggest that many cases cannot be explained by spontaneous disease. This raises a question as to where is the source of infection? A great deal of focus has been placed on the environment as a possible reservoir of infection, more specifically the soil that animals are exposed to when grazing (Leita *et al.*, 2006). Certainly there is some evidence of sheep occupying grazing land that has previously supported infected animals becoming infected themselves (Greig, 1940). In addition, more modern, controlled studies suggest that prions may persist in the environment (Hadlow *et al.*, 1982; Miller *et al.*, 2004). It is proposed that prions may enter the soil via the disposal of infected carcasses, meat products or farm effluent (Gale and Stanfield, 2001). This is supported somewhat by evidence showing that the amount of material required for oral infection is very low, in the region of 10–100 mg of infected brain (Leita *et al.*, 2006). This is clearly a concern when considering that a single bovine brain would then contain enough material to infect around 2500 animals, although this would be dependent on factors such as genetic susceptibility and cross-species infectivity barriers.

A big question that is raised by this evidence is how could a protein exist in a stable form for extended periods in a hostile environment such as soil? One fascinating study even showed that some residual infectivity remained after three years in soil that had been exposed to infected material (Brown and Gajdusek, 1991). The study left many unanswered questions, however, as no information was given on how the prions were interacting with the soil or the precise conditions of the soil. Nevertheless, the study does show that prions can exist for long periods in environments that are normally hostile to proteins. Clearly, natural processes in the soil such a bacterial activity, exposure to UV radiation and soil acidity should be deleterious for even the most resilient organic material.

Interactions with metals may be able to contribute to the protein's stability and resistance against degradation. For example, it has been shown that manganese can cause PrP to fold into a protease resistant form (Brown *et al.*, 2000). Certainly, many metals exist within soils and it may therefore be possible that these interactions contribute to PrPs longevity in the environment. The mechanism of protein adsorption on to soil particles is far from straight forward. Complicating factors include soil pH and constituents and protein PI, conformation, size, charge, solubility and flexibility (Stumm *et al.*, 1992; Norde and Giacomelli, 2000). There have been many studies of soil/protein interaction, especially with constituent clays such as kaolonite and montmorillonite (mte). One study (Servagent-Noinville *et al.,* 2000) showed the importance of electronegative interactions in the adsorbance of bovine serum albumin (BSA) onto mte and how the strength of these interactions could alter the conformation and properties of the protein. Other studies have shown specifically the very strong adsorpative nature of soil clays to proteins, especially prions. One study, (Genovesi *et al.,* 2007) for example, demonstrated the difficulty in desorbing prions from clay, especially mte and suggested that the conditions in most soils would favour an accumulation of stable prions in soils exposed to contaminated material. Another such study confirmed this and went further to suggest that mte would promote an orientation of PrP towards the soil involving elements across the entire protein in both the N and C terminus, making the adsorption almost irreversible in it's strength (Revault *et al.,* 2005). One recent study looked specifically at the disease isoform PrP^{Sc} and concluded that it was able to adsorb to many different types of soil/clay and remain infectious following desorption, even if it's removal caused cleavage of the protein (Johnson *et al.,* 2006).

While release from the soil might be necessary for transmission it does not explain how prions survive at all or why certain areas of land are "scrapie prone" and have recurring similar incidence of infection (Thorgeirsdottir *et al.,* 1999). These differences imply that some underlying environmental factor that is variable in soil could influence whether an area maintains a reservoir of prions or not. Therefore it is important to assess which factors influence PrP survival in soil. Recent studies comparing recombinant PrP and PrP^{Sc} from infected cells showed little evidence of a difference in PrP survival based on protease resistance.

Recent work looking at the survival of recombinant and native PrP in a model soil has shown that PrP can resist degradation for over two years if the soil contains manganese (Davies and Brown, 2009). The protein that survives in the soil can either be protease resistant or sensitive. There was no observed difference. In comparison PrP in the absence of manganese was almost completely degraded. The surviving protein was studied following extraction from the soil with a novel electrophoretic technique which resulted in no degradation of the protein such as truncation at the N-terminus. This implies that

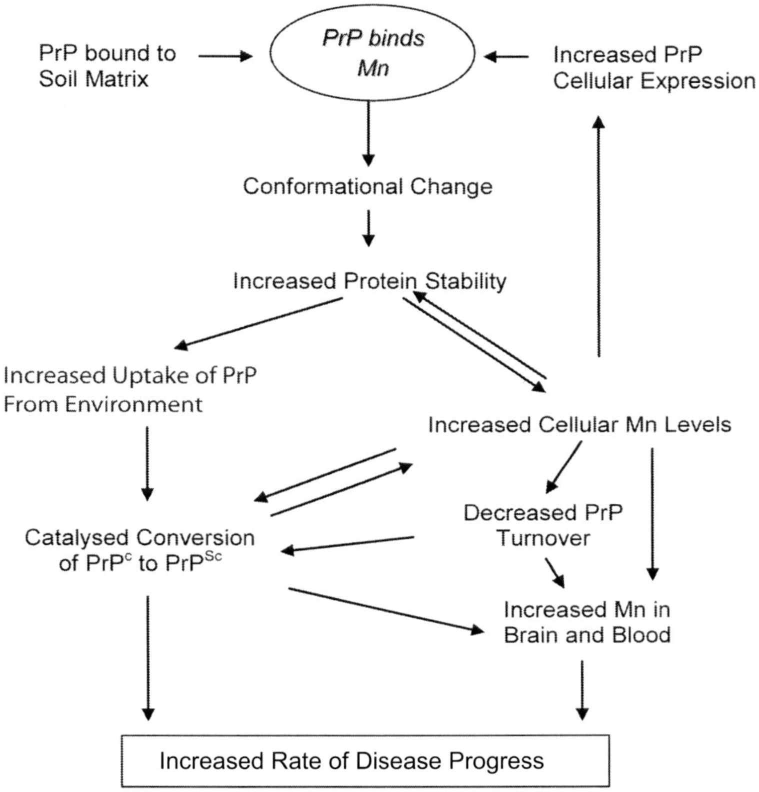

Figure 8.4 Model of Manganese – PrP interactions. Summary of the potential effects, based on published data, of manganese (Mn) binding to PrP. Mn could bind to PrP either when trapped in soil or when expressed by cells. In both cases the binding of Mn would cause a conformational change in the protein and stabilise the protein making it resistant to degradation processes. The stabilised Mn-PrP complex could then form seeds that would be ingested from the environment and catalyse protein conversion. This would generate large amounts of neurotoxic PrP and result in neurodegeneration.

under the right conditions soils can retain fully intact PrP for at least two years. While real soils may contain many other factors such as bacteria, variable pH, organic matter and other trace elements, the fact that manganese can give an almost 100 fold advantage gives genuine support to the notion that environmental manganese levels could influence incidence of disease (Fig. 8.4).

8.9 Conclusion

The prion research field has always been a controversial one. The finding that PrP is a metalloprotein did not escape such problems, but now, with hundreds of supporting publications, the truth is undeniable. What remains to be agreed is what this metal binding means, not just for the role of the normal protein in the cell, but also for protein conformation conversion and disease. The evidence presented here makes it clear that there is a very strong case to consider that PrP's metalloprotein nature is of great importance to unravelling these issues. Even more significantly, as this book testifies, PrP, as a metalloprotein belongs to a family of proteins that are both metalloproteins and proteins associated with neurodegeneration. Seeing all these proteins as metalloproteins first and disease causing proteins second might cause a revolution in thinking that will allow new breakthroughs in the understanding and treatment of all these diseases.

References

Abdelraheim, S. R., Kralovicova, S. and Brown, D. R. (2006). Hydrogen peroxide cleavage of the prion protein generates a fragment able to initiate polymerisation of full length prion protein. *Int J Biochem Cell Biol*, 38, pp. 1429–40.

Anderson, R. M., Donnelly, C. A., Ferguson, N. M., Woolhouse, M. E., Watt, C. J., Udy, H. J., MaWhinney, S., Dunstan, S. P., Southwood, T. R., Wilesmith, J. W., Ryan, J. B., Hoinville, L. J., Hillerton, J. E., Austin, A. R. and Wells, G. A. (1996). Transmission dynamics and epidemiology of BSE in British cattle. *Nature*, 382, pp. 779–88.

Andrews, N. J. (2009). "Incidence of variant Creutzfeldt-Jakob disease diagnoses and deaths in the UK." Retrieved 10/04/2009, 2009, from http://www.cjd.ed.ac.uk/cjdq60.pdf.

Aronoff-Spencer, E., Burns, C. S., Avdievich, N. I., Gerfen, G. J., Peisach, J., Antholine, W. E., Ball, H. L., Cohen, F. E., Prusiner, S. B. and Millhauser, G. L. (2000). Identification of the Cu^{2+} binding sites in the N-terminal domain of the prion protein by EPR and CD spectroscopy. *Biochemistry*, 39, pp. 13760–71.

Baskakov, I. V., Legname, G., Baldwin, M. A., Prusiner, S. B. and Cohen, F. E. (2002). Pathway complexity of prion protein assembly into amyloid. *J Biol Chem*, 277, pp. 21140–8.

Baskakov, I. V., Legname, G., Prusiner, S. B. and Cohen, F. E. (2001). Folding of prion protein to its native alpha-helical conformation is under kinetic control. *J Biol Chem*, 276, pp. 19687–90.

Basler, K., Oesch, B., Scott, M., Westaway, D., Walchli, M., Groth, D. F., McKinley, M. P., Prusiner, S. B. and Weissmann, C. (1986). Scrapie and cellular PrP isoforms are encoded by the same chromosomal gene. *Cell*, 46, pp. 417–28.

Bocharova, O. V., Breydo, L., Parfenov, A. S., Salnikov, V. V. and Baskakov, I. V. (2005a). *In vitro* conversion of full-length mammalian prion protein produces amyloid form with physical properties of PrP(Sc). *J Mol Biol*, 346, pp. 645–59.

Bocharova, O. V., Breydo, L., Salnikov, V. V. and Baskakov, I. V. (2005b). Copper(II) inhibits *in vitro* conversion of prion protein into amyloid fibrils. *Biochemistry*, 44, pp. 6776–87.

Bolton, D. C., McKinley, M. P. and Prusiner, S. B. (1982). Identification of a protein that purifies with the scrapie prion. *Science*, 218, pp. 1309–11.

Brazier, M. W., Davies, P., Player, E., Marken, F., Viles, J. H. and Brown, D. R. (2008). Manganese binding to the prion protein. *J Biol Chem*, 283, pp. 12831–9.

Brazier, M. W., Volitakis, I., Kvasnicka, M., White, A. R., Underwood, J. R., Green, J. E., Han, S., Hill, A. F., Masters, C. L. and Collins, S. J. (2010). Manganese chelation therapy extends survival in a mouse model of M1000 prion disease. *J Neurochem*, pp. 440–51.

Brown, D. R. (1999). Prion protein expression aids cellular uptake and veratridine-induced release of copper. *J Neurosci Res*, 58, pp. 717–25.

Brown, D. R. (2003). Prion protein expression modulates neuronal copper content. *J Neurochem*, 87, pp. 377–85.

Brown, D. R. (2004). Role of the prion protein in copper turnover in astrocytes. *Neurobiol Dis*, 15, pp. 534–43.

Brown, D. R., Clive, C. and Haswell, S. J. (2001). Antioxidant activity related to copper binding of native prion protein. *J Neurochem*, 76, pp. 69–76.

Brown, D. R., Hafiz, F., Glasssmith, L. L., Wong, B. S., Jones, I. M., Clive, C. and Haswell, S. J. (2000). Consequences of manganese replacement of copper for prion protein function and proteinase resistance. *Embo J*, 19, pp. 1180–6.

Brown, D. R., Qin, K., Herms, J. W., Madlung, A., Manson, J., Strome, R., Fraser, P. E., Kruck, T., von Bohlen, A., Schulz-Schaeffer, W., Giese, A., Westaway, D. and Kretzschmar, H. (1997). The cellular prion protein binds copper *in vivo*. *Nature*, 390, pp. 684–7.

Brown, D. R., Schmidt, B. and Kretzschmar, H. A. (1998). Effects of copper on survival of prion protein knockout neurons and glia. *J Neurochem*, 70, pp. 1686–93.

Brown, L. R. and Harris, D. A. (2003). Copper and zinc cause delivery of the prion protein from the plasma membrane to a subset of early endosomes and the Golgi. *J Neurochem*, 87, pp. 353–63.

Brown, P. (1998). Donor pool size and the risk of blood-borne Creutzfeldt-Jakob disease. *Transfusion*, 38, pp. 312–5.

Brown, P. and Bradley, R. (1998). 1755 and all that: a historical primer of transmissible spongiform encephalopathy. *Bmj*, 317, pp. 1688–92.

Brown, P., Preece, M. A. and Will, R. G. (1992). "Friendly fire" in medicine: hormones, homografts, and Creutzfeldt-Jakob disease. *Lancet*, 340, pp. 24–7.

Bruce, M. E., McBride, P. A., Jeffrey, M. and Scott, J. R. (1994). PrP in pathology and pathogenesis in scrapie-infected mice. *Mol Neurobiol*, 8, pp. 105–12.

Bueler, H., Aguzzi, A., Sailer, A., Greiner, R. A., Autenried, P., Aguet, M. and Weissmann, C. (1993). Mice devoid of PrP are resistant to scrapie. *Cell*, 73, pp. 1339–47.

Burns, C. S., Aronoff-Spencer, E., Dunham, C. M., Lario, P., Avdievich, N. I., Antholine, W. E., Olmstead, M. M., Vrielink, A., Gerfen, G. J., Peisach, J., Scott, W. G. and Millhauser, G. L. (2002). Molecular features of the copper binding sites in the octarepeat domain of the prion protein. *Biochemistry*, 41, pp. 3991–4001.

Burns, C. S., Aronoff-Spencer, E., Legname, G., Prusiner, S. B., Antholine, W. E., Gerfen, G. J., Peisach, J. and Millhauser, G. L. (2003). Copper coordination in the full-length, recombinant prion protein. *Biochemistry*, 42, pp. 6794–803.

Cai, T., Yao, T., Li, Y., Chen, Y., Du, K., Chen, J. and Luo, W. (2007). Proteasome inhibition is associated with manganese-induced oxidative injury in PC12 cells. *Brain Res*, 1185, pp. 359–65.

Calzolai, L., Lysek, D. A., Perez, D. R., Guntert, P. and Wuthrich, K. (2005). Prion protein NMR structures of chickens, turtles, and frogs. *Proc Natl Acad Sci USA*, 102, pp. 651–5.

Chattopadhyay, M., Walter, E. D., Newell, D. J., Jackson, P. J., Aronoff-Spencer, E., Peisach, J., Gerfen, G. J., Bennett, B., Antholine, W. E. and Millhauser, G. L. (2005). The octarepeat domain of the prion protein binds Cu(II) with three distinct coordination modes at pH 7.4. *J Am Chem Soc*, 127, pp. 12647–56.

Choi, C. J., Anantharam, V., Martin, D. P., Nicholson, E. M., Richt, J. A., Kanthasamy, A. and Kanthasamy, A. G. (2010). Manganese upregulates cellular prion protein and contributes to altered stabilization and proteolysis: relevance to role of metals in pathogenesis of prion disease. *Toxicol Sci*, 115, pp. 535–46.

Cohen, F. E. and Prusiner, S. B. (1998). Pathologic conformations of prion proteins. *Annu Rev Biochem*, 67, pp. 793–819.

Collinge, J. and Palmer, M. S. (1994). Molecular genetics of human prion diseases. *Philos Trans R Soc Lond B Biol Sci*, 343, pp. 371–8.

Collins, S., McLean, C. A. and Masters, C. L. (2001). Gerstmann-Straussler-Scheinker syndrome, fatal familial insomnia, and kuru: a review of these less common human transmissible spongiform encephalopathies. *J Clin Neurosci*, 8, pp. 387–97.

Curtis, J., Errington, M., Bliss, T., Voss, K. and MacLeod, N. (2003). Age-dependent loss of PTP and LTP in the hippocampus of PrP-null mice. *Neurobiol Dis*, 13, pp. 55–62.

Davies, P. and Brown, D. R. (2009). Manganese enhances prion protein survival in model soils and increases prion infectivity to cells. *PLoS One*, 4, pp. e7518.

Davies, P., Marken, F., Salter, S. and Brown, D. R. (2009). Thermodynamic and voltammetric characterization of the metal binding to the prion protein: insights into pH dependence and redox chemistry. *Biochemistry*, 48, pp. 2610–9.

Dupuis, L., Mbebi, C., Gonzalez de Aguilar, J. L., Rene, F., Muller, A., de Tapia, M. and Loeffler, J. P. (2002). Loss of prion protein in a transgenic model of amyotrophic lateral sclerosis. *Mol Cell Neurosci*, 19, pp. 216–24.

El Moustaine, D., Perrier, V., Smeller, L., Lange, R. and Torrent, J. (2008). Full-length prion protein aggregates to amyloid fibrils and spherical particles by distinct pathways. *FEBS J*, 275(9), pp. 2021–31.

Fitzmaurice, T. J., Burke, D. F., Hopkins, L., Yang, S., Yu, S., Sy, M. S., Thackray, A. M. and Bujdoso, R. (2008). The stability and aggregation of ovine prion protein associated with classical and atypical scrapie correlates with the ease of unwinding of helix-2. *Biochem J*, 409, pp. 367–75.

Frederikse, P. H., Zigler, S. J., Jr., Farnsworth, P. N. and Carper, D. A. (2000). Prion protein expression in mammalian lenses. *Curr Eye Res*, 20, pp. 137–43.

Furlan, S., La Penna, G., Guerrieri, F., Morante, S. and Rossi, G. C. (2007). Studying the Cu binding sites in the PrP N-terminal region: a test case for ab initio simulations. *Eur Biophys J*, 36, pp. 841–5.

Gaggelli, E., Bernardi, F., Molteni, E., Pogni, R., Valensin, D., Valensin, G., Remelli, M., Luczkowski, M. and Kozlowski, H. (2005). Interaction of the human prion PrP(106–126) sequence with copper(II), manganese(II), and zinc(II): NMR and EPR studies. *J Am Chem Soc*, 127, pp. 996–1006.

Gajdusek, D. C. (1991). The transmissible amyloidoses: genetical control of spontaneous generation of infectious amyloid proteins by nucleation of configurational change in host precursors: kuru-CJD-GSS-scrapie-BSE. *Eur J Epidemiol*, 7, pp. 567–77.

Garnett, A. P., Jones, C. E. and Viles, J. H. (2006). A survey of diamagnetic probes for copper^{2+} binding to the prion protein. 1H NMR solution structure of the palladium^{2+} bound single octarepeat. *Dalton Trans*, pp. 509–18.

Garnett, A. P. and Viles, J. H. (2003). Copper binding to the octarepeats of the prion protein. Affinity, specificity, folding, and cooperativity: insights from circular dichroism. *J Biol Chem*, 278, pp. 6795–802.

Genovesi, S., Leita, L., Sequi, P., Andrighetto, I., Sorgato, M. C. and Bertoli, A. (2007). Direct detection of soil-bound prions. *PLoS ONE*, 2, pp. e1069.

Giese, A., Levin, J., Bertsch, U. and Kretzschmar, H. (2004). Effect of metal ions on de novo aggregation of full-length prion protein. *Biochem Biophys Res Commun*, 320, pp. 1240–6.

Govaerts, C., Wille, H., Prusiner, S. B. and Cohen, F. E. (2004). Evidence for assembly of prions with left-handed beta-helices into trimers. *Proc Natl Acad Sci USA*, 101, pp. 8342–7.

Griffith, J. S. (1967). Self-replication and scrapie. *Nature*, 215, pp. 1043–4.

Haass, C. and Selkoe, D. J. (2007). Soluble protein oligomers in neurodegeneration: lessons from the Alzheimer's amyloid beta-peptide. *Nat Rev Mol Cell Biol*, 8, pp. 101–12.

Haigh, C. L., Edwards, K. and Brown, D. R. (2005). Copper binding is the governing determinant of prion protein turnover. *Mol Cell Neurosci*, 30, pp. 186–96.

Harris, D. A. (1999). Cell biological studies of the prion protein. *Curr Issues Mol Biol*, 1, pp. 65–75.

Hasnain, S. S., Murphy, L. M., Strange, R. W., Grossmann, J. G., Clarke, A. R., Jackson, G. S. and Collinge, J. (2001). XAFS study of the high-affinity copper-binding site of human PrP(91–231) and its low-resolution structure in solution. *J Mol Biol*, 311, pp. 467–73.

Hesketh, S., Sassoon, J., Knight, R. and Brown, D. R. (2008). Elevated manganese levels in blood and CNS in human prion disease. *Mol Cell Neurosci*, 37, pp. 590–8.

Hesketh, S., Sassoon, J., Knight, R., Hopkins, J. and Brown, D. R. (2007). Elevated manganese levels in blood and central nervous system occur before onset of clinical signs in scrapie and bovine spongiform encephalopathy. *J Anim Sci*, 85, pp. 1596–609.

Hope, J., Reekie, L. J., Hunter, N., Multhaup, G., Beyreuther, K., White, H., Scott, A. C., Stack, M. J., Dawson, M. and Wells, G. A. (1988). Fibrils from brains of cows with new cattle disease contain scrapie-associated protein. *Nature*, 336, pp. 390–2.

Hornshaw, M. P., McDermott, J. R. and Candy, J. M. (1995). Copper binding to the N-terminal tandem repeat regions of mammalian and avian prion protein. *Biochem Biophys Res Commun*, 207, pp. 621–9.

Hortells, P., Monleon, E., Acin, C., Vargas, A., Vasseur, V., Salomon, A., Ryffel, B., Cesbron, J. Y., Badiola, J. J. and Monzon, M. (2009). The Effect of Metal Imbalances on Scrapie Neurodegeneration. *Zoonoses Public Health*, pp. 358–66.

Hsiao, K., Dlouhy, S. R., Farlow, M. R., Cass, C., Da Costa, M., Conneally, P. M., Hodes, M. E., Ghetti, B. and Prusiner, S. B. (1992). Mutant prion proteins in Gerstmann-Straussler-Scheinker disease with neurofibrillary tangles. *Nat Genet*, 1, pp. 68–71.

Huber, R., Deboer, T. and Tobler, I. (2002). Sleep deprivation in prion protein deficient mice sleep deprivation in prion protein deficient mice and control mice: genotype dependent regional rebound. *Neuroreport*, 13, pp. 1–4.

Hureau, C., Charlet, L., Dorlet, P., Gonnet, F., Spadini, L., Anxolabehere-Mallart, E. and Girerd, J. J. (2006). A spectroscopic and voltammetric study of the pH-dependent Cu(II) coordination to the peptide GGGTH: relevance to the fifth Cu(II) site in the prion protein. *J Biol Inorg Chem*, 11, pp. 735–44.

Hutter, G., Heppner, F. L. and Aguzzi, A. (2003). No superoxide dismutase activity of cellular prion protein *in vivo*. *Biol Chem*, 384, pp. 1279–85.

Jackson, G. S., Murray, I., Hosszu, L. L., Gibbs, N., Waltho, J. P., Clarke, A. R. and Collinge, J. (2001). Location and properties of metal-binding sites on the human prion protein. *Proc Natl Acad Sci USA*, 98, pp. 8531–5.

Jeffrey, M. and Gonzalez, L. (2007). Classical sheep transmissible spongiform encephalopathies: pathogenesis, pathological phenotypes and clinical disease. *Neuropathol Appl Neurobiol*, 33, pp. 373–94.

Johnson, C. J., Phillips, K. E., Schramm, P. T., McKenzie, D., Aiken, J. M. and Pedersen, J. A. (2006). Prions adhere to soil minerals and remain infectious. *PLoS Pathog*, 2, pp. e32.

Jones, C. E., Abdelraheim, S. R., Brown, D. R. and Viles, J. H. (2004). Preferential Cu^{2+} coordination by His96 and His111 induces beta-sheet formation in the unstructured amyloidogenic region of the prion protein. *J Biol Chem*, 279, pp. 32018–27.

Jones, C. E., Klewpatinond, M., Abdelraheim, S. R., Brown, D. R. and Viles, J. H. (2005a). Probing $copper^{2+}$ binding to the prion protein using diamagnetic $nickel^{2+}$ and 1H NMR: the unstructured N terminus facilitates the coordination of six $copper^{2+}$ ions at physiological concentrations. *J Mol Biol*, 346, pp. 1393–407.

Jones, S., Batchelor, M., Bhelt, D., Clarke, A. R., Collinge, J. and Jackson, G. S. (2005b). Recombinant prion protein does not possess SOD-1 activity. *Biochem J*, 392, pp. 309–12.

Kardos, J., Kovacs, I., Hajos, F., Kalman, M. and Simonyi, M. (1989). Nerve endings from rat brain tissue release copper upon depolarization. A possible role in regulating neuronal excitability. *Neurosci Lett*, 103, pp. 139–44.

Kaufman, R. J., Swaroop, M. and Murtha-Riel, P. (1994). Depletion of manganese within the secretory pathway inhibits O-linked glycosylation in mammalian cells. *Biochemistry*, 33, pp. 9813–9.

Kim, N. H., Choi, J. K., Jeong, B. H., Kim, J. I., Kwon, M. S., Carp, R. I. and Kim, Y. S. (2005). Effect of transition metals (Mn, Cu, Fe) and deoxycholic acid (DA) on the conversion of PrPC to PrPres. *FASEB J*, 19, pp. 783–5.

Klamt, F., Dal-Pizzol, F., Conte da Frota, M. J., Walz, R., Andrades, M. E., da Silva, E. G., Brentani, R. R., Izquierdo, I. and Fonseca Moreira, J. C. (2001). Imbalance of antioxidant defense in mice lacking cellular prion protein. *Free Radic Biol Med*, 30, pp. 1137–44.

Klewpatinond, M. and Viles, J. H. (2007). Fragment length influences affinity for Cu^{2+} and Ni^{2+} binding to His96 or His111 of the prion protein and spectroscopic evidence for a multiple histidine binding only at low pH. *Biochem J*, 404, pp. 393–402.

Kralovicova, S., Fontaine, S. N., Alderton, A., Alderman, J., Ragnarsdottir, K. V., Collins, S. J. and Brown, D. R. (2009). The effects of prion protein expression on metal metabolism. *Mol Cell Neurosci*, 41, pp. 135–47.

Kramer, M. L., Kratzin, H. D., Schmidt, B., Romer, A., Windl, O., Liemann, S., Hornemann, S. and Kretzschmar, H. (2001). Prion protein binds copper within the physiological concentration range. *J Biol Chem*, 276, pp. 16711–9.

Kretzschmar, H. A., Neumann, M. and Stavrou, D. (1995). Codon 178 mutation of the human prion protein gene in a German family (Backer family): sequencing data from 72-year-old celloidin-embedded brain tissue. *Acta Neuropathol*, 89, pp. 96–8.

Kristiansen, M., Deriziotis, P., Dimcheff, D. E., Jackson, G. S., Ovaa, H., Naumann, H., Clarke, A. R., van Leeuwen, F. W., Menendez-Benito, V., Dantuma, N. P., Portis, J. L., Collinge, J. and Tabrizi, S. J. (2007). Disease-associated prion protein oligomers inhibit the 26S proteasome. *Mol Cell*, 26, pp. 175–88.

Lee, K. S., Magalhaes, A. C., Zanata, S. M., Brentani, R. R., Martins, V. R. and Prado, M. A. (2001). Internalization of mammalian fluorescent cellular prion protein and N-terminal deletion mutants in living cells. *J Neurochem*, 79, pp. 79–87.

Lekishvili, T., Sassoon, J., Thompsett, A. R., Green, A., Ironside, J. W. and Brown, D. R. (2004). BSE and vCJD cause disturbance to uric acid levels. *Exp Neurol*, 190, pp. 233–44.

Li, X. L., Dong, C. F., Wang, G. R., Zhou, R. M., Shi, Q., Tian, C., Gao, C., Mei, G. Y., Chen, C., Xu, K., Han, J. and Dong, X. P. (2009). Manganese-induced changes of the biochemical characteristics of the recombinant wild-type and mutant PrPs. *Med Microbiol Immunol*, 198, pp. 239–45.

Linder, M. C. (2001). Copper and genomic stability in mammals. *Mutat Res*, 475, pp. 141–52.

Linder, M. C. and Hazegh-Azam, M. (1996). Copper biochemistry and molecular biology. *Am J Clin Nutr*, 63, pp. 797S–811S.

Liu, M. L., Li, Y. X., Zhou, X. M. and Zhao, D. M. (2008). Copper(II) Inhibits *In vitro* Conformational Conversion of Ovine Prion Protein Triggered by Low pH. *J Biochem*, 143, pp. 333–7.

Lugaresi, E., Cirignotta, F., Coccagna, G. and Montagna, P. (1986a). Nocturnal myoclonus and restless legs syndrome. *Adv Neurol*, 43, pp. 295–307.

Lugaresi, E., Cirignotta, F. and Montagna, P. (1986b). Nocturnal paroxysmal dystonia. *J Neurol Neurosurg Psychiatry*, 49, pp. 375–80.

Marella, M., Lehmann, S., Grassi, J. and Chabry, J. (2002). Filipin prevents pathological prion protein accumulation by reducing endocytosis and inducing cellular PrP release. *J Biol Chem*, 277, pp. 25457–64.

McKintosh, E., Tabrizi, S. J. and Collinge, J. (2003). Prion diseases. *J Neurovirol*, 9, pp. 183–93.

Medori, R., Montagna, P., Tritschler, H. J., LeBlanc, A., Cortelli, P., Tinuper, P., Lugaresi, E. and Gambetti, P. (1992). Fatal familial insomnia: a second kindred with mutation of prion protein gene at codon 178. *Neurology*, 42, pp. 669–70.

Mentler, M., Weiss, A., Grantner, K., del Pino, P., Deluca, D., Fiori, S., Renner, C., Klaucke, W. M., Moroder, L., Bertsch, U., Kretzschmar, H. A., Tavan, P. and Parak, F. G. (2005). A new method to determine the structure of the metal environment in metalloproteins: investigation of the prion protein octapeptide repeat Cu(2+) complex. *Eur Biophys J*, 34, pp. 97–112.

Meyer, R. K., McKinley, M. P., Bowman, K. A., Braunfeld, M. B., Barry, R. A. and Prusiner, S. B. (1986). Separation and properties of cellular and scrapie prion proteins. *Proc Natl Acad Sci USA*, 83, pp. 2310–4.

Miele, G., Jeffrey, M., Turnbull, D., Manson, J. and Clinton, M. (2002). Ablation of cellular prion protein expression affects mitochondrial numbers and morphology. *Biochem Biophys Res Commun*, 291, pp. 372–7.

Miura, T., Sasaki, S., Toyama, A. and Takeuchi, H. (2005). Copper reduction by the octapeptide repeat region of prion protein: pH dependence and implications in cellular copper uptake. *Biochemistry*, 44, pp. 8712–20.

Moore, R. C., Lee, I. Y., Silverman, G. L., Harrison, P. M., Strome, R., Heinrich, C., Karunaratne, A., Pasternak, S. H., Chishti, M. A., Liang, Y., Mastrangelo, P., Wang, K., Smit, A. F., Katamine, S., Carlson, G. A., Cohen, F. E., Prusiner, S. B., Melton, D. W., Tremblay, P., Hood, L. E. and Westaway, D. (1999). Ataxia in prion protein (PrP)-deficient mice is associated with upregulation of the novel PrP-like protein doppel. *J Mol Biol*, 292, pp. 797–817.

Morante, S., Gonzalez-Iglesias, R., Potrich, C., Meneghini, C., Meyer-Klaucke, W., Menestrina, G. and Gasset, M. (2004). Inter- and intra-octarepeat Cu(II) site geometries in the prion protein: implications in Cu(II) binding cooperativity and Cu(II)-mediated assemblies. *J Biol Chem*, 279, pp. 11753–9.

Morillas, M., Vanik, D. L. and Surewicz, W. K. (2001). On the mechanism of alpha-helix to beta-sheet transition in the recombinant prion protein. *Biochemistry*, 40, pp. 6982–7.

Nadal, R. C., Abdelraheim, S. R., Brazier, M. W., Rigby, S. E., Brown, D. R. and Viles, J. H. (2007). Prion protein does not redox-silence Cu^{2+}, but is a sacrificial quencher of hydroxyl radicals. *Free Radic Biol Med*, 42, pp. 79–89.

Nadal, R. C., Davies, P., Brown, D. R. and Viles, J. H. (2009). Evaluation of $copper^{2+}$ affinities for the prion protein. *Biochemistry*, 48, pp. 8929–31.

Norde, W. and Giacomelli, C. E. (2000). BSA structural changes during homomolecular exchange between the adsorbed and the dissolved states. *J Biotechnol*, 79, pp. 259–68.

Orem, N. R., Geoghegan, J. C., Deleault, N. R., Kascsak, R. and Supattapone, S. (2006). Copper (II) ions potently inhibit purified PrPres amplification. *J Neurochem*, 96, pp. 1409–15.

Pan, K. M., Baldwin, M., Nguyen, J., Gasset, M., Serban, A., Groth, D., Mehlhorn, I., Huang, Z., Fletterick, R. J., Cohen, F. E. and *et al.*, (1993). Conversion of alpha-helices into beta-sheets features in the formation of the scrapie prion proteins. *Proc Natl Acad Sci USA*, 90, pp. 10962–6.

Pauly, P. C. and Harris, D. A. (1998). Copper stimulates endocytosis of the prion protein. *J Biol Chem*, 273, pp. 33107–10.

Perera, W. S. and Hooper, N. M. (2001). Ablation of the metal ion-induced endocytosis of the prion protein by disease-associated mutation of the octarepeat region. *Curr Biol*, 11, pp. 519–23.

Premzl, M., Sangiorgio, L., Strumbo, B., Marshall Graves, J. A., Simonic, T. and Gready, J. E. (2003). Shadoo, a new protein highly conserved from fish to mammals and with similarity to prion protein. *Gene*, 314, pp. 89–102.

Prusiner, S. B. (1982). Novel proteinaceous infectious particles cause scrapie. *Science*, 216, pp. 136–44.

Prusiner, S. B. (1996). Molecular biology and pathogenesis of prion diseases. *Trends Biochem Sci*, 21, pp. 482–7.

Prusiner, S. B. (1998a). The prion diseases. *Brain Pathol*, 8, pp. 499–513.

Prusiner, S. B. (1998b). Prions. *Proc Natl Acad Sci USA*, 95, pp. 13363–83.

Rachidi, W., Vilette, D., Guiraud, P., Arlotto, M., Riondel, J., Laude, H., Lehmann, S. and Favier, A. (2003). Expression of prion protein increases cellular copper binding and antioxidant enzyme activities but not copper delivery. *J Biol Chem*, 278, pp. 9064–72.

Redecke, L., Meyer-Klaucke, W., Koker, M., Clos, J., Georgieva, D., Genov, N., Echner, H., Kalbacher, H., Perbandt, M., Bredehorst, R., Voelter, W. and Betzel, C. (2005). Comparative analysis of the human and chicken prion protein copper binding regions at pH 6.5. *J Biol Chem*, 280, pp. 13987–92.

Redecke, L., von Bergen, M., Clos, J., Konarev, P. V., Svergun, D. I., Fittschen, U. E., Broekaert, J. A., Bruns, O., Georgieva, D., Mandelkow, E., Genov, N. and Betzel, C. (2007). Structural characterization of beta-sheeted oligomers formed on the pathway of oxidative prion protein aggregation *in vitro*. *J Struct Biol*, 157, pp. 308–20.

Requena, J. R., Groth, D., Legname, G., Stadtman, E. R., Prusiner, S. B. and Levine, R. L. (2001). Copper-catalyzed oxidation of the recombinant SHa(29–231) prion protein. *Proc Natl Acad Sci USA*, 98, pp. 7170–5.

Revault, M., Quiquampoix, H., Baron, M. H. and Noinville, S. (2005). Fate of prions in soil: trapped conformation of full-length ovine prion protein induced by adsorption on clays. *Biochim Biophys Acta*, 1724, pp. 367–74.

Rezaei, H., Choiset, Y., Eghiaian, F., Treguer, E., Mentre, P., Debey, P., Grosclaude, J. and Haertle, T. (2002). Amyloidogenic unfolding intermediates differentiate sheep prion protein variants. *J Mol Biol*, 322, pp. 799–814.

Rezaei, H., Eghiaian, F., Perez, J., Doublet, B., Choiset, Y., Haertle, T. and Grosclaude, J. (2005). Sequential generation of two structurally distinct ovine prion protein soluble oligomers displaying different biochemical reactivities. *J Mol Biol*, 347, pp. 665–79.

Ricchelli, F., Buggio, R., Drago, D., Salmona, M., Forloni, G., Negro, A., Tognon, G. and Zatta, P. (2006). Aggregation/fibrillogenesis of recombinant human prion protein and Gerstmann-Straussler-Scheinker disease peptides in the presence of metal ions. *Biochemistry*, 45, pp. 6724–32.

Riek, R., Hornemann, S., Wider, G., Billeter, M., Glockshuber, R. and Wuthrich, K. (1996). NMR structure of the mouse prion protein domain PrP(121–321). *Nature*, 382, pp. 180–2.

Riek, R., Hornemann, S., Wider, G., Glockshuber, R. and Wuthrich, K. (1997). NMR characterization of the full-length recombinant murine prion protein, mPrP(23–231). *FEBS Lett*, 413, pp. 282–8.

Servagent-Noinville, S., Revault, M., Quiquampoix, H. and Baron, M. (2000). Conformational Changes of Bovine Serum Albumin Induced by Adsorption on Different Clay Surfaces: FTIR Analysis. *J Colloid Interface Sci*, 221, pp. 273–83.

Shearer, J. and Soh, P. (2007a). The copper(II) adduct of the unstructured region of the amyloidogenic fragment derived from the human prion protein is redox-active at physiological pH. *Inorg Chem*, 46, pp. 710–9.

Shearer, J. and Soh, P. (2007b). Ni K-edge XAS suggests that coordination of Ni(II) to the unstructured amyloidogenic region of the human prion protein produces a Ni(2) bis-mu-hydroxo dimer. *J Inorg Biochem*, 101, pp. 370–3.

Shiraishi, N., Inai, Y., Bi, W. and Nishikimi, M. (2005). Fragmentation and dimerization of copper-loaded prion protein by copper-catalysed oxidation. *Biochem J*, 387, pp. 247–55.

Singh, A., Kong, Q., Luo, X., Petersen, R. B., Meyerson, H. and Singh, N. (2009). Prion protein (PrP) knock-out mice show altered iron metabolism: a functional role for PrP in iron uptake and transport. *PLoS One*, 4, pp. e6115.

Sokolowski, F., Modler, A. J., Masuch, R., Zirwer, D., Baier, M., Lutsch, G., Moss, D. A., Gast, K. and Naumann, D. (2003). Formation of critical oligomers is a key event during conformational transition of recombinant syrian hamster prion protein. *J Biol Chem*, 278, pp. 40481–92.

Soto, C. and Estrada, L. D. (2008). Protein misfolding and neurodegeneration. *Arch Neurol*, 65, pp. 184–9.

Stahl, N., Baldwin, M. A., Hecker, R., Pan, K. M., Burlingame, A. L. and Prusiner, S. B. (1992). Glycosylinositol phospholipid anchors of the scrapie and cellular prion proteins contain sialic acid. *Biochemistry*, 31, pp. 5043–53.

Stockel, J., Safar, J., Wallace, A. C., Cohen, F. E. and Prusiner, S. B. (1998). Prion protein selectively binds copper(II) ions. *Biochemistry*, 37, pp. 7185–93.

Stumm, W., Sigg, L. and Sulzberger, B. (1992). Chemistry of the solid-water interface: processes at the mineral-water and particle-water interface in natural systems, Wiley.

Teplow, D. B., Lazo, N. D., Bitan, G., Bernstein, S., Wyttenbach, T., Bowers, M. T., Baumketner, A., Shea, J. E., Urbanc, B., Cruz, L., Borreguero, J. and Stanley, H. E. (2006). Elucidating amyloid beta-protein folding and assembly: A multidisciplinary approach. *Acc Chem Res*, 39, pp. 635–45.

Thackray, A. M., Knight, R., Haswell, S. J., Bujdoso, R. and Brown, D. R. (2002). Metal imbalance and compromised antioxidant function are early changes in prion disease. *Biochem J*, 362, pp. 253–8.

Thompsett, A. R., Abdelraheim, S. R., Daniels, M. and Brown, D. R. (2005). High affinity binding between copper and full-length prion protein identified by two different techniques. *J Biol Chem*, 280, pp. 42750–8.

Thompsett, A. R. and Brown, D. R. (2007). Dual polarisation interferometry analysis of copper binding to the prion protein: evidence for two folding states. *Biochim Biophys Acta*, 1774, pp. 920–7.

Thorgeirsdottir, S., Sigurdarson, S., Thorisson, H. M., Georgsson, G. and Palsdottir, A. (1999). PrP gene polymorphism and natural scrapie in Icelandic sheep. *J Gen Virol*, 80 (Pt 9), pp. 2527–34.

Treiber, C., Simons, A. and Multhaup, G. (2006). Effect of copper and manganese on the de novo generation of protease-resistant prion protein in yeast cells. *Biochemistry*, 45, pp. 6674–80.

Treiber, C., Thompsett, A. R., Pipkorn, R., Brown, D. R. and Multhaup, G. (2007). Real-time kinetics of discontinuous and highly conformational metal-ion binding sites of prion protein. *J Biol Inorg Chem*, 12, pp. 711–20.

Tsenkova, R. N., Iordanova, I. K., Toyoda, K. and Brown, D. R. (2004). Prion protein fate governed by metal binding. *Biochem Biophys Res Commun*, 325, pp. 1005–12.

Tsiroulnikov, K., Rezaei, H., Dalgalarrondo, M., Chobert, J. M., Grosclaude, J. and Haertle, T. (2006). Cu(II) induces small-size aggregates with amyloid characteristics in two alleles of recombinant ovine prion proteins. *Biochim Biophys Acta*, 1764, pp. 1218–26.

Uppington, K. M. and Brown, D. R. (2008). Resistance of cell lines to prion toxicity aided by phospho-ERK expression. *J Neurochem*, 105, pp. 842–52.

Vendrely, C., Valadie, H., Bednarova, L., Cardin, L., Pasdeloup, M., Cappadoro, J., Bednar, J., Rinaudo, M. and Jamin, M. (2005). Assembly of the full-length recombinant mouse prion protein I. Formation of soluble oligomers. *Biochim Biophys Acta*, 1724, pp. 355–66.

Viles, J. H., Cohen, F. E., Prusiner, S. B., Goodin, D. B., Wright, P. E. and Dyson, H. J. (1999). Copper binding to the prion protein: structural implications of four identical cooperative binding sites. *Proc Natl Acad Sci USA*, 96, pp. 2042–7.

Vincent, B., Paitel, E., Saftig, P., Frobert, Y., Hartmann, D., De Strooper, B., Grassi, J., Lopez-Perez, E. and Checler, F. (2001). The disintegrins ADAM10 and TACE contribute to the constitutive and phorbol ester-regulated normal cleavage of the cellular prion protein. *J Biol Chem*, 276, pp. 37743–6.

Wadsworth, J. D., Hill, A. F., Joiner, S., Jackson, G. S., Clarke, A. R. and Collinge, J. (1999). Strain-specific prion-protein conformation determined by metal ions. *Nat Cell Biol*, 1, pp. 55–9.

Waggoner, D. J., Drisaldi, B., Bartnikas, T. B., Casareno, R. L., Prohaska, J. R., Gitlin, J. D. and Harris, D. A. (2000). Brain copper content and cuproenzyme activity do not vary with prion protein expression level. *J Biol Chem*, 275, pp. 7455–8.

Walmsley, A. R., Zeng, F. and Hooper, N. M. (2003). The N-terminal region of the prion protein ectodomain contains a lipid raft targeting determinant. *J Biol Chem*, 278, pp. 37241–8.

Walter, E. D., Chattopadhyay, M. and Millhauser, G. L. (2006). The affinity of copper binding to the prion protein octarepeat domain: evidence for negative cooperativity. *Biochemistry*, 45, pp. 13083–92.

Walz, R., Amaral, O. B., Rockenbach, I. C., Roesler, R., Izquierdo, I., Cavalheiro, E. A., Martins, V. R. and Brentani, R. R. (1999). Increased sensitivity to seizures in mice lacking cellular prion protein. *Epilepsia*, 40, pp. 1679–82.

Webb, S., Lekishvili, T., Loeschner, C., Sellarajah, S., Prelli, F., Wisniewski, T., Gilbert, I. H. and Brown, D. R. (2007). Mechanistic insights into the cure of prion disease by novel antiprion compounds. *J Virol*, 81, pp. 10729–41.

Weiss, A., Del Pino, P., Bertsch, U., Renner, C., Mentler, M., Grantner, K., Moroder, L., Kretzschmar, H. A. and Parak, F. G. (2007). The configuration of the Cu(2+) binding region in full-length human prion protein compared with the isolated octapeptide. *Vet Microbiol*, 123, pp. 358–66.

Weissmann, C. (1991). A 'unified theory' of prion propagation. *Nature*, 352, pp. 679–83.

Westergard, L., Christensen, H. M. and Harris, D. A. (2007). The cellular prion protein (PrP(C)): its physiological function and role in disease. *Biochim Biophys Acta*, 1772, pp. 629–44.

White, A. R., Collins, S. J., Maher, F., Jobling, M. F., Stewart, L. R., Thyer, J. M., Beyreuther, K., Masters, C. L. and Cappai, R. (1999). Prion protein-deficient neurons reveal lower glutathione reductase activity and increased susceptibility to hydrogen peroxide toxicity. *Am J Pathol*, 155, pp. 1723–30.

White, S. N., O'Rourke, K. I., Gidlewski, T., VerCauteren, K. C., Mousel, M. R., Phillips, G. E. and Spraker, T. R. (2010). Increased risk of chronic wasting disease in Rocky Mountain elk associated with decreased magnesium and increased manganese in brain tissue. *Can J Vet Res*, 74, pp. 50–3.

Will, R. G., Ironside, J. W., Zeidler, M., Cousens, S. N., Estibeiro, K., Alperovitch, A., Poser, S., Pocchiari, M., Hofman, A. and Smith, P. G. (1996). A new variant of Creutzfeldt-Jakob disease in the UK. *Lancet*, 347, pp. 921–5.

Wong, B. S., Brown, D. R., Pan, T., Whiteman, M., Liu, T., Bu, X., Li, R., Gambetti, P., Olesik, J., Rubenstein, R. and Sy, M. S. (2001a). Oxidative impairment in scrapie-infected mice is associated with brain metals perturbations and altered antioxidant activities. *J Neurochem*, 79, pp. 689–98.

Wong, B. S., Chen, S. G., Colucci, M., Xie, Z., Pan, T., Liu, T., Li, R., Gambetti, P., Sy, M. S. and Brown, D. R. (2001b). Aberrant metal binding by prion protein in human prion disease. *J Neurochem*, 78, pp. 1400–8.

Wong, B. S., Pan, T., Liu, T., Li, R., Gambetti, P. and Sy, M. S. (2000a). Differential contribution of superoxide dismutase activity by prion protein *in vivo*. *Biochem Biophys Res Commun*, 273, pp. 136–9.

Wong, B. S., Venien-Bryan, C., Williamson, R. A., Burton, D. R., Gambetti, P., Sy, M. S., Brown, D. R. and Jones, I. M. (2000b). Copper refolding of prion protein. *Biochem Biophys Res Commun*, 276, pp. 1217–24.

Wopfner, F., Weidenhofer, G., Schneider, R., von Brunn, A., Gilch, S., Schwarz, T. F., Werner, T. and Schatzl, H. M. (1999). Analysis of 27 mammalian and 9 avian PrPs reveals high conservation of flexible regions of the prion protein. *J Mol Biol*, 289, pp. 1163–78.

Zeng, F., Watt, N. T., Walmsley, A. R. and Hooper, N. M. (2003). Tethering the N-terminus of the prion protein compromises the cellular response to oxidative stress. *J Neurochem*, 84, pp. 480–90.

Zhou, Y., Shie, F. S., Piccardo, P., Montine, T. J. and Zhang, J. (2004). Proteasomal inhibition induced by manganese ethylene-bis-dithiocarbamate: relevance to Parkinson's disease. *Neuroscience*, 128, pp. 281–91.

Zhu, F., Davies, P., Thompsett, A. R., Kelly, S. M., Tranter, G. E., Hecht, L., Isaacs, N. W., Brown, D. R. and Barron, L. D. (2008). Raman optical activity and circular dichroism reveal dramatic differences in the influence of divalent copper and manganese ions on prion protein folding. *Biochemistry*, 47, pp. 2510–7.

Chapter 9

Emerging Role for Copper-Bound α-Synuclein in Parkinson's Disease Etiology

Heather R. Lucas and Jennifer C. Lee
Laboratory of Molecular Biophysics, Biochemistry and Biophysics Center National Heart, Lung, and Blood Institute, National Institutes of Health, Bethesda, MD, USA

9.1 Introduction

9.1.1 *Copper Homeostasis and the Brain*

Copper is an essential biological element that is vital for proper neurodevelopment and neurophysiology (Gaggelli *et al.*, 2006; Desai and Kaler, 2008; Culotta, 2010). Many important biological functions in humans, such as cellular respiration (cytochrome *c* oxidase), peptide amidation (peptidylglycine α-hydroxylating monoxygenase), neurotransmitter biosynthesis (dopamine β-hydroxylase), iron metabolism (ceruloplasmin), oxidative stress release (copper-zinc superoxide dismutase), and skin pigmentation (tyrosinase), are dependent on the proper delivery of copper to specific metallo-active sites. In fact, it is estimated that a third of the genome is composed of metalloproteins (Finney and O'Halloran, 2003; Hasnain,

Brain Diseases and Metalloproteins
Edited by David R. Brown

ISBN 978-981-4316-01-9 (Hardcover), 978-981-4364-07-2 (eBook)
www.panstanford.com

2004; Maret, 2010). Thus far, approximately 30 enzymes have been characterized that use copper as a cofactor (Culotta, 2010).

A common misconception in the biological and neuroscience communities is that heavy metals such as copper are found in only 'trace' quantities, however, as described above in part, this is not the case (Barnham and Bush, 2008). Various concentrations of copper have been found throughout the body providing a largely accessible pool of copper for most tissues. The total extracellular copper(II) content has been estimated as 10–25 μM in the blood, 15–30 μM in the synaptic cleft, and 0.5–2.5 μM in cerebrospinal fluid (CSF) (Brown *et al.,* 1997; Que *et al.,* 2008). Intracellular neuronal copper levels, likely in the reduced copper(I) oxidation state, can reach 2 to 3 orders of magnitude higher concentrations (Que *et al.,* 2008).

The brain houses more metal ions than any other part of the body. Actually, transition metals are 10,000 fold more prevalent in brain tissue than common neurotransmitters and neuropeptides with copper being the third most abundant biometal (Que *et al.,* 2008). It is worth noting here that copper is plentiful in the *substantia nigra* (0.4 mM), the site of dopamine production, as well as in the *locus ceruleus* (1.3 mM), the neural region that accounts for physiological stress and panic responses (Que *et al.,* 2008). For comparison, the concentration of magnesium (Mg^{2+}) in brain gray matter is found within the range 0.1–0.5 mM (Bush, 2000). Such large concentrations of metal in the brain are necessary to promote proper O_2 metabolism following respiration. Notably, the brain accommodates 20% of the body's O_2 (Gotz *et al.,* 1994).

Copper is a redox active metal that easily cycles between copper(I) and copper(II) allowing for its ability to facilitate an array of vital oxidation-reduction biochemical processes. As a result, a complex regulatory system exists to maintain proper metal storage and delivery (Lalioti *et al.,* 2009; Lutsenko, 2010). Alterations in this homeostasis, i.e., failure of copper chaperones to deliver copper to specific targets, can trigger aberrant redox chemistry leading to oxidative stress (*vide infra*) or abnormal metal-protein interactions. For example, Wilson disease is characterized by the accumulation of copper leading to copper-induced radical-mediated damage; however, Menkes disease is distinguished by an insufficient copper supply leading to inactivation of key metabolic enzymes and potentially non-enzymatic pathways (Huster and Lutsenko, 2007).

Therefore, the same characteristics of copper that make this metal essential for human health also make the metal detrimental if not controlled.

9.1.2 α-Synuclein and Parkinson's Disease

Parkinson's disease (PD) is the second most prevalent neurodegenerative disorder, affecting 0.1% of the population over 40 years of age (Siderowf and Stern, 2003). This percentage increases to 1–2% for people over 65 years old and affects as many as 4–5% of the population over 85 years of age (Uversky and Eliezer, 2009). It is currently estimated that ~1.5 million Americans have been diagnosed with PD and that on average 50,000 people throughout the world are diagnosed per year. The most common symptoms of this progressive "age-related" neurological ailment are resting tremor (trembling of hands, arms, legs, face), rigidity (stiffness of limbs), bradykinesia (slowness of movement), and postural instability (impaired balance) (Gaggelli *et al.,* 2006).

The pathophysiology of PD is characterized by the loss of dopaminergic neurons in the *substantia nigra*. The principal hallmark is the accumulation of insoluble proteinacious deposits or aggregated proteins, termed Lewy bodies, in neurons of brain cells (Dawson and Dawson, 2003). As shown in Fig. 9.1, Lewy bodies are spherical protein inclusions with a dense core surrounded by a halo of fibrillar material, also termed as amyloid, that consist primarily of the neuronal protein α-synuclein (α-syn), as well as other proteins and lipids in smaller quantities (Spillantini *et al.,* 1997; Maries *et al.,* 2003).

Various other evidence also link α-syn to Parkinson's disease (Cookson 2009). For example, three independent missense mutations, corresponding to A30P, E46K, and A53T, of the *SNCA* gene, which codes for α-syn, lead to a rare inherited form of PD (Polymeropoulos *et al.,* 1997; Kruger *et al.,* 1998; Zarranz *et al.,* 2004). Early-onset familial PD also can be caused by triplication of the *SNCA* gene locus, leading to hyperexpression of α-syn and consequently neuronal cell death (Singleton *et al.,* 2003).

α-Syn is an abundant protein that can be found throughout the central nervous system. It is especially plentiful at presynaptic terminals and can be found in both soluble and membrane-bound

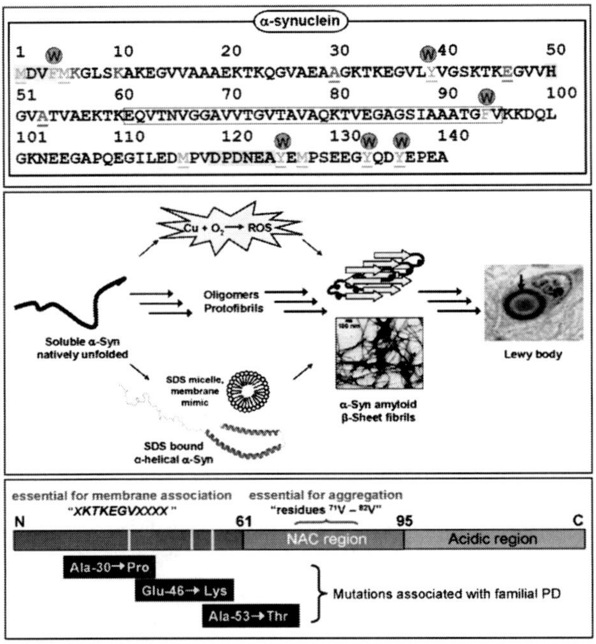

Figure 9.1 *Top:* The primary amino acid sequence of human α-syn. The minimal copper(II) binding site, i.e. first four N-terminal residues, is highlighted in yellow. Other residues involved in copper binding sites are shaded grey, specifically (i) the second N-terminal site involving His-50 and (ii) the non-specific C-terminal site, DPDNEA. The non-amyloid β component (NAC) region is boxed and the three missense mutations correlated with early-onset PD are underlined and colored red. All native aromatic residues, Tyr (underlined) and Phe, are colored orange to mark locations of available single-site Trp mutations, which are also labeled with a purple circle. All four Met residues (green) within the sequence and the two Lys residues (purple) nearest the N-terminal binding domain are labeled. *Middle:* A schematic representation of the proposed mechanism of PD pathology in regards to α-syn fibrillation and incorporation into Lewy bodies; Lewy body image was adapted from (Maries *et al.*, 2003). The electron microscopy image of β-sheet fibrils was collected under the following conditions: 100 μM α-syn, 90 μM CuII; 20 mM MOPS, 100 mM NaCl; pH 7; 5 days; 37°C. Pathways involving membrane-bound α-syn and copper-mediated ROS generation are currently active research areas; SDS-bound α-syn image was adapted from (Ulmer *et al.*, 2005). *Bottom:* A generalized diagram summarizing the characteristics that define the three regions of α-syn. See also Colour Insert.

forms. The specific function of α-syn has yet to be determined, however, various proposals with some supporting evidence exist. For example, α-syn has been implicated in the regulation of synaptic plasticity, neuronal differentiation, regulation of dopamine synthesis, and fatty acid transport (El-Agnaf *et al.,* 2003). In addition, α-syn has been shown to possess chaperone-like activity (Souza *et al.,* 2000b).

When free in solution, α-syn is natively unfolded, i.e., possessing an unstructured conformation. This intrinsically disordered nature is thought to be due to the overall high net charge at neutral pH and the low hydrophobicity (Uversky *et al.,* 2000; Fink, 2006). α-Syn is also known to adopt various other conformations. In particular, a structural transition to a helical state occurs when associated to brain-derived or synthetic lipid vesicles (Davidson *et al.,* 1998; Jo *et al.,* 2000; Eliezer *et al.,* 2001; Pfefferkorn and Lee, 2010). In the fibrillar form, α-syn assumes a cross-β sheet structure in which individual β-strands are perpendicular to the fibril axis (Tycko, 2006; Margittai and Langen, 2008).

α-Syn consists of 140 amino acids that can be categorized into three main regions (Fig. 9.1): the N-terminus (residues 1 – 60); the non-amyloid β-component or NAC region (residues 61 – 95); and the acidic C-terminus (residues 96 – 140). The N-terminus contains four imperfect 11-amino acid repeats with the conserved motif, KTKEGV, which code for amphipathic helices, making this region essential for membrane association. The NAC region, containing predominantly nonpolar side chains, is critical for aggregation and constitutes the hydrophobic amyloid core (Chen *et al.,* 2007; Qin *et al.,* 2007; Vilar *et al.,* 2008). Removal of residues 71 – 82 completely abrogates fibril assembly (Giasson *et al.,* 2001).

The role of deposition and accumulation of fibrillar structures remains widely unresolved with increasing clinical and laboratory research on various possible pathogenic mechanisms. α-Syn is generally suggested to play a central role in most mechanisms of PD; what remains a mystery is the identity of the cytotoxic species. There is emerging evidence to suggest that soluble oligomeric forms rather than the insoluble amyloid deposits are key cytotoxic agents (Caughey and Lansbury, 2003; Cookson, 2005). Specifically, several different types of protofibrillar species including spheres and annuli have been implicated as the agents responsible for cell death (Chiti and Dobson, 2006; Uversky and Eliezer, 2009). Some α-syn oligomers have been characterized, although a detailed mechanism of fibril formation has not yet been elucidated (Conway *et al.*, 2000a; Conway *et al.*, 2000b; Ding *et al.*, 2002).

Like many amyloid diseases, the role of aggregated proteinaceous materials in PD is inconclusive. On the one hand, the deposits themselves may be neurotoxic. Alternatively, these fibrillar aggregates could simply be a byproduct of neuronal death. Further, it is possible that the formation of insoluble materials could be a protective mechanism for sequestering cytotoxic species (Forno, 1996; Tompkins *et al.*, 1997; Tompkins and Hill, 1997). Considerable evidence points to fibril precursors as the pathogenic agents (Caughey and Lansbury, 2003; Cookson, 2005). Moreover, the proteins alone may not be the problem. Environmental factors such as redox-active metals and reactive oxygen species (ROS) have been variously implicated in the pathology of these and related protein misfolding diseases (*vide infra*).

9.1.3 *Copper and Parkinson's Disease*

Over the past decade, research focusing on the role of metals in the etiology of neurodegenerative diseases has increased. Most attention in the literature has been directed at the interaction of metals, namely copper, with the amyloidogenic protein amyloid-β, which is linked to Alzheimer's disease. In addition, copper-zinc superoxide dismutase and prion protein, respectively linked to Amyotrophic Lateral Sclerosis and Creutzfeldt-Jakob disease, have received a lot of attention. The focus of this report is to summarize current progress towards understanding the relationship of copper and α-syn interactions and to highlight the possible relevance to PD.

Brain metal compositions have been measured using various techniques in order to gauge the bioavailability of metal ions that could affect neurodegeneration (Que *et al.*, 2008; Spasojevic *et al.*). Mouse models have been used to examine changes in the metal concentration level in the brain with age, gender, and genetic backgrounds (Maynard *et al.*, 2006; Becker *et al.*, 2010). Recent advancements in bioimaging have been especially useful, affirming a role of copper availability in PD (Matusch *et al.*, 2010).

Interest in copper-protein interactions and their relevance to "age-related" diseases has stemmed largely from reports that copper concentrations are elevated in serum and plasma isolated from subjects over 75 years of age (Duce and Bush, 2010). In fact, patients with newly diagnosed, untreated, idiopathic PD have been reported to have raised concentrations of copper in their cerebrospinal fluid in comparison to control subjects, but there was no such difference for iron concentration (Pall *et al.*, 1987). In a recent study, the catalytic properties of copper and iron were examined in human cerebrospinal fluid (Spasojevic *et al.*, 2010). Interestingly, in contrast to iron, copper treated with H_2O_2 promoted a significant increase in hydroxyl radicals.

In a very recent study, hyperexpression of α-syn resulted in increased sensitivity of cells to copper toxicity (Wang *et al.*, 2010). Also, changes in the concentration of cellular copper resulted in changes in both α-syn aggregation and localization. For example, reduced levels of copper resulted in a decrease in intracellular aggregates and increased localization of α-syn in the plasma membrane. These trends were reversed when copper content was restored to the cells. In a separate study, new melanic pigments that appear to accumulate metals, perhaps serving a protective role, were discovered in various regions of the brain that increase with age (Zecca *et al.*, 2008).

9.2 Interaction of Copper(II) and α-Synuclein

9.2.1 *Instrumental Approaches: Stoichiometry and Dissociation Constants*

Various experimental techniques have been employed to measure the stoichiometry and dissociation constants (K_d) for copper(II) binding to α-syn under numerous solution conditions. Research from various groups have provided evidence that copper(II) can coordinate to sites within all three regions of α-syn with K_d values in the nM to mM range. Initial studies indicated that 10 copper(II) ions were bound at the C-terminus of a single α-syn monomer, K_d = 59 μM (Paik *et al.*, 1999). It was proposed that this negatively charged region would provide an electrostatic surface to promote metal-induced self-oligomerization. In other words, the metal-protein interactions were suggested to promote aggregation by suppressing the repulsive forces from the C-terminal acidic side chains, resulting in more favorable hydrophobic interactions.

Since the aforementioned work, high-resolution spectroscopic techniques have been more commonly used. In regards to the C-terminal binding sites, it is thought that they possess rather weak affinity, $K_d \sim 470$ μM (Rasia *et al.*, 2005; Sung *et al.*, 2006; Binolfi *et al.*, 2008). In addition, all C-terminal binding sites have been suggested as non-specific for copper; other divalent metals are found to bind to this region with similar affinities (Binolfi *et al.*, 2006; Sung *et al.*, 2006). This suggests that metal binding is dictated primarily by electrostatic interactions. Through experiments utilizing nuclear magnetic resonance (NMR) spectroscopy, one specific C-terminal binding motif with low selectivity for metal ions has been identified as residues 119 – 124 (DPDNEA), with Asp-121 acting as the anchoring residue (Binolfi *et al.*, 2006).

The high affinity copper(II) binding sites of α-syn are within the N-terminus. Debate over the exact nature of the site(s) has been centered on the involvement of His-50. At one point, His-50 was suggested as the anchoring residue for the primary copper(II) binding site, based primarily on NMR experiments, $K_d \sim 0.1$ μM, 100 μM protein in 20 mM MES, 100 mM NaCl, pH 6.5, 15 °C (Rasia *et al.*, 2005). This single N-terminal copper(II)-binding motif was suggested to be comprised of residues widely separated in the polypeptide sequence and involved the protein folding back on itself forming a macrochelate. In a contradictory study also using NMR spectroscopy, removal of His-50 did not prevent copper(II) binding at the N-terminus (Sung *et al.*, 2006). In addition, long-range order was not observed upon copper(II) coordination, suggesting that a macrochelate involving His-50 as a ligand is unlikely.

Fluorescence spectroscopy has also been employed to examine the involvement of His-50 (Lee *et al.*, 2008; Jackson and Lee, 2009). The high sensitivity of this technique at low concentration makes it an extremely useful tool for determining apparent K_d values, providing a physical method capable of achieving accurate measurements under physiologically relevant conditions (~μM). Moreover, the absence of intrinsic Trp residues within the native α-syn sequence provides an advantage because single-site

Trp-substitutions can then be inserted at specific locations of interest with site-directed mutagenesis. Changes in the local environment around the fluorophore can then be monitored upon addition of copper, or any other exogenous quencher (Lakowicz, 2006). Particularly, site-specific Trp substitutions replacing aromatic residues such as Phe and/or Tyr can minimize perturbation to side chain properties.

Based on all the α-syn Trp variants (F4W, Y39W, F94W, and Y125W) examined, F4W was the most sensitive to copper(II) coordination. With addition of ~1 equivalent of copper(II), Trp-4 experiences greater than 80% quenching, see Fig. 9.2 for representative Trp-4 spectra as a function of copper(II) added (Lee *et al.,* 2008). The Y39W variant exhibited little change with addition of up to 10 equivalents of copper(II), overall accounting for less than 20% fluorescence quenching. The C-terminal Trp variants showed no detectable changes with addition of comparable concentrations of copper(II). These results confirm the N-terminus as the high-affinity site and also that no significant conformational changes occur with addition of greater than 1 equivalent of copper(II).

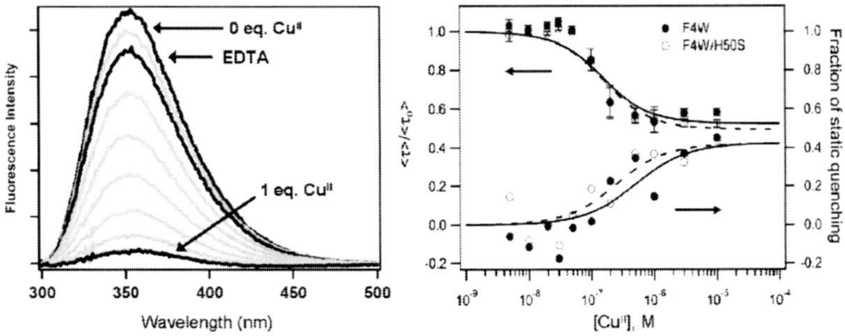

Figure 9.2 *Left*: Representative Trp fluorescence spectra as a function of added copper(II) ions; spectra corresponding to (i) F4W alone (0 eq. Cu^{II}), (ii) Cu^{II}-F4W (1 eq. Cu^{II}), and (iii) F4W following Cu^{II}-extraction *via* addition of EDTA, *i.e.* {Cu^{II}-F4W + EDTA → Cu^{II}-EDTA + F4W} (EDTA), are shown in bold. *Right*: Plot of average Trp-4 fluorescence lifetimes (upper set) and fraction of static quenching (lower set) extracted from non-negative least square (NNLS) fits of F4W (•) and F4W/H50S (o) α-syn variants ([protein] = 100 nM) as functions of added copper(II) in deoxygenated 20 mM MOPS, 100 mM NaCl, pH 7.0 buffer. Fits to a single site binding model are shown as solid (F4W) and dashed (F4W/H50S) lines.

In this same work, to obtain reliable copper(II)-α-syn dissociation constants for F4W (K_d ~100 nM) and F4W/H50S (K_d ~110 nM), titrations were performed using time-resolved fluorescence spectroscopy at submicromolar concentrations (~100 nM). Interestingly, the kinetics revealed two quenching modes: static (~60%) and dynamic (~40%) quenching. Single-site binding models (Fig. 9.2) adequately described the average Trp lifetime and static quenching data, confirming a 1:1 stoichiometry that does not involve His-50 (Lee *et al.,* 2008). The dual quenching modes have been attributed to two possible scenarios: (i) two distinct copper(II)-protein structures where one Trp-4 is within static-quenching distance and the other where Trp-4 is sufficiently close to copper(II) to make contact during its excited-state lifetime; or (ii) a broad distribution of Trp-4-copper(II) distances in a conformationally heterogeneous polypeptide.

The current consensus within the literature is that two independent, non-interacting copper-binding sites exist with very different affinities. The lower affinity site (K_d ~50 μM) is suggested to involve His-50 based on comparative analyses of circular dichroism (CD) and electronic absorption spectra (UV-vis) (Binolfi *etal.* 2008). Based on these and a variety of other measurements, the higher affinity site (K_d ~2.4 nM – 100 nM) does not involve His-50; see the following section for detailed information about the primary copper(II) binding site of α-syn (Binolfi *et al.,* 2008; Lee *et al.,* 2008; Hong and Simon, 2009; Jackson and Lee, 2009). For both sites, the N-terminal amine has unequivocally been assigned as the anchoring residue for copper(II) binding based on NMR spectroscopy and tryptophan fluorescence quenching experiments as well as unique evidence obtained from matrix-assisted laser desorption ionization mass spectrometry (MALDI MS) experiments (Binolfi *et al.,* 2008; Jackson and Lee, 2009).

The local conformation around the copper(II) ion has also been examined by electron paramagnetic resonance (EPR) spectroscopy and X-ray absorption spectroscopy (XAS) (Drew *et al.,* 2008; Lucas *et al.,* 2010). EPR measurements indicate that both N-terminal Cu^{II}-α-syn motifs are typical of type II centers with a square planar or distorted tetragonal geometry (Drew *et al.,* 2008). The coordination sphere around the high affinity N-terminal copper(II) site was predicted to involve ligation from two nitrogen and two oxygen atoms, designated

2N2O. The lower affinity site was predicted to have a 2N2O or 3N1O coordination sphere involving His-50. Extended X-ray absorption fine structure (EXAFS) data of the primary copper(II) binding site of α-syn also support a tetra-coordinate copper(II) coordination sphere involving nitrogen and oxygen containing atoms with approximate Cu-(N/O) bond distances of 1.96 Å (Lucas *et al.*, 2010).

In addition to the instrumental approaches already discussed, other physical methods have been employed to examine copper(II) binding to α-syn. For example, isothermal titration calorimetry (ITC) has suggested a 1:1 stoichiometry for copper(II) binding to wild-type α-syn and the A53T disease-related mutant, K_d ~2.4 nM (Hong and Simon, 2009). For the A30P disease-causing variant, two binding constants were reported with K_d values of 1.6 and 10 nM, respectively. The high affinity site is thought to be identical to the wild-type and A53T variants. The low affinity site for copper(II) binding to A30P is suggested to result from conformational differences that were previously reported in an NMR study of this disease-variant in the absence of metal. In the structural work, the A30P mutation was shown to disrupt a region of residual helical structure that is present in the wild-type protein (Bussell and Eliezer, 2001). An alternative explanation was that the two copper(II) binding sites may result from a small population of copper(II)-bound oligomeric species (Hong and Simon, 2009).

In separate work involving ITC, two binding constants were also reported for A30P, however different K_d values were obtained (Bharathi and Rao, 2007). ITC requires experiments to be conducted at high concentrations (≥ 70 μM), which would enhance copper(II) disproportionation in aqueous solution. Therefore, more emphasis has been placed on the results from the aforementioned ITC study because glycine, a weak copper(II) chelator, was used to stabilize the metal ion by preventing formation of insoluble copper hydroxide species (Hong and Simon, 2009).

9.2.2 *Structural Aspects of Copper(II)-α-Syn, the Primary Binding Site*

As previously discussed, copper(II) has been shown to coordinate to all three regions of α-syn with varying affinities. However, the primary, high affinity sites are within the N-terminus, as confirmed by various methodologies, including NMR, EPR, CD, UV-vis, and fluorescence spectroscopies (*vide supra*). Examining truncated forms of α-syn, i.e., N-terminal peptides, has enabled the specific nature of the primary copper(II) binding site to be determined. For example, a combined potentiometric and spectroscopic study on a series of N-terminal synthetic polypeptides first showed that sequences from 1-39, 1-28, and 1-17 are capable of binding copper(II) ions (Kowalik-Jankowska *et al.,* 2005). From this work, a 2N2O metal-peptide complex was proposed with α-amino group (NH_2), deprotonated backbone amide (N^-), carboxylate group of Asp-2 (α-COO^-), and an exogenous water molecule making up its coordination sphere. The reported CD and EPR spectra for the 1-17 peptide at pH 6.5 are reminiscent of the full length wild-type protein (Rasia *et al.,* 2005), providing further support that His-50 is not a critical ligand.

More recently, using a series of fluorescent Trp-containing synthetic peptides, the first four N-terminal residues (Syn_{1-4}) were determined as the minimal sequence necessary to fully retain copper binding (Jackson and Lee, 2009). For this minimal Cu-α-syn model (Cu^{II}-Syn_{1-4}) as well as Cu^{II}-Syn_{1-10}, the participation of the α-amino NH_2 as the essential ligand was demonstrated unequivocally by the complete abrogation of copper binding upon acetylation (Ac) of the N-termini (Fig. 9.3). The role of other potential NH_2 groups in this sequence (MDVWMKGLSK) were excluded as the single and double Lys-to-Arg mutations at positions 6 and 10 did not affect the K_d values for copper(II) binding; therefore, Lys-6 and Lys-10 do not ligate the copper ion. Interestingly, it was suggested that through posttranslational acetylation of the N-terminus, intracellular copper-α-syn chemistry could be modulated.

Effects of solution pH on copper(II)-binding were examined in detail for both the truncated and full length α-synucleins (Jackson and Lee, 2009). Nearly identical behaviors were found for Syn_{1-10}, F4W, and F4W/H50S with comparable apparent pK_a values (Fig. 9.3), consistent with potentiometric measurements (Kowalik-Jankowska *et al.,* 2005). Furthermore, data extracted from the pH titrations pointed to two nearby ionizable groups, possibly the amino and amide protons. Also, the pH dependence of copper(II) binding showed that substitution of His-50 had a negligible effect. Interestingly, Trp-4 excited-state decay kinetics measured for the minimal Syn_{1-4} sequence exhibited similar

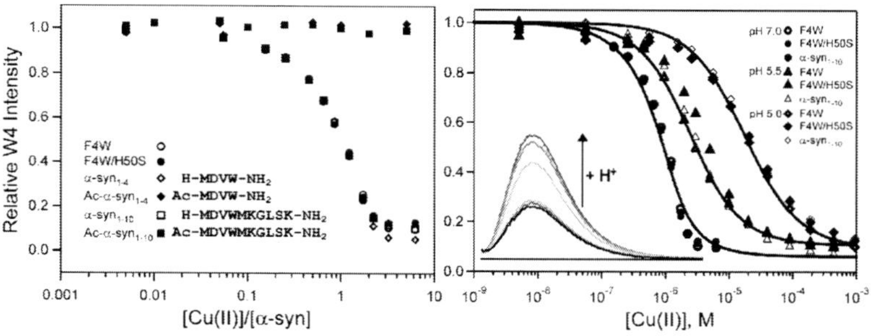

Figure 9.3 *Left*: Comparison of copper(II) binding to full length F4W α-syn, Syn_{1-4}, and Syn_{1-10} N-terminal Trp-containing peptides ([protein/peptide] = 1 μM in deoxygenated 20 mM MOPS, 100 mM NaCl, pH 7). Trp-4 quenching was not observed for peptides with acetylated (Ac) terminal amines. *Right*: Dependence of copper(II) titrations (F4W, F4W/H50S, and Syn_{1-10}) on solution pH (5.0, 5.5, and 7.0). *Inset*: Representative Trp-4 emission in copper(II)-bound F4W (1:1 Cu(II):α-syn) as a function of pH (7–5).

dual quenching modes as those observed for the F4W protein suggesting local conformational heterogeneity, arising possibly from two copper(II)-polypeptide structures.

The conformation of Cu^{II}-Syn_{1-4} was recently characterized by CD spectroscopy (Fig. 9.4) (Lucas and Lee, 2010). In this work,

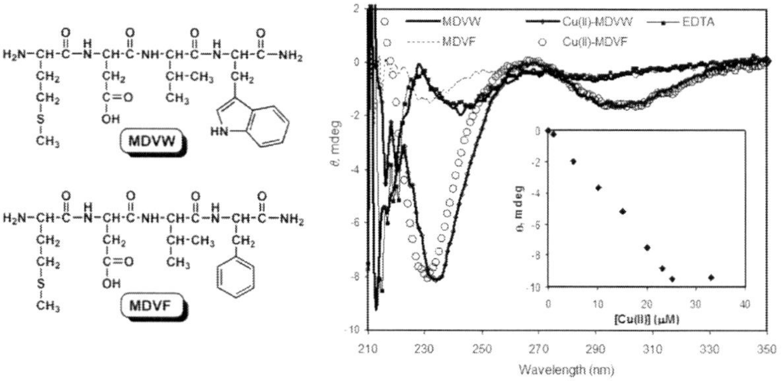

Figure 9.4 Structures of the N-terminal Syn_{1-4} peptides, MDVW and MDVF, and their corresponding circular dichroism (CD) spectra; [peptide] = 30 μM in 20 mM MOPS, 100 mM NaCl, pH 7. CD spectra corresponding to the copper bound forms, Cu(II)-MDVW and Cu(II)-MDVF, are also shown. In addition, a spectrum of MDVW following extraction of copper(II) with EDTA is plotted. The inset is a plot of Φ_{233nm} versus $[Cu^{II}]$ for MDVW indicating a 1:1 binding stoichiometry.

it was shown that copper binding to Trp-containing Syn_{1-4} has spectroscopic characteristics similar to the copper-bound wild-type sequence (MDVF). Minima at 233 and 300 nm as well as a maximum at 267 nm were observed that reached saturation at one equivalent of copper(II). The 300 and 267 nm bands are characteristic of a copper(II)-peptide complex and are assigned, respectively, as charge transfer transitions between N^- and NH_2 to the copper(II) center, analogous to other reports for α-syn (Kowalik-Jankowska *et al.*, 2005; Kowalik-Jankowska *et al.*, 2006; Binolfi *et al.*, 2008). Reversibility of metal binding was demonstrated by extraction of copper(II) with ethylenediaminetetraacetic acid (EDTA), yielding the initial, unstructured conformation. Notably, this work suggested that the spectral feature at 233 nm may indicate the presence of a copper(II)-(Phe/Trp) cation-π interaction reminiscent of the structurally characterized copper chaperone CusF and the biologically active decapeptide neuromedin-C (Loftin *et al.,* 2007; Xue *et al.,* 2008; Yorita *et al.,* 2008).

9.2.3 *Soluble versus Fibrillar Copper(II)-α-Synuclein*

Up until recently, research on the copper(II) binding site of α-syn was focused solely on soluble, monomeric forms of the protein. In a recent report, the copper(II) coordination site of α-syn was explored in the soluble and fibrillar states through XAS, revealing that only modest differences exist between soluble *versus* fibrillar Cu^{II}-α-syn (Fig. 9.5) (Lucas *et al.,* 2010). In particular, the EXAFS data support a coordination environment of three-to-four N- or O- containing ligands for both protein structures with approximate bond distances of 1.96 Å. This was interesting because CD data indicate large changes in the global polypeptide conformation with a secondary structural change from unfolded to β-sheet occurring upon fibril formation (Fig. 9.5). Based on transmission electron microscopy (TEM) imaging, the α-syn fibril morphology does not change with removal of copper by EDTA. In addition to TEM imaging and CD analyses, amyloid formation was confirmed by a thioflavin T (ThT) fluorescence assay. ThT can be used as an α-syn fibril indicator because it binds to β-sheet structure resulting in a quantum yield increase.

ThT has also been used to measure the aggregation kinetics. By utilizing this probe, α-syn fibrillation has been shown to

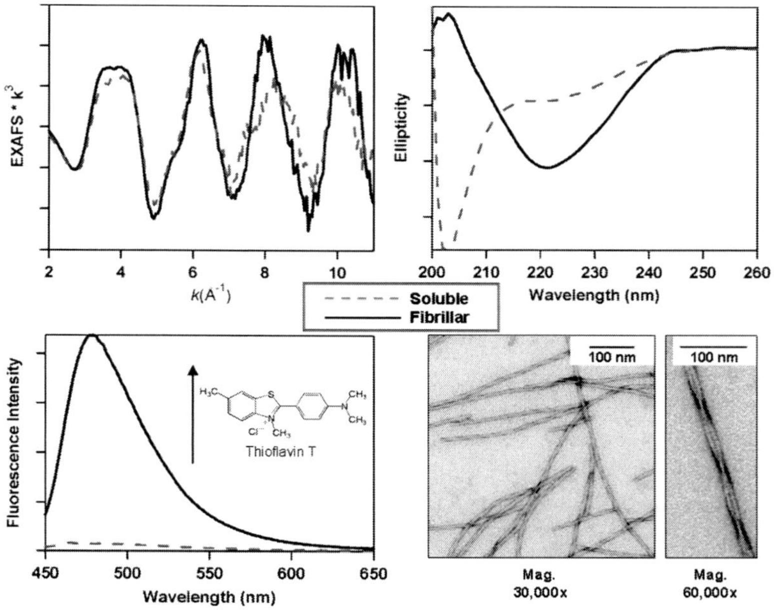

Figure 9.5 Extended X-ray absorption fine structure (EXAFS; *top, left*) and circular dichroism (CD; *top, right*) data of soluble versus fibrillar Cu^{II}-α-syn. *Bottom, right:* ThT fluorescence assay indicating fibril formation. *Bottom, left:* Transmission electron microscopy images of fibrillar Cu^{II}-α-syn.

accelerate in the presence of copper(II) in comparison to various other divalent metals examined (Binolfi *et al.,* 2006). More specifically, the midpoint transition ($t_{1/2}$) for α-syn aggregation time decreased from ~65 hours to ~30 hours in the presence of copper(II).

Interestingly, at this same time point (30 h), no α-syn fibrils had formed for protein aged in the absence of metal, nor for protein aged in the presence of manganese(II), iron(II), cobalt(II), or nickel(II). A closer look at the metal binding affinities of α-syn, revealed a specificity for copper(II) with an apparent K_d value much higher (< 1 μM) than the other metals, which were in the 1–2 mM range. In this work, however, the possibility of metal-dioxygen side reactions that could affect the fibrillation process were not addressed.

9.3 Metal-Catalysed Protein Oxidation

9.3.1 *Generation of Reactive Oxygen Species*

The generation of reactive oxygen species (ROS) and subsequent chemistry is commonly thought to be the basis for irreversible protein modifications and/or changes in native structure. As a result, proteins may be less susceptible to proteolytic degradation, which can result in accumulation and/or aggregation. Protein oxidation is known to be more prevalent with age, leading to dysfunctional enzymes or altered catalytic pathways, as well as oxidative stress, as is relevant here (Stadtman, 1992, 2001; Stadtman *et al.,* 2005). The covalent modification of proteins as a result of oxidative chemistry has been implicated in various physiological and pathological pathways, including neurodegenerative disease etiology.

Parkinson's disease can be induced in mice and monkeys by injection of the pro-toxin 1-methyl-4-phenyl-1,2,3,6-tetrahydropyridine (MPTP), which has been a gold standard for PD studies (Maries *et al.,* 2003). MPTP was discovered as a contaminant in illicit opiates following the sporadic onset of PD in a small community of drug-users (Langston *et al.,* 1983). Following intravenous injection of the pro-drug, MPTP is oxidized by monoamine oxidase B (MAO-B) to 1-methyl-4-phenyl-2,3-dihydropyridinium (MPHP), then localized in glia, where spontaneous conversion to the active metabolite 1-methyl-4-phenyl-pyridinium (MPP^+) occurs (Fig. 9.6). MPP^+ is then selectively incorporated into dopaminergic neurons through a membrane bound dopaminergic transporter. The Parkinsonian effects of MPTP are due primarily to the loss of ATP in nigrostriatal neurons and to a lesser extent, the inhibition of mitochondrial complex I. Overall, this leads to elevated levels of oxidative stress due to ROS production, which can lead to cell death.

Copper-mediated generation of ROS *via* Fenton chemistry or Haber-Weiss reactions (Fig. 9.6) has been suggested to promote α-syn fibrillation (Fig. 9.1). A crucial point that is often ignored outside of the bioinorganic science community is that in order for metal-catalyzed oxidation of proteins to occur, the metal must bind to the protein and form a transient metal-dioxygen species. In regards to copper-dioxygen chemistry, much is known about copper metalloenzymes in terms of both structure and function (see introduction). A wealth

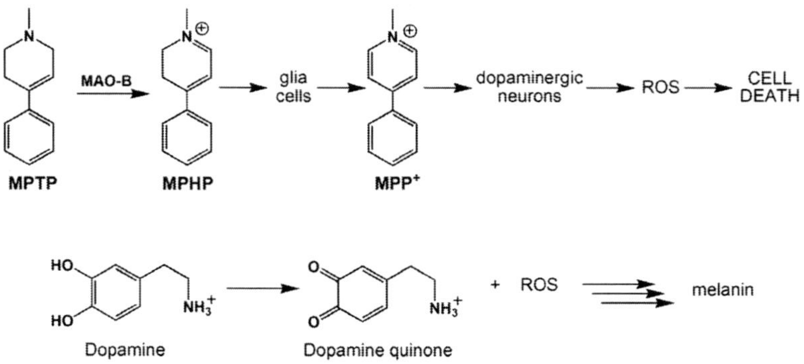

$$\begin{array}{rl} \textbf{Haber-Weiss Reaction:} & Cu^{2+} + O_2^{\bullet-} \longrightarrow Cu^{+} + O_2 \\ \textbf{Fenton Reaction:} & Cu^{+} + H_2O_2 \longrightarrow Cu^{2+} + OH^{-} + {}^{\bullet}OH \\ \hline \textbf{Net Reaction:} & O_2^{\bullet-} + H_2O_2 \xrightarrow[Cu^{2+}]{} OH^{-} + {}^{\bullet}OH + O_2 \end{array}$$

Figure 9.6 *Top:* Schematic representation of 1-methyl-4-phenyl-1,2,3,6-tetrahydropyridine (MPTP) activation by monoamine oxidase B (MAO-B), forming 1-methyl-4-phenyl-pyridinium (MPP^+), and the pathway to cytotoxicity. *Middle:* Dopamine oxidation to dopamine quinone and reactive oxygen species (ROS), prior to forming melanin. Bottom: Reaction schemes for copper-mediated generation of ROS.

of information concerning the nature of potentially relevant copper-dioxygen synthetic species is also available (Solomon *et al.,* 1996; Lewis and Tolman 2004; Mirica *et al.,* 2004; Solomon *et al.,* 2004). Moreover, it is important to note that hydrogen peroxide (H_2O_2) is an inactive molecule that can transform into highly reactive products, such as hydroxyl radicals; however, unlike superoxide ($O_2^{-\bullet}$), H_2O_2 can cross cell membranes.

Recently, evidence of copper-bound α-syn mediated generation of ROS studied through XAS has been reported (Lucas *et al.,* 2010). Interestingly, the Cu K-edge spectrum (Fig. 9.7) of soluble copper-bound α-syn exhibited an estimated 20% metal reduction with the appearance of a characteristic copper(I) absorption feature at 8983 eV; the estimation of copper(I) content was based on comparisons to a four coordinate copper(I) model. The Cu K-edge measurements also were conducted on samples prepared under anaerobic conditions, which resulted in a slightly higher concentration of copper(I)-bound α-syn species. To assess the role of O_2-chemistry

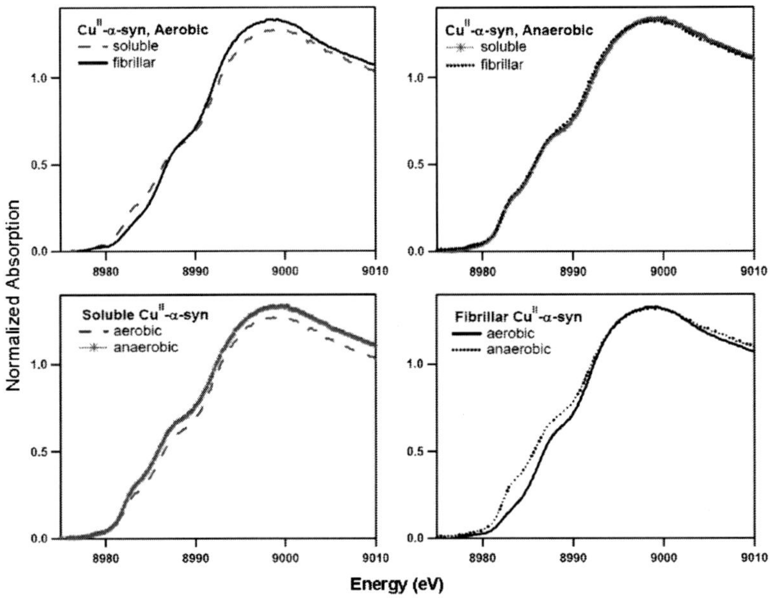

Figure 9.7 Comparison of Cu K-edge X-ray absorption spectroscopy (XAS) data of α-syn aged under aerobic (*top, left*) versus anaerobic conditions (*top, right*). Different plots of the same spectra are also shown for clarity at the beginning (soluble; *bottom, left*) and end (fibrillar; *bottom, right*) of the aggregation process.

in the aging process, fibrillar samples were examined by Cu K-edge XAS indicating that metal reduction occurs in the absence of O_2 and that re-oxidation occurs in the presence of O_2 (Fig. 9.7). The direct observation of copper(I) was in line with a previous study examining changes in the K_d value for copper(II) binding in the presence of O_2 based on Trp-4 fluorescence quenching experiments (Lucas and Lee, 2010). In this latter study, copper(II) binding was enhanced by an order of magnitude in the presence of dioxygen, $K_{d(aerobic)}$ ~10 nM and $K_{d(anaerobic)}$ ~100 nM. Overall the results confirm that coordination of copper to α-syn promotes oxidative stress.

9.3.2 *Methionine Oxidation*

The oxidation of methionine residues to methionine sulfide radicals, methionine sulfoxide (MetO), and/or methionine sulfone is a key feature of oxidative stress and is also known to decrease the biological activity of proteins. In fact, the cyclic interconversion of Met and MetO within proteins is involved in several different biological processes such as: (i) ROS scavenging; (ii) regulation of enzyme activity; (iii) cell signaling; as well as (iv) targeting proteins for proteolytic degradation (Stadtman *et al.,* 2003). Several oxidants, including metals, can directly oxidize Met to MetO *via* a formal oxygen transfer.

Various research groups have shown that methionine-oxidized α-syn prevents formation of β-sheet (Cole *et al.,* 2005; Fink, 2006; Leong *et al.,* 2009a). Instead, the intermediate oligomeric form of α-syn is stabilized; note that although Lewy bodies are mainly composed of fibrillar α-syn, the specific toxic form is under debate (see introduction). It also has been shown that a four-fold molar excess of fully oxidized α-syn (there are four native Mets, Fig. 9.1) is sufficient to completely inhibit the aggregation of wild-type α-syn (Uversky *et al.,* 2002; Fink, 2006). This finding suggests that oligomeric α-syn can incorporate unmodified protein, preventing fibrillation. In addition, this same research group has shown that the presence of certain metals (Ti^{3+}, Zn^{2+}, Al^{3+}, and Pb^{2+}) can overcome this inhibition, which was proposed to relate to the epidemiological correlation between chronic metal exposure and PD (Yamin *et al.,* 2003).

Methionine-oxidized α-syn is typically prepared by addition of H_2O_2 to the protein (Uversky *et al.,* 2002). Recently, dopamine-mediated formation of oxidized Met and soluble α-syn oligomers has been demonstrated. Specifically, the C-terminal ^{125}YEMPS^{129} motif is identified to modulate the ability of dopamine to inhibit α-syn fibrillation (Leong *et al.,* 2009b). The data strongly indicated that dopamine promoted ROS generation, resulting in Met oxidation. Because dopamine can be readily oxidized to dopamine quinone, $O_2^{-\bullet}$, H_2O_2, and free radicals in the presynaptic terminals of dopaminergic neurons, these experiments highlight a physiological oxidative pathway for generating soluble oligomers, potentially neurotoxic agents (LaVoie and Hastings, 1999; Sulzer, 2001). Similar results can be obtained using a Cu/H_2O_2 oxidizing system, however it appears to be less efficient (Uversky *et al.,* 2002; Leong *et al.,* 2009b).

In studies examining the effect of dioxygen on copper(II) binding and structure (*vide supra*), an autoreduction event yielding copper(I)

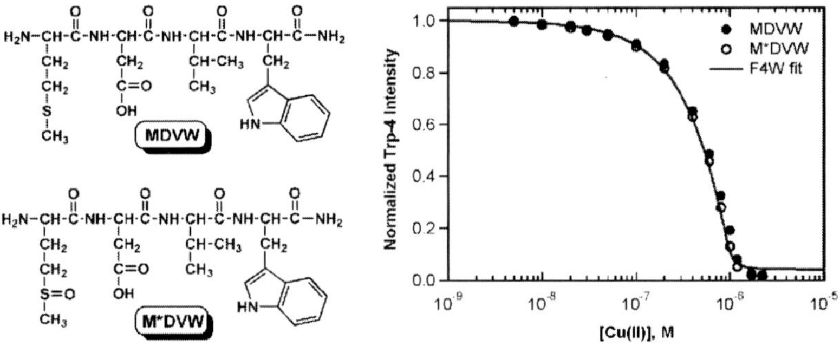

Figure 9.8 Structures of the N-terminal Syn_{1-4} peptides, MDVW and M*DVW (M*=methionine sulfoxide), and their corresponding copper(II) binding curves monitored by Trp-4 fluorescence in comparison to the full length F4W α-syn variant; [peptide/protein] = 1 μM in 20 mM MOPS, 100 mM NaCl, pH 7.

was reported (Lucas *et al.*, 2010; Lucas and Lee, 2010). Based on the fundamental principle that when one molecule is reduced another is oxidized, it was hypothesized that Met oxidation was occurring. This process seemed plausible because two Met residues (Met-1 and Met-5) are in close proximity to the primary N-terminal copper(II) binding site and therefore susceptible to oxidation. Copper(II) binding to Syn_{1-4} peptide with one oxidized Met (MetO-1) was examined in order to probe possible effects of protein oxidation within the copper(II) binding motif (Fig. 9.8). The results revealed that Met-1 oxidation does not prevent nor alter copper(II) binding. Currently, no direct evidence of copper-mediated Met-oxidation of α-syn exists. Nevertheless, a recent report examining oxidation of Mets by H_2O_2, that is perhaps relevant, suggested that Met-5 is the most reactive amongst all four α-syn Met residues (Met-1, 5, 116 and 127) (Zhou *et al.*, 2010).

9.3.3 *Protein Crosslinking*

Covalent post-translational modifications of a protein can inhibit intracellular proteolysis or cause a soluble protein to become insoluble (Halliwell, 1992; Benzi and Moretti, 1995). As a result, oxygen derived free radical damage has been thought to be involved in neurodegeneration. Interestingly, the oxidative stress marker dityrosine has been detected in the brains of mice treated with MPTP (*vide supra*) (Pennathur *et al.,* 1999). α-Syn contains four tyrosine residues, three of which are found in the C-terminus (Fig. 9.1). Formation of such crosslinks could lead to the stabilization and/or polymerization of α-syn. In fact, dityrosine formation as a result of tyrosine-tyrosine coupling has been reported for α-syn in the presence of copper following exposure to ROS.

Early work suggested that copper(II) binding at the C-terminus facilitated oxidative crosslinking of the Tyr residues in the presence of H_2O_2 (Paik *et al.,* 2000). C-terminal truncation mutants, α-syn114 and α-syn97, containing a single Tyr residue at position 39, were unable to form dityrosines suggesting that the C-terminal Tyr residues (Tyr-125, Tyr-133, and Tyr-136) were responsible. In a separate report from this same group, Tyr crosslinking was enhanced in the presence of lipids (Lee *et al.,* 2003). Independently, another report suggested that Tyr-39 was essential for covalent protein crosslinking (Ruf *et al.,* 2008). This is consistent with a recent report that suggests long-range intramolecular interactions exist between Tyr-39 and Tyr-133, based on Tyr-to-Ala α-syn variants (Ulrih *et al.,* 2008).

In a contradictory report, copper(II)-mediated dityrosine formation was examined in the presence of nitrating agents (Souza *et al.,* 2000a). Again the C-terminal Tyr residues were suggested as responsible, however Tyr-39 was not thought to be involved because this residue was nitrated rather than crosslinked. Additionally, only one out of the four Tyr residues of α-syn was suggested to form a crosslink.

Recently, evidence of ROS formation as a result of copper(I)/O_2 reactivity was demonstrated (Lucas *et al.,* 2010). In this work, a new emission feature was observed at 395 nm following fibril formation in samples aged aerobically in the presence of copper(I) (Fig. 9.9). In the absence of copper, a red-shifted and less pronounced feature was observed at 425 nm (Fig. 9.9). Both emission bands were attributed to dityrosine formation and the large difference was proposed to result from the formation of inter- *versus* intra-molecular crosslinks. In the absence of dioxygen, neither emission band was observed.

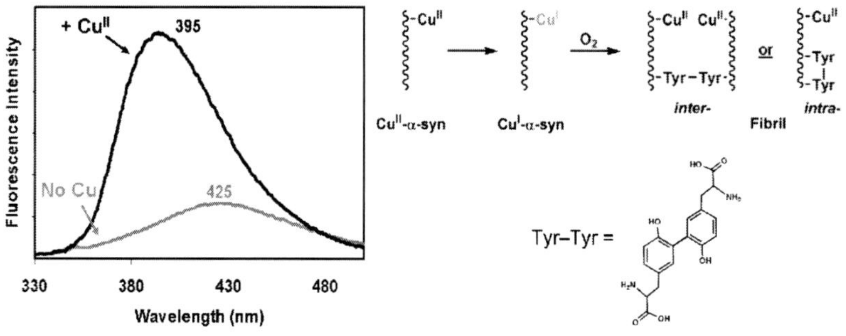

Figure 9.9 *Left*: Fluorescence detection of dityrosine (Tyr-Tyr) crosslinks in fibrillar α-syn both in the presence (black) and absence (grey) of copper under aerobic conditions; 180 μM in 20 mM MOPS, 100 mM NaCl, pH 7. *Right*: Schematic representation of the possible reaction mechanism for copper/dioxygen-mediated fibril assembly.

9.4 Conclusion

As discussed in this chapter, α-synuclein (α-syn) conformational studies and copper-mediated oxidative chemistry are of great importance for understanding the molecular mechanisms involved in Parkinson's disease (PD) progression. In the Introduction (Section 9.1), the bioavailability of copper is discussed, underlining this biometal as a vital nutrient that is crucial to human health (Section 9.1.1). In addition, the factors that relate α-syn, a neuronal protein, to PD are reviewed (Section 9.1.2). Moreover, alterations in metal homeostasis with age are highlighted, which could be involved in PD etiology, possibly due to changes in α-syn conformation and/or cellular localization (Section 9.1.3).

Since copper must bind to α-syn in order to influence protein structure and/or modulate metal-polypeptide chemistry, numerous spectroscopic and biochemical studies have been directed at elucidating the specifics of the metal binding sites in all three regions of α-syn (Section 9.2). The current consensus within the field appear to point to the N-terminus (residues 1 – 10) as the physiologically relevant metal-binding site, as it contains the strongest copper(II)-affinity (submicromolar apparent dissociation constant, $K_d \sim 1$ nM – 100 nM, pH 6.5 – 7.4). The copper(II)-α-syn coordination sphere is thought to be composed of 2 nitrogen and 2 oxygen-containing ligands with the α-amino group as the critical anchor. N-terminal peptides have been especially useful for determining the minimal site for copper(II) binding, MDVW.

While these findings have provided some structural and chemical insights, the existence and exact nature of intracellular copper-α-syn remain to be elucidated. Several studies point to the involvement of post-translational, covalent modifications of α-syn, namely methionine oxidation and dityrosine formation (Section 9.3), hat result from metal-mediated oxidative chemistry. Recent studies have provided evidence that a considerable amount of α-syn is N-terminally acetylated (Li *et al.,* 2005; Anderson *et al.,* 2006). As a result, it is difficult to reconcile and assess the biological significance of the well-characterized copper(II) binding site, since acetylation was shown to abolish copper binding to N-terminal α-syn peptides.

The ultimate challenge for this field is to define and understand metal-α-syn chemistry *in vivo.* Sophisticated techniques continue to be developed, such as X-ray fluorescence and infrared imaging (Miller *et al.,* 2006), which should make this achievable in the future. In the meantime, new insights into the role of metals in PD will be gained by examining copper-α-syn interactions in the presence of other biomolecules, for example, phospholipid bilayers. Furthermore, future studies aimed at the copper(I) binding site of α-syn will be especially important since reduced metal oxidation states are generally considered to be more physiologically relevant. In addition, copper(I) has a high potential for promoting ROS chemistry and therefore may lead to a more cytotoxic oligomeric or fibrillar species.

References

Anderson, J. P., Walker, D. E., Goldstein, J. M., de Laat, R., Banducci, K., Caccavello, R. J., Barbour, R., Huang, J., Kling, K., Lee, M., Diep, L., Keim, P. S., Shen, X., Chataway, T., Schlossmacher, M. G., Seubert, P., Schenk, D., Sinha, S., Gai, W. P. and Chilcote, T. J. (2006). Phosphorylation of Ser-129 is the dominant pathological modification of α-synuclein in familial and sporadic Lewy body disease. *J. Biol. Chem.*, 281, pp. 29739–52.

Barnham, K. J. and Bush, A. I. (2008). Metals in Alzheimer's and Parkinson's diseases. *Curr. Opin. Chem. Biol.*, 12, pp. 222–8.

Becker, J. S., Zoriy, M., Matusch, A., Wu, B., Salber, D., Palm, C. and Becker, J. S. (2010). Bioimaging of metals by laser ablation inductively coupled plasma mass spectrometry (LA-ICP-MS). *Mass Spectrom. Rev.*, 29, pp. 156–75.

Benzi, G. and Moretti, A. (1995). Are reactive oxygen species involved in Alzheimer's disease? *Neurobiol. Aging*, 16, pp. 661–74.

Bharathi and Rao, K. S. (2007). Thermodynamics imprinting reveals differential binding of metals to α-synuclein: relevance to Parkinson's disease. *Biochem. Biophys. Res. Commun.*, 359, pp. 115–20.

Binolfi, A., Lamberto, G. R., Duran, R., Quintanar, L., Bertoncini, C. W., Souza, J. M., Cervenansky, C., Zweckstetter, M., Griesinger, C. and Fernandez, C. O. (2008). Site-specific interactions of Cu(II) with α- and β-synuclein: Bridging the molecular gap between metal binding and aggregation. *J. Am. Chem. Soc.*, 130, pp. 11801–12.

Binolfi, A., Rasia, R. M., Bertoncini, C. W., Ceolin, M., Zweckstetter, M., Griesinger, C., Jovin, T. M. and Fernandez, C. O. (2006). Interaction of α-synuclein with divalent metal ions reveals key differences: a link between structure, binding specificity and fibrillation enhancement. *J. Am. Chem. Soc.*, 128, pp. 9893–901.

Brown, D. R., Qin, K., Herms, J. W., Madlung, A., Manson, J., Strome, R., Fraser, P. E., Kruck, T., von Bohlen, A., Schulz-Schaeffer, W., Giese, A., Westaway, D. and Kretzschmar, H. (1997). The cellular prion protein binds copper *in vivo*. *Nature*, 390, pp. 684–7.

Bush, A. I. (2000). Metals and neuroscience. *Curr. Opin. Chem. Biol.*, 4, pp. 184–91.

Bussell, R., Jr. and Eliezer, D. (2001). Residual structure and dynamics in Parkinson's disease-associated mutants of α-synuclein. *J. Biol. Chem.*, 276, pp. 45996–6003.

Caughey, B. and Lansbury, P. T. (2003). Protofibrils, pores, fibrils, and neurodegeneration: separating the responsible protein aggregates from the innocent bystanders. *Annu. Rev. Neurosci.*, 26, pp. 267–98.

Chen, M., Margittai, M., Chen, J. and Langen, R. (2007). Investigation of α-synuclein fibril structure by site-directed spin labeling. *J. Biol. Chem.*, 282, pp. 24970–9.

Chiti, F. and Dobson, C. M. (2006). Protein misfolding, functional amyloid, and human disease. *Annu. Rev. Biochem.*, 75, pp. 333–66.

Cole, N. B., Murphy, D. D., Lebowitz, J., Di Noto, L., Levine, R. L. and Nussbaum, R. L. (2005). Metal-catalyzed oxidation of α-synuclein: helping to define the relationship between oligomers, protofibrils, and filaments. *J. Biol. Chem.*, 280, pp. 9678–90.

Conway, K. A., Harper, J. D. and Lansbury, P. T., Jr. (2000a). Fibrils formed *in vitro* from α-synuclein and two mutant forms linked to Parkinson's disease are typical amyloid. *Biochemistry*, 39, pp. 2552–63.

Conway, K. A., Lee, S. J., Rochet, J. C., Ding, T. T., Williamson, R. E. and Lansbury, P. T., Jr. (2000b). Acceleration of oligomerization, not fibrillization, is a shared property of both α-synuclein mutations linked to early-onset Parkinson's disease: Implications for pathogenesis and therapy. *Proc. Nat. Acad. Sci. USA*, 97, pp. 571–6.

Cookson, M. R. (2005). The biochemistry of Parkinson's disease. *Annu. Rev. Biochem.*, 74, pp. 29–52.

Cookson, M. R. (2009). α-Synuclein and neuronal cell death. *Mol. Neurodegener.*, 4, p. 9.

Culotta, V. (2010). Cell biology of copper. *J. Biol. Inorg. Chem.*, 15, pp. 1–2.

Davidson, W. S., Jonas, A., Clayton, D. F. and George, J. M. (1998). Stabilization of α-synuclein secondary structure upon binding to synthetic membranes. *J. Biol. Chem.*, 273, pp. 9443–9.

Dawson, T. M. and Dawson, V. L. (2003). Molecular pathways of neurodegeneration in Parkinson's disease. *Science*, 302, pp. 819–22.

Desai, V. and Kaler, S. G. (2008). Role of copper in human neurological disorders. *Am. J. Clin. Nutr.*, 88, pp. 855S–8S.

Ding, T. T., Lee, S. J., Rochet, J. C. and Lansbury, P. T., Jr. (2002). Annular α-synuclein protofibrils are produced when spherical protofibrils are incubated in solution or bound to brain-derived membranes. *Biochemistry*, 41, pp. 10209–17.

Drew, S. C., Leong, S. L., Pham, C. L., Tew, D. J., Masters, C. L., Miles, L. A., Cappai, R. and Barnham, K. J. (2008). Cu^{2+} binding modes of recombinant α-synuclein--insights from EPR spectroscopy. *J. Am. Chem. Soc.*, 130, pp. 7766–73.

Duce, J. A. and Bush, A. I. (2010). Biological metals and Alzheimer's disease: Implications for therapeutics and diagnostics. *Prog. Neurobiol.*, p. 18.

El-Agnaf, O. M., Salem, S. A., Paleologou, K. E., Cooper, L. J., Fullwood, N. J., Gibson, M. J., Curran, M. D., Court, J. A., Mann, D. M., Ikeda, S., Cookson, M. R., Hardy, J. and Allsop, D. (2003). α-Synuclein implicated in Parkinson's disease is present in extracellular biological fluids, including human plasma. *FASEB J.*, 17, pp. 1945–7.

Eliezer, D., Kutluay, E., Bussell, R., Jr. and Browne, G. (2001). Conformational properties of α-synuclein in its free and lipid-associated states. *J. Mol. Biol.*, 307, pp. 1061–73.

Fink, A. L. (2006). The aggregation and fibrillation of α-synuclein. *Acc. Chem. Res.*, 39, pp. 628–34.

Finney, L. A. and O'Halloran, T. V. (2003). Transition metal speciation in the cell: Insights from the chemistry of metal ion receptors. *Science*, 300, pp. 931–936.

Forno, L. S. (1996). Neuropathology of Parkinson's disease. *J. Neuropathol. Exp. Neurol.*, 55, pp. 259–72.

Gaggelli, E., Kozlowski, H., Valensin, D. and Valensin, G. (2006). Copper homeostasis and neurodegenerative disorders (Alzheimer's, prion, and Parkinson's diseases and amyotrophic lateral sclerosis). *Chem. Rev.*, 106, pp. 1995–2044.

Giasson, B. I., Murray, I. V., Trojanowski, J. Q. and Lee, V. M. (2001). A hydrophobic stretch of 12 amino acid residues in the middle of α-synuclein is essential for filament assembly. *J. Biol. Chem.*, 276, pp. 2380–6.

Gotz, M. E., Kunig, G., Riederer, P. and Youdim, M. B. (1994). Oxidative stress: free radical production in neural degeneration. *Pharmacol. Ther.*, 63, pp. 37–122.

Halliwell, B. (1992). Reactive oxygen species and the central nervous system. *J. Neurochem.*, 59, pp. 1609–23.

Hasnain, S. S. (2004). Synchrotron techniques for metalloproteins and human disease in post genome era. *J. Synchrotron Radiat.*, 11, pp. 7–11.

Hong, L. and Simon, J. D. (2009). Binding of Cu(II) to human α-synucleins: Comparison of wild type and the point mutations associated with the familial Parkinson's disease. *J. Phys. Chem. B*, 113, pp. 9551–61.

Huster, D. and Lutsenko, S. (2007). Wilson disease: not just a copper disorder. Analysis of a Wilson disease model demonstrates the link between copper and lipid metabolism. *Mol. Biosyst.*, 3, pp. 816–24.

Jackson, M. S. and Lee, J. C. (2009). Identification of the minimal copper(II)-binding α-synuclein sequence. *Inorg. Chem.*, 48, pp. 9303–7.

Jo, E., McLaurin, J., Yip, C. M., St George-Hyslop, P. and Fraser, P. E. (2000). α-Synuclein membrane interactions and lipid specificity. *J. Biol. Chem.*, 275, pp. 34328–34.

Kowalik-Jankowska, T., Rajewska, A., Jankowska, E. and Grzonka, Z. (2006). Copper(II) binding by fragments of α-synuclein containing M1-D2- and -H50-residues; a combined potentiometric and spectroscopic study. *Dalton Trans.*, pp. 5068–76.

Kowalik-Jankowska, T., Rajewska, A., Wisniewska, K., Grzonka, Z. and Jezierska, J. (2005). Coordination abilities of N-terminal fragments of α-synuclein towards copper(II) ions: A combined potentiometric and spectroscopic study. *J. Inorg. Biochem.*, 99, pp. 2282–91.

Kruger, R., Kuhn, W., Muller, T., Woitalla, D., Graeber, M., Kosel, S., Przuntek, H., Epplen, J. T., Schols, L. and Riess, O. (1998). Ala30Pro mutation in the gene encoding α-synuclein in Parkinson's disease. *Nat. Genet.*, 18, pp. 106–8.

Lakowicz, J. R. (2006). Protein Fluorescence. In Principles of Fluorescence Spectroscopy. New York, Springer: pp. 529–75.

Lalioti, V., Muruais, G., Tsuchiya, Y., Pulido, D. and Sandoval, I. V. (2009). Molecular mechanisms of copper homeostasis. *Front. Biosci.*, 14, pp. 4878–903.

Langston, J. W., Ballard, P., Tetrud, J. W. and Irwin, I. (1983). Chronic Parkinsonism in humans due to a product of meperidine-analog synthesis. *Science*, 219, pp. 979–80.

LaVoie, M. J. and Hastings, T. G. (1999). Dopamine quinone formation and protein modification associated with the striatal neurotoxicity of methamphetamine: evidence against a role for extracellular dopamine. *J. Neurosci.*, 19, pp. 1484–91.

Lee, E. N., Lee, S. Y., Lee, D., Kim, J. and Paik, S. R. (2003). Lipid interaction of α-synuclein during the metal-catalyzed oxidation in the presence of Cu^{2+} and H_2O_2. *J. Neurochem.*, 84, pp. 1128–42.

Lee, J. C., Gray, H. B. and Winkler, J. R. (2008). Copper(II) binding to α-synuclein, the Parkinson's protein. *J. Am. Chem. Soc.*, 130, pp. 6898–9.

Leong, S. L., Cappai, R., Barnham, K. J. and Pham, C. L. (2009a). Modulation of α-synuclein aggregation by dopamine: a review. *Neurochem. Res.*, 34, pp. 1838–46.

Leong, S. L., Pham, C. L., Galatis, D., Fodero-Tavoletti, M. T., Perez, K., Hill, A. F., Masters, C. L., Ali, F. E., Barnham, K. J. and Cappai, R. (2009b). Formation of dopamine-mediated α-synuclein-soluble oligomers requires methionine oxidation. *Free Radic. Biol. Med.*, 46, pp. 1328–37.

Lewis, E. A. and Tolman, W. B. (2004). Reactivity of dioxygen-copper systems. *Chem. Rev.*, 104, pp. 1047–76.

Li, W., West, N., Colla, E., Pletnikova, O., Troncoso, J. C., Marsh, L., Dawson, T. M., Jakala, P., Hartmann, T., Price, D. L. and Lee, M. K. (2005). Aggregation promoting C-terminal truncation of α-synuclein is a normal cellular process and is enhanced by the familial Parkinson's disease-linked mutations. *Proc. Nat. Acad. Sci. USA*, 102, pp. 2162–7.

Loftin, I. R., Franke, S., Blackburn, N. J. and McEvoy, M. M. (2007). Unusual Cu(I)/Ag(I) coordination of E. coli CusF as revealed by atomic resolution crystallography and X-ray absorption spectroscopy. *Protein Sci.*, 16, pp. 2287–93.

Lucas, H. R., Debeer, S., Hong, M. S. and Lee, J. C. (2010). Evidence for copper-dioxygen reactivity during α-synuclein fibril formation. *J. Am. Chem. Soc.*, 132, pp. 6636–7.

Lucas, H. R. and Lee, J. C. (2010). Effect of dioxygen on copper(II) binding to α-synuclein. *J. Inorg. Biochem.*, 104, pp. 245–9.

Lutsenko, S. (2010). Human copper homeostasis: a network of interconnected pathways. *Curr. Opin. Chem. Biol.*, 14, pp. 211–17.

Maret, W. (2010). Metalloproteomics, metalloproteomes, and the annotation of metalloproteins. *Metallomics*, 2, pp. 117–25.

Margittai, M. and Langen, R. (2008). Fibrils with parallel in-register structure constitute a major class of amyloid fibrils: molecular insights from electron paramagnetic resonance spectroscopy. *Q. Rev. Biophys.*, 41, pp. 265–97.

Maries, E., Dass, B., Collier, T. J., Kordower, J. H. and Steece-Collier, K. (2003). The role of α-synuclein in Parkinson's disease: insights from animal models. *Nat. Rev. Neurosci.*, 4, pp. 727–38.

Matusch, A., Depboylu, C., Palm, C., Wu, B., Hoglinger, G. U., Schafer, M. K. and Becker, J. S. (2010). Cerebral bioimaging of Cu, Fe, Zn, and Mn in the MPTP mouse model of Parkinson's disease using laser ablation inductively coupled plasma mass spectrometry (LA-ICP-MS). *J. Am. Soc. Mass Spectrom.*, 21, pp. 161–71.

Maynard, C. J., Cappai, R., Volitakis, I., Cherny, R. A., Masters, C. L., Li, Q. X. and Bush, A. I. (2006). Gender and genetic background effects on brain metal levels in APP transgenic and normal mice: implications for Alzheimer β-amyloid pathology. *J. Inorg. Biochem.*, 100, pp. 952–62.

Miller, L. M., Wang, Q., Telivala, T. P., Smith, R. J., Lanzirotti, A. and Miklossy, J. (2006). Synchrotron-based infrared and X-ray imaging shows focalized accumulation of Cu and Zn co-localized with beta-amyloid deposits in Alzheimer's disease. *J. Struct. Biol.*, 155, pp. 30–7.

Mirica, L. M., Ottenwaelder, X. and Stack, T. D. P. (2004). Structure and spectroscopy of copper-dioxygen complexes. *Chem. Rev.*, 104, pp. 1013–45.

Paik, S. R., Shin, H. J. and Lee, J. H. (2000). Metal-catalyzed oxidation of α-synuclein in the presence of Copper(II) and hydrogen peroxide. *Arch. Biochem. Biophys.*, 378, pp. 269–77.

Paik, S. R., Shin, H. J., Lee, J. H., Chang, C. S. and Kim, J. (1999). Copper(II)-induced self-oligomerization of α-synuclein. *Biochem. J.*, 340 (Pt 3), pp. 821–8.

Pall, H. S., Williams, A. C., Blake, D. R., Lunec, J., Gutteridge, J. M., Hall, M. and Taylor, A. (1987). Raised cerebrospinal-fluid copper concentration in Parkinson's disease. *Lancet*, 2, pp. 238–41.

Pennathur, S., Jackson-Lewis, V., Przedborski, S. and Heinecke, J. W. (1999). Mass spectrometric quantification of 3-nitrotyrosine, ortho-tyrosine, and o,o'-dityrosine in brain tissue of 1-methyl-4-phenyl-1,2,3,6-tetrahydropyridine-treated mice, a model of oxidative stress in Parkinson's disease. *J. Biol. Chem.*, 274, pp. 34621–8.

Pfefferkorn, C. M. and Lee, J. C. (2010). Tryptophan probes at the α-synuclein and membrane interface. *J. Phys. Chem. B*, 114, pp. 4615–22.

Polymeropoulos, M. H., Lavedan, C., Leroy, E., Ide, S. E., Dehejia, A., Dutra, A., Pike, B., Root, H., Rubenstein, J., Boyer, R., Stenroos, E. S., Chandrasekharappa, S., Athanassiadou, A., Papapetropoulos, T., Johnson, W. G., Lazzarini, A. M., Duvoisin, R. C., Di Iorio, G., Golbe, L. I. and Nussbaum, R. L. (1997). Mutation in the α-synuclein gene identified in families with Parkinson's disease. *Science*, 276, pp. 2045–7.

Qin, Z., Hu, D., Han, S., Hong, D. P. and Fink, A. L. (2007). Role of different regions of α-synuclein in the assembly of fibrils. *Biochemistry*, 46, pp. 13322–30.

Que, E. L., Domaille, D. W. and Chang, C. J. (2008). Metals in neurobiology: probing their chemistry and biology with molecular imaging. *Chem. Rev.*, 108, pp. 1517–49.

Rasia, R. M., Bertoncini, C. W., Marsh, D., Hoyer, W., Cherny, D., Zweckstetter, M., Griesinger, C., Jovin, T. M. and Fernandez, C. O. (2005). Structural characterization of copper(II) binding to α-synuclein: Insights into the bioinorganic chemistry of Parkinson's disease. *Proc. Nat. Acad. Sci. USA*, 102, pp. 4294–9.

Ruf, R. A., Lutz, E. A., Zigoneanu, I. G. and Pielak, G. J. (2008). α-Synuclein conformation affects its tyrosine-dependent oxidative aggregation. *Biochemistry*, 47, pp. 13604–9.

Siderowf, A. and Stern, M. (2003). Update on Parkinson disease. *Ann. Intern. Med.*, 138, pp. 651–8.

Singleton, A. B., Farrer, M., Johnson, J., Singleton, A., Hague, S., Kachergus, J., Hulihan, M., Peuralinna, T., Dutra, A., Nussbaum, R., Lincoln, S., Crawley, A., Hanson, M., Maraganore, D., Adler, C., Cookson, M. R., Muenter, M., Baptista, M., Miller, D., Blancato, J., Hardy, J. and Gwinn-Hardy, K. (2003). α-Synuclein locus triplication causes Parkinson's disease. *Science*, 302, pp. 841.

Solomon, E. I., Sundaram, U. M. and Machonkin, T. E. (1996). Multicopper oxidases and oxygenases. *Chem. Rev.*, 96, pp. 2563–605.

Solomon, E. I., Szilagyi, R. K., George, S. D. and Basumallick, L. (2004). Electronic structures of metal sites in proteins and models: Contributions to function in blue copper proteins. *Chem. Rev.*, 104, pp. 419–58.

Souza, J. M., Giasson, B. I., Chen, Q., Lee, V. M. and Ischiropoulos, H. (2000a). Dityrosine cross-linking promotes formation of stable α-synuclein polymers. Implication of nitrative and oxidative stress in the pathogenesis of neurodegenerative synucleinopathies. *J. Biol. Chem.*, 275, pp. 18344–9.

Souza, J. M., Giasson, B. I., Lee, V. M. and Ischiropoulos, H. (2000b). Chaperone-like activity of synucleins. *FEBS Lett.*, 474, pp. 116–9.

Spasojevic, I., Mojovic, M., Stevic, Z., Spasic, S. D., Jones, D. R., Morina, A. and Spasic, M. B. (2010). Bioavailability and catalytic properties of copper and iron for Fenton chemistry in human cerebrospinal fluid. *Redox Rep.*, 15, pp. 29–35.

Spillantini, M. G., Schmidt, M. L., Lee, V. M., Trojanowski, J. Q., Jakes, R. and Goedert, M. (1997). α-Synuclein in Lewy bodies. *Nature*, 388, pp. 839–40.

Stadtman, E. R. (1992). Protein oxidation and aging. *Science*, 257, pp. 1220–4.

Stadtman, E. R. (2001). Protein oxidation in aging and age-related diseases. *Ann. N. Y. Acad. Sci.*, 928, pp. 22–38.

Stadtman, E. R., Moskovitz, J. and Levine, R. L. (2003). Oxidation of methionine residues of proteins: Biological consequences. *Antioxid. Redox Sign.*, 5, pp. 577–582.

Stadtman, E. R., Van Remmen, H., Richardson, A., Wehr, N. B. and Levine, R. L. (2005). Methionine oxidation and aging. *Biochim. Biophys. Acta*, 1703, pp. 135–40.

Sulzer, D. (2001). α-Synuclein and cytosolic dopamine: stabilizing a bad situation. *Nat. Med.*, 7, pp. 1280–2.

Sung, Y. H., Rospigliosi, C. and Eliezer, D. (2006). NMR mapping of copper binding sites in α-synuclein. *Biochim. Biophys. Acta*, 1764, pp. 5–12.

Tompkins, M. M., Basgall, E. J., Zamrini, E. and Hill, W. D. (1997). Apoptotic-like changes in Lewy-body-associated disorders and normal aging in *substantia nigral* neurons. *Am. J. Pathol.*, 150, pp. 119–31.

Tompkins, M. M. and Hill, W. D. (1997). Contribution of somal Lewy bodies to neuronal death. *Brain Res.*, 775, pp. 24–9.

Tycko, R. (2006). Molecular structure of amyloid fibrils: insights from solid-state NMR. *Q. Rev. Biophys.*, 39, pp. 1–55.

Ulmer, T. S., Bax, A., Cole, N. B. and Nussbaum, R. L. (2005). Structure and dynamics of micelle-bound human α-synuclein. *J. Biol. Chem.*, 280, pp. 9595–603.

Ulrih, N. P., Barry, C. H. and Fink, A. L. (2008). Impact of Tyr to Ala mutations on α-synuclein fibrillation and structural properties. *Biochim. Biophys. Acta*, 1782, pp. 581–5.

Uversky, V. N. and Eliezer, D. (2009). Biophysics of Parkinson's disease: structure and aggregation of α-synuclein. *Curr. Protein Pept. Sci.*, 10, pp. 483–99.

Uversky, V. N., Gillespie, J. R. and Fink, A. L. (2000). Why are "natively unfolded" proteins unstructured under physiologic conditions? *Proteins*, 41, pp. 415–27.

Uversky, V. N., Yamin, G., Souillac, P. O., Goers, J., Glaser, C. B. and Fink, A. L. (2002). Methionine oxidation inhibits fibrillation of human α-synuclein *in vitro*. *FEBS Lett.*, 517, pp. 239–44.

Vilar, M., Chou, H. T., Luhrs, T., Maji, S. K., Riek-Loher, D., Verel, R., Manning, G., Stahlberg, H. and Riek, R. (2008). The fold of α-synuclein fibrils. *Proc. Nat. Acad. Sci. USA*, 105, pp. 8637–42.

Wang, X., Moualla, D., Wright, J. A. and Brown, D. R. (2010). Copper binding regulates intracellular α-synuclein localisation, aggregation and toxicity. *J. Neurochem.*, 113, pp. 704–14.

Xue, Y., Davis, A. V., Balakrishnan, G., Stasser, J. P., Staehlin, B. M., Focia, P., Spiro, T. G., Penner-Hahn, J. E. and O'Halloran, T. V. (2008). Cu(I) recognition via cation-π and methionine interactions in CusF. *Nat. Chem. Biol.*, 4, pp. 107–9.

Yamin, G., Glaser, C. B., Uversky, V. N. and Fink, A. L. (2003). Certain metals trigger fibrillation of methionine-oxidized α-synuclein. *J. Biol. Chem.*, 278, pp. 27630–5.

Yorita, H., Otomo, K., Hiramatsu, H., Toyama, A., Miura, T. and Takeuchi, H. (2008). Evidence for the cation-π interaction between Cu^{2+} and tryptophan. *J. Am. Chem. Soc.*, 130, pp. 15266–7.

Zarranz, J. J., Alegre, J., Gomez-Esteban, J. C., Lezcano, E., Ros, R., Ampuero, I., Vidal, L., Hoenicka, J., Rodriguez, O., Atares, B., Llorens, V., Gomez Tortosa, E., del Ser, T., Munoz, D. G. and de Yebenes, J. G. (2004). The new mutation, E46K, of α-synuclein causes Parkinson and Lewy body dementia. *Ann. Neurol.*, 55, pp. 164–73.

Zecca, L., Bellei, C., Costi, P., Albertini, A., Monzani, E., Casella, L., Gallorini, M., Bergamaschi, L., Moscatelli, A., Turro, N. J., Eisner, M., Crippa, P. R., Ito, S., Wakamatsu, K., Bush, W. D., Ward, W. C., Simon, J. D. and Zucca, F. A. (2008). New melanic pigments in the human brain that accumulate in aging and block environmental toxic metals. *Proc. Nat. Acad. Sci. USA*, 105, pp. 17567–72.

Zhou, W., Long, C., Reaney, S. H., Di Monte, D. A., Fink, A. L. and Uversky, V. N. (2010). Methionine oxidation stabilizes non-toxic oligomers of α-synuclein through strengthening the auto-inhibitory intra-molecular long-range interactions. *Biochim. Biophys. Acta*, 1802, pp. 322–30.

Chapter 10

Interactions of α-Synuclein with Metal Ions: New Insights into the Structural Biology and Bioinorganic Chemistry of Parkinson's Disease

Andrés Binolfi and Claudio O. Fernández
Institute of Molecular and Cell Biology of Rosario, CONICET, S2002LRK, Rosario, Santa Fe, Argentina

10.1 Introduction

10.1.1 *α-Synuclein and Neurodegenerative Diseases*

The misfolding of proteins into a toxic conformation and its deposition as amyloid fibrils is proposed to be at the molecular foundation of a number of neurodegenerative disorders including Creutzfeldt-Jacob's disease, Alzheimer's disease (AD) and Parkinson's disease (PD) (Table 10.1) (Dobson, 2004).

The clinical features of PD were first described by James Parkinson in 1817, although, until recently, little was known about its etiology and the molecular composition of its defining neuropathological characteristic, the Lewy body (Lewy, 1912; Forno, 1996). This situation changed a few years ago, when two findings brought the small protein α-synuclein (AS) to the fore. First, a missense mutation in the AS gene was found to cause a rare, familial form of

Brain Diseases and Metalloproteins
Edited by David R. Brown

ISBN 978-981-4316-01-9 (Hardcover), 978-981-4364-07-2 (eBook)
www.panstanford.com

Table 10.1 Protein misfolding diseases. Selected disorders and proteins linked to their pathogenesis

Disease	Protein
Alzheimer's disease	Amyloid β-peptide/Tau
Parkinson's disease	Alpha-Synuclein
Creutzfeldt-Jacobs disease	Prion
Type-2 diabetes mellitus	IAPP (Amylin peptide)
Huntington's disease	Huntingtin
Dyalisis related amyloidosis	β2 Microglobulin
Familial amiloidosis	Transthyretin/Lysozyme
Cataract	Cristallins
Amyothropic lateral sclerosis	Cu, Zn, Superoxide Dismutase
Systemic AL amyloidosis	Immunoglobin Light Chain AL

PD (Polymeropoulos *et al.,* 1997). Second, the Lewy bodies present in cases of sporadic PD were found to be strongly immunoreactive for AS (Spillantini *et al.,* 1997).

The implication of AS in neurodegenerative diseases is supported actually by 3 major lines of evidence: (i) it is the main component of neuronal and glial cytoplasmatic inclusions, the pathological hallmarks of PD, dementia with Lewy bodies (DLB), multiple system atrophy (MSA) and other neurodegenerative disorders collectively referred to as synucleinopathies (Spillantini *et al.,* 1997; Goedert, 2001); (ii) three point mutations in the AS gene (A53T, A30P, E46K) are linked to early onset familial PD (Polymeropoulos *et al.,* 1997; Kruger *et al.,* 1998; Zarranz *et al.,* 2004); and (iii) transgenic animal models involving overexpression of AS develop cytoplasmatic inclusions and motor deficiencies (Masliah *et al.,* 2000).

The discovery that AS accounts for the filamentous deposits of PD, DLB and MSA has provided a general underlying theme to the study of these disorders. Therefore, delineating the mechanism of AS aggregation and its pathophysiological role in neurodegeneration has been the focus of many investigations during the last years.

10.1.2 *The Physiological Role of AS*

Synucleins are proteins that are abundant in the brain and whose physiological functions are poorly understood. In humans, the synuclein family consists at least of three members (α-synuclein (AS), β-synuclein (BS) and γ-synuclein (GS)) that range from 127–140 amino acids and are 55–62% identical in sequence, with a similar domain organization (Fig. 10.1(a)) (Clayton and George, 1999).

By immunohistochemistry, AS and BS were shown to be colocalized in nerve terminals, in close proximity to synaptic vesicles. Instead, GS seems to be present predominantly throughout nerve cells (Buchman *et al.*, 1998). Interestingly, Lewy bodies do not stain for BS or GS, a very puzzling result given the sequence homology, and particularly the overlapping cellular localization of AS and BS (Buchman *et al.*, 1998; Clayton and George, 1999)

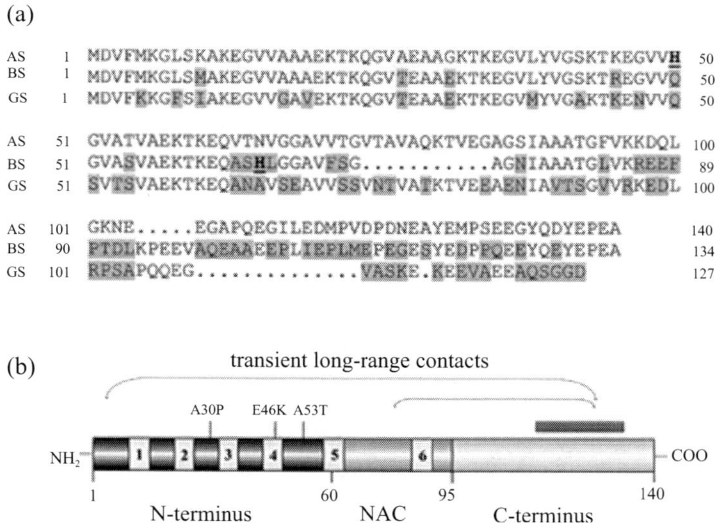

Figure 10.1 (a) Alignment of amino acid sequences of AS, BS and GS. Letters shaded in grey indicate amino acids that are different in BS and GS sequences compared to AS. Histidine residues are highlighted. (b) Schematic representation of AS. The N-terminal, NAC and C-terminal regions are identified as well as the inherited pathogenic point mutations. Numbered boxes in grey indicate the location of the imperfect KTKEGV repeats. Lines in black indicate the transient long-range contacts between the C-terminus and the NAC region, and the C-terminus and N-terminus. The bar at the C-terminus identifies the region with intrinsic residual structure.

The incapability of BS and GS to spontaneously aggregate *in vitro*, is the other feature distinguishing AS from its two close relatives (Giasson *et al.,* 2001; Uversky *et al.,* 2002(a)). An interesting ing finding reported recently showed that molar excesses of BS effectively inhibit AS fibrillation *in vitro* (Hashimoto *et al.,* 2001; Uversky *et al.,* 2002a), and studies on transgenic mice brain showed a comparable inhibitory effect *in vivo* (Hashimoto *et al.,* 2004). These evidences indicate that BS might regulate AS fibrillation. On the basis of these observations, it has been concluded that a decrease in the levels of BS should be considered as a possible factor in PD etiology. However, the mechanisms through which BS might inhibit AS aggregation are still unclear.

Although the normal physiological function of AS remains unknown, several lines of evidence suggest a role in membrane-associated processes at the presynaptic terminals (Clayton and George, 1999). In particular, AS was proposed to regulate dopamine neurotransmission by modulation of vesicular dopamine storage (Lotharius and Brundin, 2002). More recently, new emerging evidences pointed towards the involvement of AS in rescuing lethality associated with the lack of CSPα protein, a co-chaperone associated with synaptic vesicles and implicated in the folding pathways of SNARE proteins, suggesting that AS might act as an auxiliary chaperone preserving the function and integrity of synapses (Chandra *et al.*, 2005).

10.1.3 *Structural Properties of AS*

The protein AS comprises 140 residues distributed in three different regions (Fig. 10.1b): (i) the amphipatic N-terminus (residues 1–60), showing the consensus repeats KTKEGV and involved in lipid binding (Davidson *et al.,* 1998; Bussell and Eliezer, 2003; Chandra *et al.,* 2003; Jao *et al.,* 2004), (ii) the highly hydrophobic self-aggregating sequence known as NAC (non-Aβ component, residues 61–95), which initiates fibrillation (Giasson *et al.,* 2001), and (iii) the acidic C-terminal region (residues 96–140), rich in Pro, Asp and Glu residues and critical for blocking rapid AS filament assembly (Crowther *et al.,* 1998; Fernandez *et al.,* 2004; Hoyer *et al.,* 2004).

In its native monomeric state AS adopts an ensemble of conformations with no rigid secondary structure although long-range interactions have been shown to stabilize an aggregation-autoinhibited conformation (Lee *et al.*, 2004; Bertoncini *et al.*, 2005; Dedmon *et al.*, 2005; Lee *et al.*, 2005; Lee *et al.*, 2007). The protein undergoes dramatic conformational transitions from its natively unstructured state to an α-helical conformation upon interaction with lipid membranes (Bussell and Eliezer, 2003; Chandra *et al.*, 2003; Jao *et al.*, 2004), or to the characteristic crossed β-conformation in highly organized amyloid-like fibrils under conditions that trigger aggregation (Serpell *et al.*, 2000; Der-Sarkissian *et al.*, 2003). Whereas the structural basis behind the α-helical transition is well understood, the structural changes involved in the fibrillization pathway of the protein are still matte r of intense debate.

10.1.4 *Mechanism of AS Amyloid Assembly*

The aggregation kinetics of AS proceeds according to two processes; the first is a nucleus formation step (lag-phase), followed by an extension of fibrils from the nucleus (growth phase), until a stationary phase is reached (Fig. 10.2). Usually, the nucleation process is very slow and rate-determining, and the nucleus-dependent extension process is much faster.

The mechanism of AS aggregation is characterized by the formation of prefibrillar aggregates of heterogeneous size and morphology (protofibrils), that includes spheres, chain-like structures, annular pore-like structures, and large granular structures (Ding *et al.*, 2002; Lashuel *et al.*, 2002a; Lashuel *et al.*, 2002b). Initially, the amyloid hypothesis supported by strong genetic, pathologic and biochemical evidences implicated amyloid fibrils as the main cause for neurodegeneration and/or organ dysfunction in several human diseases including PD and other amyloidoses. The identification and characterization of potential neurotoxic quaternary structure intermediates, referred to as protofibrils, that precede fibril formation and the finding that several pathogenic mutations promote protofibril formation suggested that the protofibril rather than the fibrils are the pathogenic species (Lansbury, 1999; Haass and Steiner, 2001; Lashuel *et al.*, 2002b; Caughey and Lansbury, 2003). Thus, development of methods for stabilizing AS protofibrils of defined size and morphology is critical to investigate the relationship between monomer, protofibrillar intermediates and

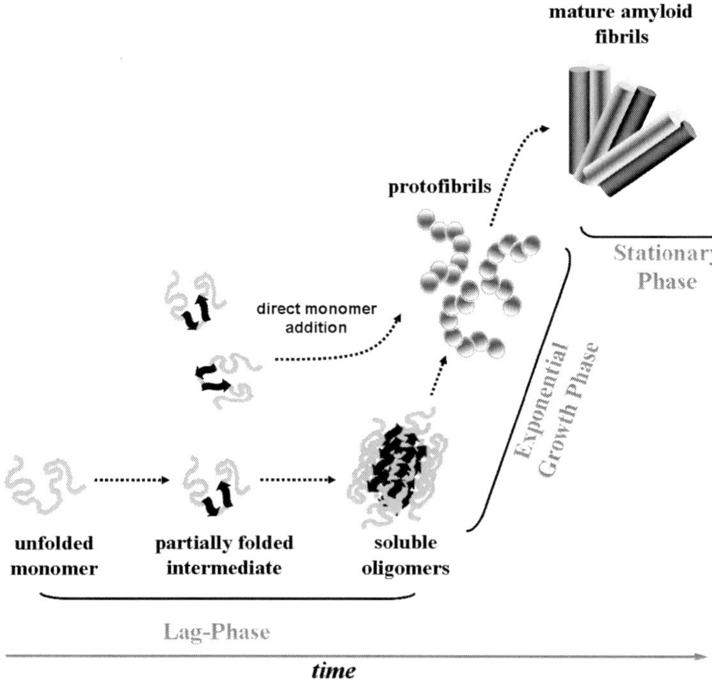

Figure 10.2 Schematic representation of the time course of aggregation of AS and the molecular species involved in the fibrillation pathway.

fibrils in the pathogenesis of PD and the identification of targets for therapeutic intervention.

10.1.5 *Role of Metal Ions in AS Fibril Formation*

There is now increasing evidence that altered metal homeostasis may be involved in the progression of neurodegenerative diseases (Sayre *et al.*, 1999; Bush, 2000; Gaeta and Hider, 2005; Gaggelli *et al.*, 2006; Kozlowski *et al.*, 2006; Molina-Holgado *et al.*, 2007). Protein-metal interactions appear to play a critical role in protein aggregation and are therefore likely to provide a link between the accumulation of aggregated proteins, oxidative damage of the brain and neuronal cell loss (Atwood *et al.*, 1998; Paik *et al.*, 1999; Paik *et al.*, 2000; Requena *et al.*, 2001; Uversky *et al.*, 2001). This data support the hypothesis that metal interactions with the target protein in several of these age-dependent degenerative diseases might constitute one of the major factors contributing to their etiology. Interestingly, copper and manganese have been implicated in Creutzfeldt-Jakob disease (Brown and Kozlowski, 2004; Gaggelli *et al.*, 2005; Kim *et al.*, 2005; Karr and Szalai, 2008; Klewpatinond *et al.*, 2008), whereas recent studies emphasize the role of copper, iron and zinc as contributors to both amyloid Aβ assembly *in vitro* and to the neuropathology of Alzheimer's disease (AD) (Bush *et al.*, 1993; Garzon-Rodriguez *et al.*, 1999; Huang *et al.*, 1999; Miura *et al.*, 2000; Syme and Viles, 2006). High levels of copper, zinc and iron were also found in and around amyloid plaques of AD brains (Lovell *et al.*, 1998). Furthermore, coordination environments for Cu(II) complexes in the amyloid precursor protein, the amyloid Aβ-peptide and the prion protein have been very well characterized by several biophysical and structural studies (Aronoff-Spencer *et al.*, 2000; Kowalik-Jankowska *et al.*, 2001, 2002; Barnham *et al.*, 2003; Garnett and Viles, 2003; Kowalik-Jankowska *et al.*, 2003; Belosi *et al.*, 2004; Karr *et al.*, 2004a; Karr *et al.*, 2004b; Valensin *et al.*, 2004a; Valensin *et al.*, 2004b; Chattopadhyay *et al.*, 2005; Karr *et al.*, 2005; Klewpatinond and Viles, 2007; Karr and Szalai, 2008; Klewpatinond *et al.*, 2008). Detailed knowledge of the structural and binding features of relevant metal ions as well as the mechanism by which these metal ions might participate in the fibrillization of these proteins have contributed to the design of a new therapeutical scheme based on the development of metal ion chelators (Gaeta and Hider, 2005; Molina-Holgado *et al.*, 2007).

The role of metal ions in AS amyloid assembly and neurodegeneration is also becoming a central question in the pathophysiology of PD. Iron deposits have been identified in Lewy bodies in the substantia nigra (Castellani *et al.*, 2000), and elevated copper concentrations have been reported in the cerebrospinal fluid of PD patients (Pall *et al.*, 1987). In addition, individuals with chronic industrial exposure to copper, manganese or iron have an increased rate of PD (Gorell *et al.*, 1999), whereas a systematic analysis of the effect of various metal ions revealed that copper is the most effective ion in promoting AS oligomerization (Uversky *et al.*, 2001; Golts *et al.*, 2002). However, the reported effects of the transition metal ions on AS fibrillation were examined at concentrations 0.5–5.0 mM, far greater than those normally occurring in tissues, such that the physiological relevance of these effects remained open to question.

Advances in the bioinorganic chemistry of Parkinson's disease then required reevaluation of AS-metal interactions under more physiologically relevant conditions as a first step towards the understanding of the molecular mechanism by which metal ions accelerate AS filament assembly. This knowledge results crucial to establish the role of metal ions in synucleinopathies at the molecular resolution currently available for other amyloidoses and to address the central question of whether these agents constitute a common denominator underlying the amyloid-related disorders known as AD, PD, and prion disease.

The main objective of this work is to present the significant advances obtained during the last years into the structural characterization of AS-metal interactions and to discuss their implications on the mechanism of AS amyloid formation.

10.2 Interaction of AS with Cu(II) Ions

10.2.1 *Cu(II) Levels in the Micromolar Range are Effective in Inducing AS Aggregation*

The enhanced fluorescence emission of the dye thioflavin-T (ThioT) is frequently used for monitoring the kinetics of amyloid formation. Thioflavin-T is a weakly fluorescent dye that exhibits a pronounced intensity increase upon binding to amyloid fibrils, becoming a specific marker for the β-pleated sheet conformation of amyloid structures (LeVine, 1999). The chemical structure of ThioT and the fluorescence changes accompanying its binding to the aggregated state of an amyloidogenic protein are depicted in Figs. 10.3(a, b).

The time course of aggregation of AS, as monitored by the ThioT assay (Fig. 10.3(c)), is consistent with a nucleation-dependent mechanism that might be defined by the equation (10.1) (Fernandez *et al.*, 2004; Binolfi *et al.*, 2006).

$$\alpha[t] = (1 - e^{-k_{app}t})/(1 + e^{-k_{app}(t - t_{1/2})}) \qquad (10.1)$$

The quantities k_{app} and $t_{1/2}$ are related to the aggregation reaction as follows:

$$k_{app} = k\varepsilon[AS] \qquad (10.2)$$

$$t_{1/2} = \ln(\varepsilon[AS]/[nc])/k_{app} \qquad (10.3)$$

where k is the rate constant for incorporation of monomers into growth points located in aggregates, [AS] is the total concentration of AS monomers, [nc] is the concentration of nucleation centers, ε denotes the fraction of addition-competent monomer sites in already formed aggregates and $t_{1/2}$ represents the lag time to the midpoint of the fractional transition (α[t]) from monomer to aggregate.

An study focused on the time course of AS aggregation in the absence and presence of Cu(II) revealed that micromolar levels of the metal ion were effective in accelerating the aggregation of AS. Addition of Cu(II) decreased the characteristic lag time for aggregation of the unligated protein to a degree that was dependent on Cu(II) concentration (Fig. 10.3(d)) (Rasia *et al.*, 2005). An inverse relationship was found between $t_{1/2}$ and Cu(II) concentration, but no

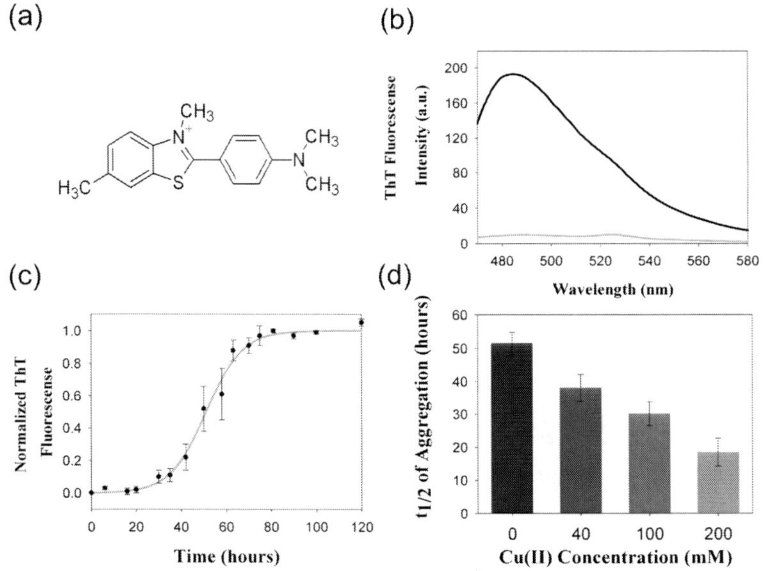

Figure 10.3 (a) Chemical structure of the fluorescent dye ThioT. (b) Fluorescence emission spectra of ThioT in presence of native (grey) and aggregated AS (black). (c) Variation of fluorescence emission intensity at 480 nm with incubation time. 100 μM AS was incubated at 37°C for fluorescence measurements and the experimental data was fitted according to equation 10.1 to obtain $t_{1/2}$ values. Error bars correspond to standard deviation obtained from independent repetitions of the experiment. (d) Aggregation $t_{1/2}$ values (hours) of 100 μM AS in the presence of increasing concentrations of Cu(II) (Rasia *et al.*, 2005).

major changes were observed in the rate of aggregate growth (k_{app}), indicating that Cu(II) promotes nucleation but not the fibril growth phase. Interestingly, 70–90% of the Cu(II) added in the aggregation assay was found to be incorporated into the AS aggregates. Altogether, these results indicated that the protein aggregates in its copper-bound form and that metal-bound AS is more prone to nucleate that the unliganded protein (Rasia *et al.*, 2005).

10.2.2 *Cu(II) Binds Preferentially to the N-Terminal Region of AS*

The structural and affinity features of AS-Cu(II) complexes were addressed by several studies. Working with N- and C-terminal truncated forms of the protein it was reported that the enhancement of fibrillation mediated by Cu(II) was a C-terminus dependent phenomenon. By using an indirect approach, the authors concluded that the metal binding interface for Cu(II) was located in the sequence 97–140 of the C-terminus, where five or more copper ions might bind with a K_d of 45–60 μM (Paik *et al.*, 1999).

In another work, the Cu(II) binding properties of AS were studied by CD and UV-Vis spectroscopy (Rasia *et al.*, 2005). The location of ligand field (d-d) and ligand-to-metal charge transfer transitions (LTM c.t.) in the electronic spectra of the metal complex provides diagnostic information about the structural features of AS-Cu(II) interactions. Direct information about the coordination geometry of the metal complex, the nature of the ligands in the coordination environment, and the affinity features of the interaction can be obtained (Bryce and Gurd, 1966; Sigel and Martin, 1982). Indeed, CD and UV-Vis spectroscopy has been successfully applied to characterize the complexes of Cu(II) with other proteins involved in neurodegenerative diseases, such as the amyloid precursor protein, the Aβ peptide and the prion protein (Viles *et al.*, 1999; Kowalik-Jankowska *et al.*, 2002; Kowalik-Jankowska *et al.*, 2003; Syme *et al.*, 2004; Syme and Viles, 2006).

The UV-Vis spectra of AS complexed with Cu(II) showed a single visible absorption band at ~ 620 nm ($\varepsilon = 64\ M^{-1}\ cm^{-1}$), characteristic of two- or three-nitrogen coordination in a type (II) square-planar or distorted tetragonal arrangement (Rasia *et al.*, 2005). The CD spectra of the AS-Cu(II) complex showed a positive band in the d-d region (≈600 nm) and a negative band at 300 nm. The latter band was assigned tentatively to a LTM c.t. between the metal center and an imidazol group (π_1 N_{im}–Cu LTM c.t., 280–345 nm) or a deprotonated peptide nitrogen (π_2 N_{im}–Cu LTM c.t., 295–315 nm) (Sigel and Martin 1982; Daniele *et al.,* 1996). From UV-Vis experiments performed at increasing concentrations of Cu(II) it was shown that AS tightly bound only two Cu(II) ions per monomer, with dissociation constants in the 0.1 to 50 μM range. It was also concluded that more Cu(II) ions might be ligated by the protein, but with significantly lower affinity and probably via nonspecific electrostatic interactions with charged amino acids side-chains (Rasia *et al.,* 2005).

Contributions of different regions of the protein to Cu(II) binding were evaluated by following the association of the metal ion with distinct domain truncated species of AS. These studies demonstrated that the UV-Vis and CD spectra of a C-terminal truncated variant of AS (1–108 AS) complexed with Cu(II) were similar to that of the full-length protein, whereas the spectroscopic features of the Cu(II) complex formed with the N-terminal truncated species 95–137 AS was considerably different (Rasia *et al.,* 2005). The results from this work constituted the first experimental evidence supporting a role for the N-terminal region of AS in Cu(II) binding.

10.2.3 *Identification of Cu(II) Binding Sites in AS*

NMR spectroscopy is an excellent technique to study protein-ligand interactions processes, providing maps of the interaction interfaces and binding constants. In particular, the ^{1}H-^{15}N heteronuclear single quantum correlation spectrum (^{1}H-^{15}N HSQC) (Bodenhausen and Rubern, 1980) of proteins contain one cross-peak for each amide group in the molecule (except those involving proline residues) and thus provide sequence-specific probes for locating metal binding sites in the protein. Metal ions with unpaired electrons in its electronic structure, like Cu(II), modifies dramatically the relaxation properties of resonances from nuclei linked to the metal center by through-bond (scalar) or through-space (dipolar) interactions (Bertini and Luchinat, 1996). In consequence, a differential broadening (lower intensities) is observed for the amide resonances in the NMR experiment, providing the means for mapping the metal-binding interfaces in the protein (Fig. 10.4(a)).

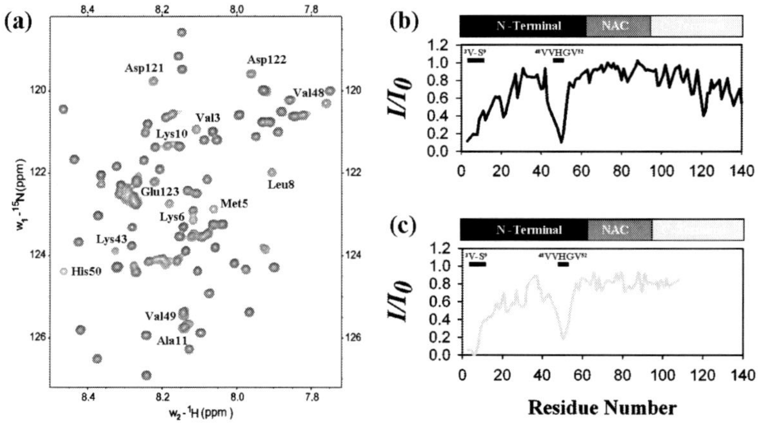

Figure 10.4 (a) ^{1}H-^{15}N HSQC spectrum of AS in the presence of Cu(II). Overlaid contour plots of the ^{1}H-^{15}N HSQC spectra of 100 μM AS at pH 6.5, 15 °C, in the absence (red) and presence (blue) of 50 μM Cu(II). Amino acid residues broadened significantly or beyond detection are identified. (b, c) Intensity profiles of the backbone amide groups of AS (b) and the C-terminal truncated variant (1–108 AS) in the presence of 0.5 equivalents of Cu(II). I and I_0 correspond to the intensities of cross-peaks in the presence and absence of the metal ion, respectively. See also Colour Insert.

Details of Cu(II) binding to AS by NMR spectroscopy revealed significant changes in well defined regions of the protein structure. The Cu(II) titration experiments showed that residues 3–9 (the amide resonances of residues 1 and 2 are not detected because of solvent-exchange effects) and 48–52, were strongly perturbed (Fig. 10.4(b)) (Rasia *et al.,* 2005). Identical results were obtained from experiments performed on C-terminal truncated variants of AS (Fig. 10.4(c)), indicating that at substoichiometric levels Cu(II) mainly interacts with residues located in the N-terminal region of AS. The amide resonances assigned to the C-terminal region were perturbed only at higher Cu(II) concentrations, with the strongest effect centered in the 119–124 region. In all cases, NMR spectral changes induced by Cu(II) were abolished upon EDTA addition, demonstrating the reversibility of the protein-metal interactions (Rasia *et al.,* 2005).

The NMR titration experiments discussed here identified two primary binding sites for Cu(II) in AS. The effects observed on the 3–9 and 48–52 regions attributed a potential role to the amino terminal group of Met-1 and the imidazol ring of His-50 as metal anchoring residues for Cu(II) binding to the N-terminal region of AS. Interestingly, whereas the C-terminal region was shown to bind Cu(II) with low affinity, the NAC region remained insensitive to the presence of the metal ion (Rasia *et al.*, 2005).

That Cu(II) can promote AS aggregation efficiently at physiologically relevant concentrations and the unequivocal identification of binding interfaces in the N-terminal region for which Cu(II) has the highest affinity were the most important findings of the works discussed in this section, that supported strongly the notion of PD as a metal-associated neurodegenerative disorder.

10.3 Interaction of AS with Other Divalent Metal Ions

10.3.1 *Binding Affinity and Effects on AS Aggregation*

In addition to the finding of AS as a possible copper-binding protein, other metal ions such as Fe(II) and Mn(II) have been linked to the etiology of PD (Verity, 1999; Castellani *et al.*, 2000; Golts *et al.*, 2002). Some studies reported that Cu(II), Fe(II) and Mn(II) quenched tyrosine fluorescence and these metal ions were presumed to form stable metal-protein complexes (Uversky *et al.*, 2001; Golts *et al.*, 2002). *In vitro* studies demonstrated that millimolar (mM) levels of these divalent cations caused a significant acceleration of AS fibril formation (Uversky *et al.*, 2001).

However, when the action of some of these metal ions on AS aggregation was evaluated in the micromolar (μM) range of metal concentrations, they showed minimal or no stimulatory effects (Binolfi *et al.*, 2006). As discussed previously, Cu(II) was very effective on promoting AS aggregation even in the micromolar range of metal concentrations, a property that was not shared by the other divalent metal ions analyzed (Rasia *et al.*, 2005; Binolfi *et al.*, 2006). Thus, a primary difference between Cu(II) and other metal ions studied related to the capability of accelerating the kinetics of AS aggregation (Figs. 10.5(a, b)).

The dissociation constants of various AS-metal(II) complexes were then estimated by different techniques. By equilibrium dialysis and NMR spectroscopy, the dissociation constants estimated for Mn(II), Fe(II), Co(II) and Ni(II) were all in the 1–2 mM range,

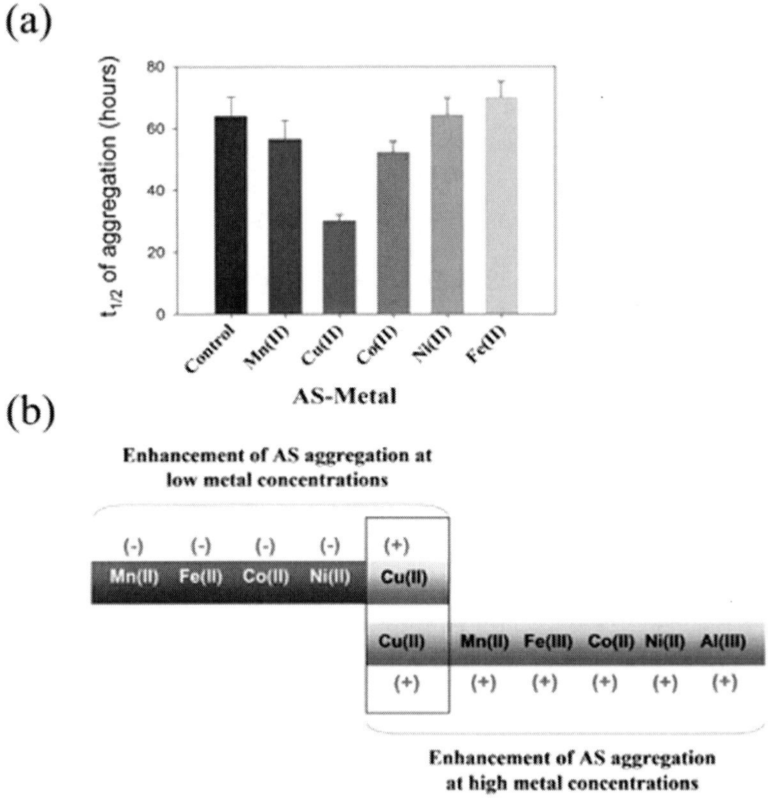

Figure 10.5 (a) Plot of $t_{1/2}$ values (hours) obtained for each metal ion from the fitting of aggregation kinetics data of AS to equation 10.1. The concentration of AS and each metal ion was 100 μM. (b) Comparative analysis of the effects of different divalent metal ions on AS (100 μM) aggregation kinetics. Metal concentrations ranged between 40–100 μM and 2.0–5.0 mM in the low and high- metal concentration experiments, respectively.

indicating that at the low concentrations of metal ions used in the aggregation experiments the degree of occupancy of the metal binding sites in AS was significant only for Cu(II) (Rasia *et al.*, 2005; Binolfi *et al.*, 2006). In conclusion, these studies demonstrated that the different effects of Cu(II) and the divalent metal ions Mn(II), Fe(II), Co(II) and Ni(II) on AS aggregation were linked to a reduced affinity of the latter cations for the protein.

10.3.2 *Identification of the AS-Metal(II) Binding Sites*

The specific regions of metal binding to AS were mapped by two-dimensional heteronuclear NMR spectroscopy, as described previously. Although scarcely populated in the micromolar range of metal concentrations, the AS-metal complexes could be extensively characterized by NMR due the paramagnetic nature of the metal ions studied. The range of estimated AS-metal(II) affinities indicates that the metal complexes at the C-terminus must have lifetimes substantially shorter than that of Cu(II) at the N-terminus. This also implies that the resonances of the nuclei close to the paramagnetic ions can be more dramatically affected than those in a tight complex (i.e., AS-Cu(II) complex), due to a phenomenon described as paramagnetic exchange broadening (Bertini and Luchinat, 1996; Gaggelli *et al.,* 2005). Thus, added to the biological relevance of Fe(II) and Mn(II), other divalent metal ions such as Co(II) and Ni(II), displaying different degrees of paramagnetism were used to achieve a detailed characterization of metal binding sites in AS (Binolfi *et al.,* 2006).

The Mn(II) titration experiments revealed the presence of significant changes in cross-peak intensities, all restricted to residues located in the C-terminal region of AS (Figs. 10.6(a–c)). The strongest broadening effects at low levels of Mn(II) corresponded to amide groups of residues of Asp-121, Asn-122 and Glu-123, indicating that this site was the most populated under these conditions. The paramagnetic effect was further pronounced and generalized at higher Mn(II) concentrations, reflecting an increasing fraction of the primary metal complex and the transient population of secondary binding sites at the C-terminus (Binolfi *et al.,* 2006).

Titration of AS with the metal ions Fe(II), Co(II) and Ni(II), which are less paramagnetic than Mn(II), showed effects qualitatively similar to those caused by low levels of Mn(II) (Figs. 10.6(d–f)). In contrast, no generalized line broadening was observed at the C-terminus by higher concentrations of these divalent metal ions. Since the affinities for the AS-metal(II) complexes were all in the 1–2 mM range, the different degree of broadening reflected the magnitude of the electron spin relaxation times (τ_S) of each paramagnetic metal ion, whose values were reported to be 10^{-8} s for Mn(II) and spanning between 10^{-11}-10^{-13} s for Co(II), Ni(II) and Fe(II) (Bertini and Luchinat, 1996; Binolfi *et al.,* 2006).

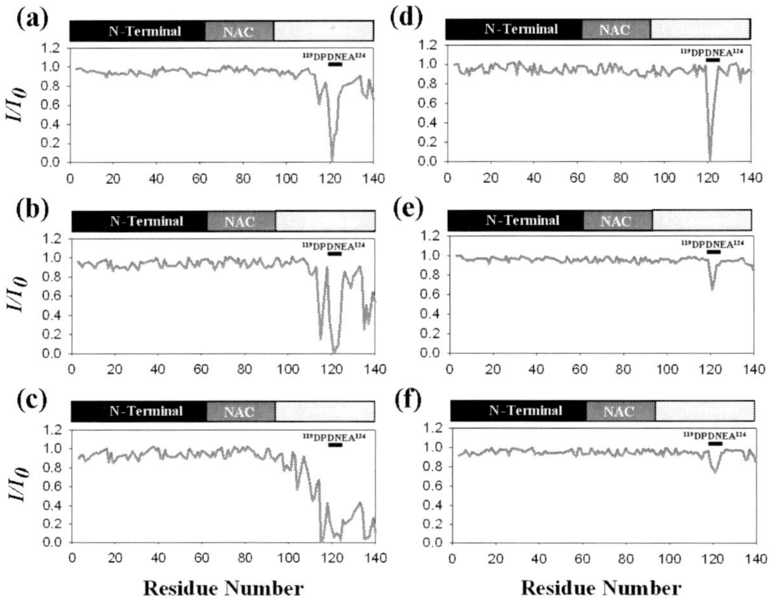

Figure 10.6 Intensity profiles of the backbone amide groups of AS in the presence of divalent metal ions. ^{1}H-^{15}N HSQC spectra of 100 μM AS at pH 6.5 and 15°C were registered upon addition of (a) 15 μM Mn(II), (b) 40 μM Mn(II), (c) 100 μM Mn(II), (d) 100 μM Co(II), (e) 100 μM Ni(II) or (f) 100 μM Fe(II). I and I_0 correspond to the intensities of cross-peaks in the presence and absence of the metal ion, respectively.

The NMR analysis of the AS-metal(II) complexes indicated that the divalent metal ions Mn(II), Fe(II), Co(II) and Ni(II) bind preferentially to the C-terminal region of AS in its native state. The work revealed the existence of a common, multiple binding site(s) for metal ions in the region comprising residues 110–140, which was shown to constitute also the binding interface for polycations such as the natural occurring polyamines putrescine, spermidine and spermine (Fernandez *et al.*, 2004). By exploiting the different degree of paramagnetism of the metal ions or using substoichiometric metal-to-protein ratios the primary site for metal ion coordination was identified. The metal interaction was localized on residues ^{119}DPDNEA^{124}, the spectral features of Asp-121 being the most affected. The picture reported by these studies was similar to that determined previously for the

AS-Cu(II) complex located at the C-terminus (Rasia *et al.*, 2005), proving conclusively that: (i) the divalent metal ions Mn(II) and Fe(II) bind preferentially to the C-terminus of AS; (ii) the binding takes place primarily in a well-defined region involving Asp-121 as the main anchoring residue; and (iii) AS binds metal ions to C-terminus with very low selectivity and affinity.

10.3.3 *Structural Determinants of Metal(II) Binding to AS*

The identification of a similar binding interface for divalent metal ions in the highly acidic C-terminus of AS suggested a common mode of binding dictated by electrostatic interactions. Under these circumstances, the metal ions would more likely interact with carboxylate side-chains clustered in that region. This might be the case with the Asp-121 binding site, which is surrounded by the aspartic and glutamic acids Asp-119, Glu-123 and Glu-126. A similar effect would then be expected for the cluster of residues around Asp-135, comprising Glu-130, Glu-131, Glu-137 and Glu-139 (see Fig. 10.1(a)). However, a noticeable difference in the metal binding capabilities of the two regions was observed.

As reported by another study, residual dipolar couplings (RDC) were shown to be an excellent tool to characterize the slow-dynamics and transient long-range interactions of the AS monomer (Bernado *et al.*, 2005; Bertoncini *et al.*, 2005). The RDC profile was characterized by predominantly positive couplings that become exceptionally large for residues 115–119 and 125–129 in the C-terminus, indicative of the higher degree of restricted motions in that region (Fig. 10.7(a)). Interestingly, it was noted that the residues constituting the primary metal binding site in the C-terminus, Asp-121, Asn-122 and Glu-123, corresponded to the linker sequence showing couplings close to zero and connecting the two major peaks of the RDC profile (115–119 and 125–129) (Figs. 10.7(b,c)). The strong correlation established by this study between the location of the primary metal binding site and the dynamic and structural properties inherent to the C-terminal region suggested that the presence of a specific spatial organization about residues 121–123 might result in a particular orientation of the coordination moieties favoring metal binding to this region (Binolfi *et al.*, 2006). According to this, the binding of metal ions to the C-terminus of AS was proposed not to be driven exclusively

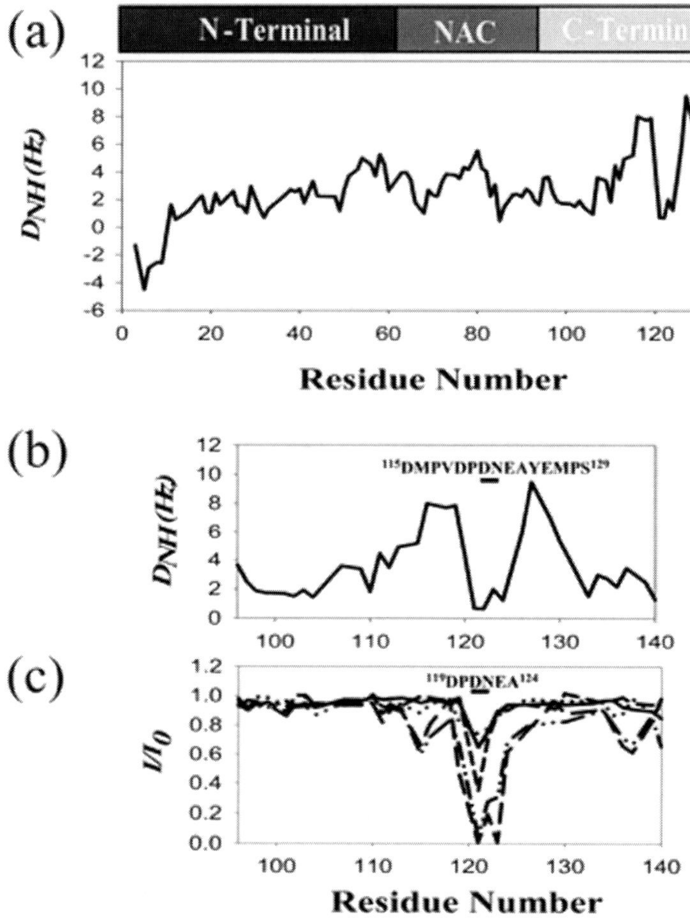

Figure 10.7 (a) RDC profile of backbone amide groups of AS. (b) Enlargement of the RDC profile of backbone amide groups in the C-terminal region of AS. (c) Intensity profiles of the backbone amide groups of AS in the presence of sub-stoichiometric or stoichimetric concentrations of Fe(II) (solid line), Ni(II) (dotted line), Co(II) (dashed line), Mn(II) (double dotted-dashed line) and Cu(II) (large dashed line). I and I_0 correspond to the intensities of cross-peaks in the presence and absence of the metal ion, respectively.

by electrostatic interactions but mostly modulated by the intrinsic conformation of that region (Binolfi *et al.*, 2006).

These results provided compelling evidence for a common low-affinity metal binding interface at the C-terminus. On the other hand, it has been discussed in previous sections that AS is able to bind Cu(II) with high affinity at the N-terminal region. These key differences in the affinity features are clearly dictated by structural factors corresponding to different protein regions and reveal a strong link between the specificity of metal binding to AS and their effectiveness in accelerating AS aggregation. These findings led to a new conceptual scheme in the metallobiology of PD, according to which the hierarchy of AS-metal interactions reflects both biological and structural effects, the latter resulting from the nature of the coordinating moieties of the protein (Binolfi *et al.*, 2006). The impact of these structural-affinity differences on the mechanistic basis behind the metal induced AS fibrillation process is discussed in the next section.

10.3.4 *Mechanistic Basis for the Aggregation of AS Mediated by Metal Ions*

A coordination site formed mostly by carboxylate side-chains is in agreement with the modest affinity constants (mM range) observed for metal binding to the C-terminus and thus the high levels of metal ions required to induce the aggregation of AS. A similar weak binding was determined for the interaction between AS and natural polyamines that bind exclusively to the C-terminus of AS and trigger its aggregation. As in that case, the existence of multiple metal binding sites at the C-terminus might be involved in the aggregation pathway by forming transient asymmetrically arranged dimers, in which the metal ions would serve to bridge multiple interfaces. Indeed, these interactions would be significantly stabilized only at high metal:protein ratios, such as those necessary to induce the aggregation of AS (Binolfi *et al.*, 2006).

On the other hand, other studies have proposed that the binding of metal ions to the negatively charged carboxylates in the C-terminal region would lead to masking of the electrostatic repulsion and the collapse to a partially folded conformation. The formation of a partially folded intermediate, which would be more prone to aggregate than the protein in its native state, has been suggested to occur under conditions of low pH, elevated temperatures, or millimolar concentrations of metal ions (Uversky *et al.*, 2002a; Uversky *et al.*, 2002b). Although the invariance of NMR parameters such as the chemical shifts of backbone amide resonances and the hydrodynamic radius of the protein indicated the absence of substantial conformational changes or the induction of a partially misfolded species even in the presence of mM levels of metal ions (Binolfi *et al.*, 2006), the established structural-affinity relationship supported a mechanism of metal-induced aggregation sharing common features for the divalent metal ions Mn(II), Fe(II), Co(II) and Ni(II), yet indicating that this process differs significantly from that induced by Cu(II).

In the case of Cu(II), the formation of highly specific complexes at the N-terminal region of the protein would represent the critical step in the early stage of fibrillation of AS. As reported for the amyloid β-peptide and prion protein, AS is also highly susceptible to metal catalyzed oxidation, a reaction that induced extensive oligomerization and precipitation of these proteins (Atwood *et al.*, 1998; Paik *et al.*, 2000; Requena *et al.*, 2001). Since metal-catalyzed oxidation of proteins is a highly selective, site-specific process that occurs primarily at protein sites with transition metal-binding capacity, it is possible to hypothesize that copper binding to the N-terminus of AS renders the protein a relative easy target for oxidative damage, a reaction that might lead to a cascade of structural alterations promoting the generation of a pool of AS molecules more prone to aggregate.

10.4 Structural Details Behind the Specificity of AS-Cu(II) Interactions

As discussed in other sections, the lack of experimental evidences proving a relationship between metal binding affinity and aggregation enhancement prompted several researchers to perform a detailed structural characterization of AS-metal interactions. These studies proved conclusively that Cu(II) binds specifically to AS and is effective in accelerating aggregation at physiologically relevant concentrations (Rasia *et al.*, 2005). NMR mapping revealed that the most affected regions were located at the N-terminus, comprising the stretch of residues ^{3}VFMKGLS^{9} and the His within the sequence ^{48}VAHGV^{52}, in line with several studies supporting a role for His residues as metal anchoring sites in other amyloidogenic proteins (Viles *et al.*, 1999; Aronoff-Spencer *et al.*, 2000; Burns *et al.*, 2002; Chattopadhyay *et al.*, 2004; Karr *et al.*, 2004b; Valensin *et al.*, 2004a; Valensin *et al.*, 2004b; Gaggelli *et al.*, 2005; Guilloreau *et al.*, 2006;

Klewpatinond and Viles, 2007; Calabrese and Miranker, 2009). A clear effect was also observed at the C-terminus of AS, [119]DPDNEA[124], but only at higher Cu(II) concentrations. Finally, a comparative analysis with other divalent metal ions revealed a selective effect of AS-metal interactions on AS aggregation kinetics, dictated by structural factors corresponding to different protein regions.

Elucidation of structural factors dictating Cu(II) binding specificity to AS was then the subject of multidisciplinary studies that involved the combined application of NMR, EPR, UV-Vis, CD spectroscopy and MALDI-MS. Site-directed and domain-truncated mutants of AS and its natural homologue β-synuclein (BS) were analyzed by this approach (Binolfi *et al.,* 2008). The evidences emerging form these works constitute the main body of information discussed in the following sections.

10.4.1 *The Nature of Cu(II) Anchoring Residues at the N-Terminal Region of AS*

Wild-type AS and its His mutant H50A were used to explore the location of specific copper binding sites by protection from diethyl pyrocarbonate (DEPC) modification and subsequent MALDI-MS analysis (Fig. 10.8) (Binolfi *et al.,* 2008).

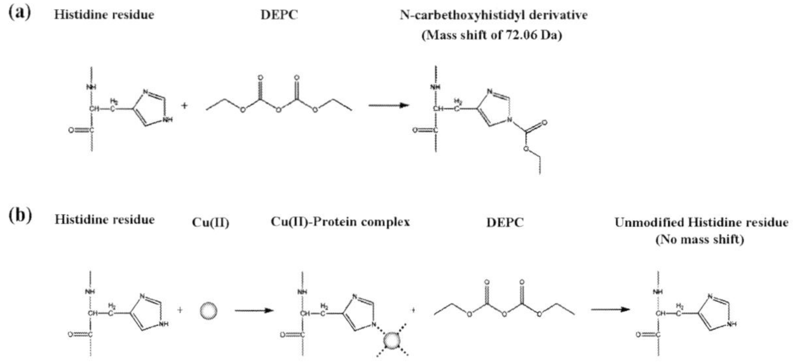

Figure 10.8 Scheme of the DEPC modification and Cu(II) protection reactions. (a) DEPC reacts with the imidazol ring of histidine to produce the N-carbethoxyhistidyl derivative. A mass shift of 72.06 Da is predicted per DEPC adduct formed. (b) Copper coordination to the imidazol ring of histidine protects this residue from DEPC modification.

DEPC is a chemical compound that has been widely used as a protein modification reagent capable of reacting with histidine residues to produce an N-carbethoxy-histidyl derivative (Miles, 1977). This molecule can also react with other nucleophilic residues including cysteinyl, arginyl, and tyrosyl residues, as well with α- and ε-amino groups. However, in contrast to arginine and tyrosine (AS lacks Cys residues), the imidazole ring and the α-amino groups are both known as copper anchoring sites. Many studies demonstrated that metal coordination to peptides or proteins protects the involved amino acids from DEPC modification, constituting an excellent approach to locate metal binding sites in biomolecules (Figs. 10.8(a,b)) (Qin *et al.*, 2002; Binolfi *et al.*, 2008).

On the basis of a predicted mass shift of 72 Da per DEPC molecule incorporated into the AS monomer, the comparative analysis of the modification patterns before and after preincubation with Cu(II) ions revealed that two residues within the AS protein structure were protected from DEPC modification by Cu(II) (Binolfi *et al.*, 2008). The observation of similar copper-dependent changes on the C-terminal truncated variant of AS (1–108 AS) indicated that the residues protected by Cu(II) coordination were located at the N-terminal region of the protein, confirming the picture obtained previously by NMR mapping experiments (Binolfi *et al.*, 2008).

Whereas two residues in AS were protected from DEPC modification by Cu(II), only one residue was reported to be protected by metal coordination on the H50A AS species, demonstrating that His was one of the anchoring residues for Cu(II) in the N-terminal region of the protein. To delimit the position of the nonhistidine copper binding site at the N-terminus of AS, tryptic digestion followed by MS was performed on DEPC-modified samples before and after preincubation with Cu(II) ions. The sequence ^{1}MDVFMK^{6} was identified as another Cu(II) binding motif and Met-1 as the other residue directly involved in Cu(II) coordination (Binolfi *et al.*, 2008). Although amide resonances of Met-1 and Asp-2 were not detectable by ^{1}H-^{15}N HSQC NMR experiments because of their fast exchange with the solvent, MALDI-MS provided unequivocal evidence for the direct involvement of Met-1 as the primary anchoring residue for Cu(II) in the protein.

10.4.2 *Deconvolution of Cu(II) Binding Sites in AS*

To evaluate the structural factors modulating Cu(II) binding to AS, the metal binding capability of BS was also studied. This homologue of AS lacks the long-range interactions involving N- and C-terminus, and NAC and C-terminus contacts that are present in AS (Bertoncini *et al.*, 2007). In addition, BS shows a high degree of homology with AS at the N-terminal region, where specific Cu(II) binding occurs, and the sole His residue is shifted from position 50 in AS to 65 in BS, constituting a useful and naturally occurring model for understanding the structural basis behind Cu(II) binding specificity.

The details of Cu(II) binding to the N-terminal region of BS and the mutants H65A BS and H50A AS were explored at single residue resolution by NMR spectroscopy. The interaction of Cu(II) with BS was identical to that reported for AS, whereas the interaction of BS with other divalent metal ions revealed that they bind exclusively to the C-terminal region (Binolfi *et al.*, 2008). Comparison with the His mutant species of AS demonstrated that the changes induced by Cu(II) in the region containing the His residues were the only ones abolished upon mutation of that site (Figs. 10.9(a, b)).

In agreement, another study based on tryptophan fluorescence measurements and conducted to probe Cu(II) protein binding in AS mutants with single W substitutions (F4W, Y39W, F94W and Y125W) reported that the fluorophore at position 4 was the most strongly quenched by the presence of the metal ion. Interestingly, the metal binding site revealed by W4 fluorescence quenching did not involve His-50 nor global polypeptide arrangement that would bring W39, W94, or W125 into proximity (Lee *et al.*, 2008). Finally, spectroscopic studies focused on the interaction of Cu(II) with peptides containing the sequences 1–17, 1–28 and 1–39 of AS demonstrated that the metal binding capability of the full-length protein were preserved in the synthetic species (Kowalik-Jankowska *et al.*, 2005).

Overall, the results discussed in this section confirm that: (i) the N-terminal region of AS is the high-affinity interface for Cu(II) binding; (ii) the free amino group at the N-terminus and the imidazol ring of the His residue are the primary anchoring groups for Cu(II) binding to the protein; (iii) the Cu(II)-binding motifs in the N-terminal region constitute separate, independent metal-binding sites; (iv) the transient long-range interactions in AS do not influence the binding

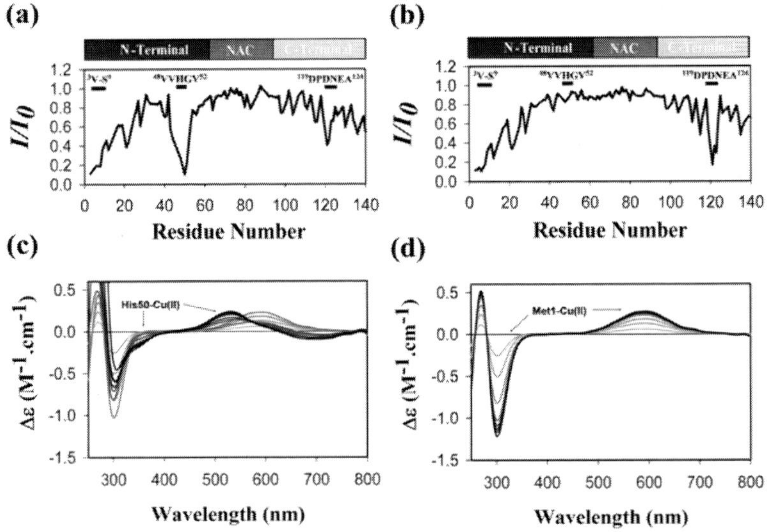

Figure 10.9 Intensity profiles of the backbone amide groups of 100 μM WT AS (a) and H50A AS (b) in the presence of 50 μM Cu(II). I and I_0 correspond to the intensities of cross-peaks in the presence and absence of the metal ion, respectively. Panels (c) and (d) shows the CD spectra of different AS-Cu(II) complexes. Direct titration of AS (c) and H50A AS (d) with Cu(II). Additions of 0.2 equivalents of Cu(II) were made as the concentration was incremented from 0 to 3 mol equivalents (from grey to blak). The metal complexes assigned to different protein regions are identified.

preferences of Cu(II) at each site; and (v) the binding of Cu(II) to the C-terminal region represents a very low affinity (K_d in the millimolar range), nonspecific process.

10.4.3 *Coordination Environment of AS-Cu(II) Complexes*

The non-interacting nature of the Cu(II) binding sites in AS was also confirmed by CD and EPR spectroscopy (Binolfi *et al.*, 2008). In addition, these techniques provided valuable information related to the electronic structure and the coordination geometry of the metal binding sites.

As discussed in Section 10.2.2, the CD spectra of AS complexed with 1 equiv. of Cu(II) was characterized by the presence of a positive d-d transition band at 600 nm and a negative LTM c.t. band at 300 nm.

Beyond 1 equiv. of Cu(II), two additional bands were evident at 520 and 340 nm, both of which saturated with 2 equiv. of added Cu(II) (Fig. 10.9(c)). Interestingly, the latter bands were not detected in the CD spectra of the Cu(II) complexes of H50A AS, demonstrating that the bands at 520 and 340 nm corresponded to the Cu(II) complex in which His acts as main anchoring residue (Fig. 10.9(d)) (Binolfi *et al.*, 2008).

The band at 340 nm was attributed to a LTM c.t. band between the metal center and an imidazol group, which is typical of His-Cu(II) complexes (Sigel and Martin, 1982; Daniele *et al.*, 1996). The positive band at 600 nm and the negative band at 300 nm reflected then the coordination of Cu(II) to the binding site with the highest affinity for the metal ion. The band at 300 nm in this Cu(II) complex was assigned to a LTM c.t. transition between the metal center and a deprotonated peptide nitrogen (Sigel and Martin, 1982; Daniele *et al.*, 1996). It is known that when the metal binds to the terminal amino group it favors the successive deprotonation of amide groups of the second and third residues (Sigel and Martin, 1982). Accordingly, it was postulated that Cu(II) coordination to the amino terminal group of Met-1 would favor deprotonation of the amide nitrogen of Asp-2, which once complexed to Cu(II) give rise to the LTM c.t. band at 300 nm in the CD spectrum (Binolfi *et al.*, 2008). Supporting this hypothesis, several studies on model peptides with aspartyl residues in position 2 showed the formation of highly stable Cu(II) complexes through direct Cu(II) coordination to the amino group of residue 1 and deprotonated amide nitrogen and carboxylate side-chains of residue at position 2 (Kallay *et al.*, 2005). The enhanced stability of the (NH_2, N^-, side-chain-COO^-)-Cu(II) coordination environment was attributed to the presence of an extra negative charge and the formation of (5,6)-member joined chelate rings (Sanna *et al.*, 2001; Kallay *et al.*, 2005).

From EPR experiments at least three different Cu(II) species could be discerned from the spectra of AS and BS on the basis of their parallel g values and hyperfine splitting (Rasia *et al.*, 2005; Binolfi *et al.*, 2008; Drew *et al.*, 2008). Two of these binding modes were located at the N-terminal region of the proteins and were characteristic of type-2 Cu(II) centers with a coordination environment composed of a mixture of N- and O-based ligands in a planar or distorted tetragonal arrangement. A comparative analysis with the CD and

EPR spectra of the His to Ala mutant species made possible to assign the EPR bands to specific Cu(II) complexes in the protein. The highest-affinity binding site or binding mode 1 was characterized by a 2N2O coordination environment, where the amino terminal group of Met-1 and the amide backbone of Asp-2 act as nitrogen ligands (Fig. 10.10) (Rasia *et al.*, 2005; Binolfi *et al.*, 2008; Drew *et al.*, 2008, Binolfi *et al.*, 2010). The oxygen ligands are provided by the carboxylate side-chain of Asp-2 and a water molecule (Binolfi *et al.*, 2010). A 1N3O or 2N2O arrangement might be implicated in binding mode 2, assigned to the His-Cu(II) complex (Binolfi *et al.*, 2008; Drew *et al.*, 2008). Histidine provides one of the nitrogen ligands, whereas the oxygen ligands could be provided by water molecules and/or backbone carbonyls from the peptide backbone. Whereas binding modes 1 and 2 corresponded to the N-terminal Cu(II) complexes involving Met-1 and His-50, respectively, an oxygen-rich (O_4) coordination environment was assigned to binding mode 3 (Fig. 10.10), in agreement with the binding site mapped by NMR for Cu(II) binding to the C-terminal region, where carboxylates from Asp-119, Asp-121 and Glu-123 would act as the major contributors to metal binding in this region (Rasia *et al.*, 2005; Binolfi *et al.*, 2008; Drew *et al.*, 2008).

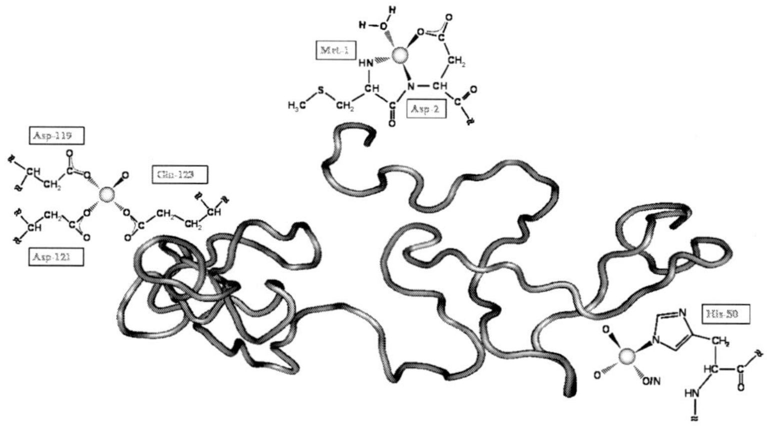

Figure 10.10 Schematic representation of the location of Cu(II) complexes in AS. Structure and ligands involved in metal binding sites are those derived from the spectroscopic data discussed throughout the chapter. The AS-Cu(II) complexes are drawn in close proximity to the regions containing the metal coordinating residues.

Finally, a recent work has suggested the possible formation of a Cu(II) mediated macrochelate in AS, involving Met-1 and His-50 residues in a unique 3N1O coordination arrangement (Drew *et al.,* 2008). However, there are several evidences arguing against the occurrence of such a metal-complex conformation under physiologically relevant conditions (Sung *et al.,* 2006; Binolfi *et al.,* 2008; Lee *et al.,* 2008; Drew *et al.,* 2009).

10.4.4 *Cu(II) Binding Affinity Features at the N-Terminal Region*

The affinity features for the Cu(II) complexes in the N-terminal region of AS were determined by different techniques. CD titration experiments reported dissociation constants $K_{d1} = 0.7 \pm 0.6$ μM and $K_{d2} = 60 \pm 20$ μM for the high and low-affinity sites, respectively, in agreement with the data reported previously by UV-Vis spectroscopy (see section 10.2.2). Despite the large numerical uncertainties associated with these quantities, which are due to the high concentration of the protein required to achieve accurately measurable signals, their relative order of magnitude was well-established.

More reliable K_{d1} values for the strongly bound Cu(II) ion were obtained by competition experiments involving the chromophoric metal ligand Mag-fura-2 (MF), whose spectroscopic features change upon Cu(II) binding. This procedure was extensively used to monitor Zn(II), Cd(II) and Co(II) binding to metalloproteins (de Seny *et al.*, 2001; Llarrull *et al.*, 2007). By using this approach, a submicromolar dissociation constant ($K_{d1} = 0.20 \pm 0.02$ μM) was determined for the first equivalent of Cu(II) bound to AS, consistent with the values reported in other studies by tryptophan fluorescence and isothermal titration calorimetry measurements (Binolfi *et al.*, 2008; Lee *et al.*, 2008; Binolfi *et al.*, 2010).

10.5 Conclusions and Future Perspectives

In the preceding sections a detailed structural description of the interactions of AS with divalent metal ions has been presented. The major findings of these studies are summarized in Table 10.2.

Table 10.2 Affinity and structural features AS-metal(II) interactions

Metal ion	Binding region[a]	Binding sequence	Ligand donor set	Primary anchoring groups	Disscociation constants
Cu(II)	NT	[1]MDVFMK[6] (AS and BS)	2N2O	Terminal α-NH_2 of Met-1	0.1 μM
		[48]VVHGV[52] (AS)	2N2O/ 1N3O	His-50 imidazol ring	35 μM
		[62]QASHLGG[68] (BS)	1N3O	His-65 imidazol ring	
Other metal ionsb	CT	[119]DPDNEA[124] (AS)	4O	Carboxylate group of Asp-121	~ 1 mM
		[120]EDPPQEE[126] (BS)	4O	Carboxylate groups of Asp-121 and Glu-126	

[a]NT and CT correspond to N-terminal and C-terminal regions, respectively.
[b]Other metal ions: Mn(II), Fe(II), Ni(II) and Co(II). Copper complexes with BS are included for comparative purposes.

From the binding features of the Cu(II) complexes studied it can be concluded that: (i) the high-affinity Cu(II) binding site in AS (K_d = 0.11 ± 0.01 μM) corresponds to that in which the N-terminal amino nitrogen of Met-1 acts as anchoring group and where Asp-2 and a water molecule are also involved in the ligand donor set (Binolfi *et al.*, 2010); (ii) a second lower-affinity binding motif for Cu(II) (K_d = 35 μM) is centered at position 50 in AS (Binolfi *et al.*, 2010) corresponding to the location of the sole His residue in the primary sequence of the protein; (iii) AS bind metal ions to the C-terminus with very low selectivity (K_d ~1 mM), where carboxylate side-chain of Asp-121 acts as major contributor to the binding process.

The comparative analysis between Cu(II) and other divalent metal ions revealed a hierarchal effect of AS-metal interactions on AS aggregation kinetics, dictated largely by structural factors corresponding to different protein regions. As mentioned before,

Cu(II) has been implicated in the pathogenesis of age-dependent degenerative diseases characterized by deposition of amyloid material. The affinity of Cu(II) for the target proteins involved in several of these diseases is in the low or submicromolar range. The fact that AS binds Cu(II) with an affinity in the submicromolar range and that Cu(II) can promote AS aggregation efficiently at physiologically relevant concentrations establish a tight link with other amyloid-related disorders such as Alzheimer's disease and prion disease and support the notion of PD as a metal-associated neurodegenerative disorder.

Focusing on the possible role played by His in the Cu(II) mediated aggregation of AS, the evidences indicate that the effect of Cu(II) concentrations on AS aggregation *in vitro*, at least at the earliest aggregational events, seem not to be related neither to specific histidine binding nor to the general factor usually invoked for AS aggregation such as the interaction with clusters of negatively charged residues at the C-terminal region.

Related to the mechanism of Cu(II)-induced AS aggregation, elucidation of the structures of the different AS-Cu(II) complexes and the role played by the metal ion on the oligomeric and fibrillar states AS are crucial to understand the mechanistic rules that direct the Cu(II)-mediated aggregation of AS. The structural properties of Cu(I) binding to AS and detection of redox active species need to be also investigated to advance in the hypothesis of an aggregation mechanism based on metal-catalyzed oxidation of AS.

Although more studies are clearly needed to explore the biological consequences of AS-Cu(II) interactions, the advances in the structural biology and bioinorganic chemistry of AS-metal interactions during the last years indicate that perturbations in copper metabolism might constitute a more widespread element in neurodegenerative disorders than has been recognized previously.

References

Aronoff-Spencer, E., Burns, C. S., Avdievich, N. I., Gerfen, G. J., Peisach, J., Antholine, W. E., Ball, H. L., Cohen, F. E., Prusiner, S. B. and Millhauser, G. L. (2000). Identification of the Cu^{2+} binding sites in the N-terminal domain of the prion protein by EPR and CD spectroscopy. *Biochemistry*, 39, pp. 13760–71.

Atwood, C. S., Moir, R. D., Huang, X. D., Scarpa, R. C., Bacarra, N. M. E., Romano, D. M., Hartshorn, M. K., Tanzi, R. E. and Bush, A. I. (1998). Dramatic aggregation of Alzheimer A beta by Cu(II) is induced by conditions representing physiological acidosis. *J. Biol. Chem.*, 273, pp. 12817–26.

Barnham, K. J., McKinstry, W. J., Multhaup, G., Galatis, D., Morton, C. J., Curtain, C. C., Williamson, N. A., White, A. R., Hinds, M. G., Norton, R. S., Beyreuther, K., Masters, C. L., Parker, M. W. and Cappai, R. (2003). Structure of the Alzheimer's disease amyloid precursor protein copper binding domain. A regulator of neuronal copper homeostasis. *J. Biol. Chem.*, 278, pp. 17401–7.

Belosi, B., Gaggelli, E., Guerrini, R., Kozlowski, H., Luczkowski, M., Mancini, F. M., Remelli, M., Valensin, D. and Valensin, G. (2004). Copper binding to the neurotoxic peptide PrP106–126: Thermodynamic and structural studies. *Chembiochem.*, 5, pp. 349–59.

Bernado, P., Bertoncini, C. W., Griesinger, C., Zweckstetter, M. and Blackledge, M. (2005). Defining long-range order and local disorder in native alpha-synuclein using residual dipolar couplings. *J. Am. Chem. Soc.*, 127, pp. 17968–9.

Bertini, I. and Luchinat, C. (1996). NMR of paramagnetic substances. *Coord. Chem. Rev.*, 150, pp. 1–296.

Bertoncini, C. W., Jung, Y. S., Fernandez, C. O., Hoyer, W., Griesinger, C., Jovin, T. M. and Zweckstetter, M. (2005). Release of long-range tertiary interactions potentiates aggregation of natively unstructured alpha-synuclein. *Proc. Natl. Acad. Sci. USA*, 102, pp. 1430–5.

Bertoncini, C. W., Rasia, R. M., Lamberto, G. R., Binolfi, A., Zweckstetter, M., Griesinger, C. and Fernandez, C. O. (2007). Structural characterization of the intrinsically unfolded protein beta-synuclein, a natural negative regulator of alpha-synuclein aggregation. *J. Mol. Biol.*, 372, pp. 708–22.

Binolfi, A., Lamberto, G. R., Duran, R., Quintanar, L., Bertoncini, C. W., Souza, J. M., Cervenansky, C., Zweckstetter, M., Griesinger, C. and Fernandez, C. O. (2008). Site-specific interactions of Cu(II) with alpha and beta-synuclein: bridging the molecular gap between metal binding and aggregation. *J. Am. Chem. Soc.*, 130, pp. 11801–12.

Binolfi, A., Rasia, R. M., Bertoncini, C. W., Ceolin, M., Zweckstetter, M., Griesinger, C., Jovin, T. M. and Fernandez, C. O. (2006). Interaction of alpha-synuclein with divalent metal ions reveals key differences: a link between structure, binding specificity and fibrillation enhancement. *J. Am. Chem. Soc.*, 128, pp. 9893–901.

Binolfi, A., Rodriguez, E. E., Valensin, D., D'Amelio, N., Ippoliti, E., Obal, G., Duran, R., Magistrato, A., Pritsch, O., Zweckstetter, M., Valensin, G., Carloni, P., Quintanar, L., Griesinger, C. and Fernández, C.O (2010). Bioinorganic chemistry of Parkinson's disease: structural determinants for the copper-mediated amyloid formation of alpha-synuclein. *Inorg. Chem.*, 49, pp. 10668–79.

Bodenhausen, G. and Rubern, D. J. (1980). Natural abundance nitrogen-15 NMR by enhanced heteronuclear spectroscopy *Chem. Phys. Lett.*, 69, pp. 185–9.

Brown, D. R. and Kozlowski, H. (2004). Biological inorganic and bioinorganic chemistry of neurodegeneration based on prion and Alzheimer diseases. *Dalton Trans.*, pp. 1907–17.

Bryce, G. F. and Gurd, F. R. (1966). Optical rotatory dispersion and circular dichroism spectra of copper(II)- and nickel(II)-peptide complexes. *J. Biol. Chem.*, 241, pp. 1439–48.

Buchman, V. L., Hunter, H. J., Pinon, L. G., Thompson, J., Privalova, E. M., Ninkina, N. N. and Davies, A. M. (1998). Persyn, a member of the synuclein family, has a distinct pattern of expression in the developing nervous system. *J. Neurosci. Res.*, 18, pp. 9335–41.

Burns, C. S., Aronoff-Spencer, E., Dunham, C. M., Lario, P., Avdievich, N. I., Antholine, W. E., Olmstead, M. M., Vrielink, A., Gerfen, G. J., Peisach, J., Scott, W. G. and Millhauser, G. L. (2002). Molecular features of the copper binding sites in the octarepeat domain of the prion protein. *Biochemistry*, 41, pp. 3991–4001.

Bush, A. I. (2000). Metals and neuroscience. *Curr. Opin. Chem. Biol.*, 4, pp. 184–91.

Bush, A. I., Multhaup, G., Moir, R. D., Williamson, T. G., Small, D. H., Rumble, B., Pollwein, P., Beyreuther, K. and Masters, C. L. (1993). A novel zinc(II) binding site modulates the function of the beta A4 amyloid protein precursor of Alzheimer's disease. *J. Biol. Chem.*, 268, pp. 16109–12.

Bussell, R., Jr. and Eliezer, D. (2003). A structural and functional role for 11-mer repeats in alpha-synuclein and other exchangeable lipid binding proteins. *J. Mol. Biol.*, 329, pp. 763–78.

Calabrese, M. F. and Miranker, A. D. (2009). Metal binding sheds light on mechanisms of amyloid assembly. *Prion*, 3, pp. 1–4.

Castellani, R. J., Siedlak, S. L., Perry, G. and Smith, M. A. (2000). Sequestration of iron by Lewy bodies in Parkinson's disease. *Acta Neuropathol.*, 100, pp. 111–4.

Caughey, B. and Lansbury, P. T. (2003). Protofibrils, pores, fibrils, and neurodegeneration: separating the responsible protein aggregates from the innocent bystanders. *Annu. Rev. Neurosci.*, 26, pp. 267–98.

Chandra, S., Chen, X., Rizo, J., Jahn, R. and Sudhof, T. C. (2003). A broken α-helix in folded α-Synuclein. *J. Biol. Chem.*, 278, pp. 15313–8.

Chandra, S., Gallardo, G., Fernandez-Chacon, R., Schluter, O. M. and Sudhof, T. C. (2005). Alpha-synuclein cooperates with CSPalpha in preventing neurodegeneration. *Cell*, 123, pp. 383–96.

Chattopadhyay, M., Aronoff-Spencer, E., Burns, C. S., Antholine, W., Scott, W., Peisach, J., Gerfen, G. J. and Millhauser, G. (2004). Cu(II) binding in the prion protein. *Abstracts of Papers of the American Chemical Society*, 227, pp. U1503–U1503.

Chattopadhyay, M., Walter, E. D., Newell, D. J., Jackson, P. J., Aronoff-Spencer, E., Peisach, J., Gerfen, G. J., Bennett, B., Antholine, W. E. and Millhauser, G. L. (2005). The Octarepeat Domain of the Prion Protein Binds Cu(II) with Three Distinct Coordination Modes at pH 7.4. *J. Am. Chem. Soc.*, 127, pp. 12647–56.

Clayton, D. F. and George, J. M. (1999). Synucleins in synaptic plasticity and neurodegenerative disorders. *J. Neurosci. Res.*, 58, pp. 120–9.

Crowther, R. A., Jakes, R., Spillantini, M. G. and Goedert, M. (1998). Synthetic filaments assembled from C-terminally truncated α-synuclein. *FEBS Lett.*, 436, pp. 309–12.

Daniele, P. G., Prenesti, E. and Ostacoli, G. (1996). Ultraviolet-circular dichroism spectra for structural analysis of copper(II) complexes with aliphatic and aromatic ligands in aqueous solution. *J. Chem. Soc. Dalton Trans.*, pp. 3269–75.

Davidson, W. S., Jonas, A., Clayton, D. F. and George, J. M. (1998). Stabilization of alpha-synuclein secondary structure upon binding to synthetic membranes. *J. Biol. Chem.*, 273, pp. 9443–9.

de Seny, D., Heinz, U., Wommer, S., Kiefer, M., Meyer-Klaucke, W., Galleni, M., Frere, J. M., Bauer, R. and Adolph, H. W. (2001). Metal ion binding and coordination geometry for wild type and mutants of metallo-beta-lactamase from Bacillus cereus 569/H/9 (BcII): a combined thermodynamic, kinetic, and spectroscopic approach. *J. Biol. Chem.*, 276, pp. 45065–78.

Dedmon, M. M., Lindorff-Larsen, K., Christodoulou, J., Vendruscolo, M. and Dobson, C. M. (2005). Mapping long-range interactions in α-synuclein using spin-label NMR and ensemble molecular dynamics simulations. *J. Am. Chem. Soc.*, 127, pp. 476–7.

Der-Sarkissian, A., Jao, C. C., Chen, J. and Langen, R. (2003). Structural organization of alpha-synuclein fibrils studied by site-directed spin labeling. *J. Biol. Chem.*, 278, pp. 37530–5.

Ding, T. T., Lee, S. J., Rochet, J. C. and Lansbury, P. T., Jr. (2002). Annular alpha-synuclein protofibrils are produced when spherical protofibrils are incubated in solution or bound to brain-derived membranes. *Biochemistry*, 41, pp. 10209–17.

Dobson, C. M. (2004). Principles of protein folding, misfolding and aggregation. *Semin. Cell. Dev. Biol.*, 15, pp. 3–16.

Drew, S. C., Leong, S. L., Pham, C. L., Tew, D. J., Masters, C. L., Miles, L. A., Cappai, R. and Barnham, K. J. (2008). Cu^{2+} binding modes of recombinant alpha-synuclein--insights from EPR spectroscopy. *J. Am. Chem. Soc.*, 130, pp. 7766–73.

Drew, S. C., Tew, D. J., Masters, C. L., Cappai, R. and Barnham, K. J. (2009). Copper coordination by familial mutants of Parkinson's disease-associated alpha-synuclein. *Appl. Magn. Res.*, 36, pp. 223–9.

Fernandez, C. O., Hoyer, W., Zweckstetter, M., Jares-Erijman, E. A., Subramaniam, V., Griesinger, C. and Jovin, T. M. (2004). NMR of α-synuclein-polyamine complexes elucidates the mechanism and kinetics of induced aggregation. *EMBO J.*, 23, pp. 2039–46.

Forno, L. S. (1996). Neuropathology of Parkinson's disease. *J. Neuropathol. Exp. Neurol.*, 55, pp. 259–72.

Gaeta, A. and Hider, R. C. (2005). The crucial role of metal ions in neurodegeneration: the basis for a promising therapeutic strategy. *Br. J. Pharmacol.*, 146, pp. 1041–59.

Gaggelli, E., Bernardi, F., Molteni, E., Pogni, R., Valensin, D., Valensin, G., Remelli, M., Luczkowski, M. and Kozlowski, H. (2005). Interaction of the Human Prion PrP(106–126) Sequence with Copper(II), Manganese(II), and Zinc(II): NMR and EPR Studies. *J. Am. Chem. Soc.*, 127, pp. 996–1006.

Gaggelli, E., Kozlowski, H., Valensin, D. and Valensin, G. (2006). Copper homeostasis and neurodegenerative disorders (Alzheimer's, prion, and Parkinson's diseases and amyotrophic lateral sclerosis). *Chem. Rev.*, 106, pp. 1995–2044.

Garnett, A. P. and Viles, J. H. (2003). Copper binding to the octarepeats of the prion protein. Affinity, specificity, folding, and cooperativity: insights from circular dichroism. *J. Biol. Chem.*, 278, pp. 6795–802.

Garzon-Rodriguez, W., Yatsimirsky, A. K. and Glabe, C. G. (1999). Binding of Zn(II), Cu(II), and Fe(II) ions to Alzheimer's A beta peptide studied by fluorescence. *Bioorg. Med. Chem. Lett.*, 9, pp. 2243–8.

Giasson, B. I., Murray, I. V., Trojanowski, J. Q. and Lee, V. M. (2001). A hydrophobic stretch of 12 amino acid residues in the middle of α-synuclein is essential for filament assembly. *J. Biol. Chem.*, 276, pp. 2380–6.

Goedert, M. (2001). α-synuclein and neurodegenerative diseases. *Nat. Rev. Neurosci.*, 2, pp. 492–501.

Golts, N., Snyder, H., Frasier, M., Theisler, C., Choi, P. and Wolozin, B. (2002). Magnesium inhibits spontaneous and iron-induced aggregation of α-synuclein. *J. Biol. Chem.*, 277, pp. 16116–23.

Gorell, J. M., Johnson, C. C., Rybicki, B. A., Peterson, E. L., Kortsha, G. X., Brown, G. G. and Richardson, R. J. (1999). Occupational exposure to manganese, copper, lead, iron, mercury and zinc and the risk of Parkinson's disease. *Neurotoxicology*, 20, pp. 239–47.

Guilloreau, L., Damian, L., Coppel, Y., Mazarguil, H., Winterhalter, M. and Faller, P. (2006). Structural and thermodynamical properties of Cu^{II} amyloid-beta16/28 complexes associated with Alzheimer's disease. *J. Biol. Inorg. Chem.*, 11, pp. 1024–38.

Haass, C. and Steiner, H. (2001). Protofibrils, the unifying toxic molecule of neurodegenerative disorders? *Nat. Neurosci.*, 4, pp. 859–60.

Hashimoto, M., Rockenstein, E., Mante, M., Crews, L., Bar-On, P., Gage, F. H., Marr, R. and Masliah, E. (2004). An antiaggregation gene therapy strategy for Lewy body disease utilizing beta-synuclein lentivirus in a transgenic model. *Gene Ther*, 11, pp. 1713–23.

Hashimoto, M., Rockenstein, E., Mante, M., Mallory, M. and Masliah, E. (2001). beta-Synuclein inhibits alpha-synuclein aggregation: a possible role as an anti-parkinsonian factor. *Neuron*, 32, pp. 213–23.

Hoyer, W., Cherny, D., Subramaniam, V. and Jovin, T. M. (2004). Impact of the Acidic C-Terminal Region Comprising Amino Acids 109–140 on α-Synuclein Aggregation *in vitro*. *Biochemistry*, 43, pp. 16233–42.

Huang, X., Cuajungco, M. P., Atwood, C. S., Hartshorn, M. A., Tyndall, J. D., Hanson, G. R., Stokes, K. C., Leopold, M., Multhaup, G., Goldstein, L. E., Scarpa, R. C., Saunders, A. J., Lim, J., Moir, R. D., Glabe, C., Bowden, E. F., Masters, C. L., Fairlie, D. P., Tanzi, R. E. and Bush, A. I. (1999). Cu(II) potentiation of alzheimer abeta neurotoxicity. Correlation with cell-free hydrogen peroxide production and metal reduction. *J. Biol. Chem.*, 274, pp. 37111–6.

Jao, C. C., Der-Sarkissian, A., Chen, J. and Langen, R. (2004). Structure of membrane-bound alpha-synuclein studied by site-directed spin labeling. *Proc. Natl. Acad. Sci. USA*, 101, pp. 8331–6.

Kallay, C., Varnagy, K., Micera, G., Sanna, D. and Sovago, I. (2005). Copper(II) complexes of oligopeptides containing aspartyl and glutamyl residues. Potentiometric and spectroscopic studies. *J. Inorg. Biochem.*, 99, pp. 1514–25.

Karr, J. W., Akintoye, H., Kaupp, L. J. and Szalai, V. A. (2004a). N-Terminal Deletions Modify the Cu^{2+} Binding Site in Amyloid-b. *Biochemistry*, 44, pp. 5478–87.

Karr, J. W., Akintoye, H., Kaupp, L. J. and Szalai, V. A. (2005). N-Terminal deletions modify the Cu^{2+} binding site in amyloid-beta. *Biochemistry*, 44, pp. 5478–87.

Karr, J. W., Kaupp, L. J. and Szalai, V. A. (2004b). Amyloid-beta binds Cu^{2+} in a mononuclear metal ion binding site. *J. Am. Chem. Soc.*, 126, pp. 13534–8.

Karr, J. W. and Szalai, V. A. (2008). Cu(II) Binding to Monomeric, Oligomeric, and Fibrillar Forms of the Alzheimer's Disease Amyloid-beta Peptide. *Biochemistry*, 47, pp. 5006–16.

Kim, N. H., Choi, J. K., Jeong, B. H., Kim, J. I., Kwon, M. S., Carp, R. I. and Kim, Y. S. (2005). Effect of transition metals (Mn, Cu, Fe) and deoxycholic acid (DA) on the conversion of PrPC to PrPres. *FASEB J.*, 19, pp. 783–5.

Klewpatinond, M., Davies, P., Bowen, S., Brown, D. R. and Viles, J. H. (2008). Deconvoluting the Cu^{2+} binding modes of full-length prion protein. *J Biol Chem*, 283, pp. 1870–81.

Klewpatinond, M. and Viles, J. H. (2007). Fragment length influences affinity for Cu^{2+} and Ni^{2+} binding to His96 or His111 of the prion protein and spectroscopic evidence for a multiple histidine binding only at low pH. *Biochem J.*, 404, pp. 393–402.

Kowalik-Jankowska, T., Rajewska, A., Wisniewska, K., Grzonka, Z. and Jezierska, J. (2005). Coordination abilities of N-terminal fragments of alpha-synuclein towards copper(II) ions: a combined potentiometric and spectroscopic study. *J. Inorg. Biochem.*, 99, pp. 2282–91.

Kowalik-Jankowska, T., Ruta-Dolejsz, M., Wisniewska, K. and Lankiewicz, L. (2001). Cu(II) interaction with N-terminal fragments of human and mouse beta-amyloid peptide. *J. Inorg. Biochem.*, 86, pp. 535–45.

Kowalik-Jankowska, T., Ruta-Dolejsz, M., Wisniewska, K. and Lankiewicz, L. (2002). Coordination of copper(II) ions by the 11-20 and 11-28 fragments of human and mouse beta-amyloid peptide. *J. Inorg. Biochem.*, 92, pp. 1–10.

Kowalik-Jankowska, T., Ruta, M., Wisniewska, K. and Lankiewicz, L. (2003). Coordination abilities of the 1-16 and 1-28 fragments of beta-amyloid peptide towards copper(II) ions: a combined potentiometric and spectroscopic study. *J. Inorg. Biochem.*, 95, pp. 270–82.

Kozlowski, H., Brown, D. and Valensin, G., Eds. (2006). Metallochemistry of Neurodegeneration: Biological, Chemical, and Genetic Aspects, (RSC Publishing: Cambridge, UK).

Kruger, R., Kuhn, W., Muller, T., Woitalla, D., Graeber, M., Kosel, S., Przuntek, H., Epplen, J. T., Schols, L. and Riess, O. (1998). Ala30Pro mutation in the gene encoding alpha-synuclein in Parkinson's disease. *Nat. Genet.*, 18, pp. 106–108.

Lansbury, P. T., Jr. (1999). Evolution of amyloid: what normal protein folding may tell us about fibrillogenesis and disease. *Proc. Natl. Acad. Sci. USA*, 96, pp. 3342–4.

Lashuel, H. A., Hartley, D., Petre, B. M., Walz, T. and Lansbury, P. T., Jr. (2002a). Neurodegenerative disease: amyloid pores from pathogenic mutations. *Nature*, 418, pp. 291.

Lashuel, H. A., Petre, B. M., Wall, J., Simon, M., Nowak, R. J., Walz, T. and Lansbury, P. T., Jr. (2002b). Alpha-synuclein, especially the Parkinson's disease-associated mutants, forms pore-like annular and tubular protofibrils. *J. Mol. Biol.*, 322, pp. 1089–102.

Lee, J. C., Gray, H. B. and Winkler, J. R. (2005). Tertiary contact formation in alpha-synuclein probed by electron transfer. *J. Am. Chem. Soc.*, 127, pp. 16388–9.

Lee, J. C., Gray, H. B. and Winkler, J. R. (2008). Copper(II) binding to alpha-synuclein, the Parkinson's protein. *J. Am. Chem. Soc.*, 130, pp. 6898–9.

Lee, J. C., Lai, B. T., Kozak, J. J., Gray, H. B. and Winkler, J. R. (2007). α-synuclein tertiary contact dynamics. *J. Phys. Chem. B*, 111, pp. 2107–12.

Lee, J. C., Langen, R., Hummel, P. A., Gray, H. B. and Winkler, J. R. (2004). α-synuclein structures from fluorescence energy-transfer kinetics: implications for the role of the protein in Parkinson's disease. *Proc. Natl. Acad. Sci. USA*, 101, pp. 16466–71.

LeVine, H., 3rd (1999). Quantification of beta-sheet amyloid fibril structures with thioflavin T. *Methods Enzymol.*, 309, pp. 274–84.

Lewy, F. H. (1912). In Handbuch der Neurologie. M. Lewandowsky, Abelsdorff, G. Berlin, Springer-Verlag. 3: 920–933.

Llarrull, L. I., Tioni, M. F., Kowalski, J., Bennett, B. and Vila, A. J. (2007). Evidence for a dinuclear active site in the metallo-beta-lactamase BcII with substoichiometric Co(II). A new model for metal uptake. *J. Biol. Chem.*, 282, pp. 30586–95.

Lotharius, J. and Brundin, P. (2002). Pathogenesis of Parkinson's disease: dopamine, vesicles and alpha-synuclein. *Nat. Rev. Neurosci.*, 3, pp. 932–42.

Lovell, M. A., Robertson, J. D., Teesdale, W. J., Campbell, J. L. and Markesbery, W. R. (1998). Copper, iron and zinc in Alzheimer's disease senile plaques. *J. Neurol. Sci.*, 158, pp. 47–52.

Masliah, E., Rockenstein, E., Veinbergs, I., Mallory, M., Hashimoto, M., Takeda, A., Sagara, Y., Sisk, A. and Mucke, L. (2000). Dopaminergic loss and inclusion body formation in α-synuclein mice: implications for neurodegenerative disorders. *Science*, 287, pp. 1265–9.

Miles, E. W. (1977). Modification of histidyl residues in proteins by diethylpyrocarbonate. *Methods Enzymol.*, 47, pp. 431–42.

Miura, T., Suzuki, K., Kohata, N. and Takeuchi, H. (2000). Metal binding modes of Alzheimer's amyloid β-peptide in insoluble aggregates and soluble complexes. *Biochemistry*, 39, pp. 7024–31.

Molina-Holgado, F., Hider, R. C., Gaeta, A., Williams, R. and Francis, P. (2007). Metals ions and neurodegeneration. *Biometals*, 20, pp. 639–54.

Paik, S. R., Shin, H. J. and Lee, J. H. (2000). Metal-catalyzed oxidation of α-synuclein in the presence of Copper(II) and hydrogen peroxide. *Arch. Biochem. Biophys.*, 378, pp. 269–77.

Paik, S. R., Shin, H. J., Lee, J. H., Chang, C. S. and Kim, J. (1999). Copper(II)-induced self-oligomerization of alpha-synuclein. *Biochem. J.*, 340, pp. 821–8.

Pall, H. S., Blake, D. R., Gutteridge, J. M., Williams, A. C., Lunec, J., Hall, M. and Taylor, A. (1987). Raised Cerebrospinal-Fluid Copper Concentration in Parkinsons-Disease. *Lancet*, 330, pp. 238–41.

Polymeropoulos, M. H., Lavedan, C., Leroy, E., Ide, S. E., Dehejia, A., Dutra, A., Pike, B., Root, H., Rubenstein, J., Boyer, R., Stenroos, E. S., Chandrasekharappa, S., Athanassiadou, A., Papapetropoulos, T., Johnson, W. G., Lazzarini, A. M., Duvoisin, R. C., Di Iorio, G., Golbe, L. I. and Nussbaum, R. L. (1997). Mutation in the alpha-Synuclein Gene Identified in Families with Parkinson's Disease. *Science*, 276, pp. 2045–2047.

Qin, K., Yang, Y., Mastrangelo, P. and Westaway, D. (2002). Mapping Cu(II) binding sites in prion proteins by diethyl pyrocarbonate modification and matrix-assisted laser desorption ionization-time of flight (MALDI-TOF) mass spectrometric footprinting. *J. Biol. Chem.*, 277, pp. 1981–90.

Rasia, R. M., Bertoncini, C. W., Marsh, D., Hoyer, W., Cherny, D., Zweckstetter, M., Griesinger, C., Jovin, T. and Fernández, C. O. (2005). Structural characterization of copper(II) binding to a-synuclein: Insights into the bioinorganic chemistry of Parkinson's disease. *Proc. Natl. Acad. Sci. USA*, 102, pp. 4294–4299.

Requena, J. R., Groth, D., Legname, G., Stadtman, E. R., Prusiner, S. B. and Levine, R. L. (2001). Copper-catalyzed oxidation of the recombinant SHa(29-231) prion protein. *Proc. Natl. Acad. Sci. USA*, 98, pp. 7170–5.

Sanna, D., Agoston, C. G., Micera, G. and Sovago, I. (2001). The effect of the ring size of fused chelates on the thermodynamic and spectroscopic properties of peptide complexes of Cu(II). *Polyhedron*, 20, pp. 3079–90.

Sayre, L. M., Perry, G. and Smith, M. A. (1999). Redox metals and neurodegenerative disease. *Curr. Opin. Chem. Biol.*, 3, pp. 220–5.

Serpell, L. C., Berriman, J., Jakes, R., Goedert, M. and Crowther, R. A. (2000). Fiber diffraction of synthetic alpha-synuclein filaments shows amyloid-like cross-beta conformation. *Proc. Natl. Acad. Sci. USA*, 97, pp. 4897–902.

Sigel, H. and Martin, R. B. (1982). Coordinating properties of the amide bond. Stability and structure of metal ion complexes of peptides and related ligands *Chem. Rev.*, 82, pp. 385–426.

Spillantini, M. G., Schmidt, M. L., Lee, V. M., Trojanowski, J. Q., Jakes, R. and Goedert, M. (1997). Alpha-synuclein in Lewy bodies. *Nature*, 388, pp. 839–40.

Sung, Y. H., Rospigliosi, C. and Eliezer, D. (2006). NMR mapping of copper binding sites in alpha-synuclein. *Biochim. Biophys. Acta.*, 1764, pp. 5–12.

Syme, C. D., Nadal, R. C., Rigby, S. E. and Viles, J. H. (2004). Copper binding to the amyloid-beta (Abeta) peptide associated with Alzheimer's disease: folding, coordination geometry, pH dependence, stoichiometry, and affinity of Abeta-(1-28): insights from a range of complementary spectroscopic techniques. *J. Biol. Chem.*, 279, pp. 18169–77.

Syme, C. D. and Viles, J. H. (2006). Solution 1H NMR investigation of Zn^{2+} and Cd^{2+} binding to amyloid-beta peptide (Abeta) of Alzheimer's disease. *Biochim. Biophys. Acta*, 1764, pp. 246–56.

Uversky, V. N., Li, J. and Fink, A. L. (2001). Metal-triggered structural transformations, aggregation, and fibrillation of human alpha-synuclein. A possible molecular NK between Parkinson's disease and heavy metal exposure. *J. Biol. Chem.*, 276, pp. 44284–96.

Uversky, V. N., Li, J., Souillac, P., Millett, I. S., Doniach, S., Jakes, R., Goedert, M. and Fink, A. L. (2002a). Biophysical properties of the synucleins and their propensities to fibrillate: inhibition of alpha-synuclein assembly by beta- and gamma-synucleins. *J. Biol. Chem.*, 277, pp. 11970–8.

Uversky, V. N., Yamin, G., Souillac, P. O., Goers, J., Glaser, C. B. and Fink, A. L. (2002b). Methionine oxidation inhibits fibrillation of human α-synuclein in vitro. *FEBS Lett*, 517, pp. 239–44.

Valensin, D., Luczkowski, M., Mancini, F. M., Legowska, A., Gaggelli, E., Valensin, G., Rolka, K. and Kozlowski, H. (2004a). The dimeric and tetrameric octarepeat fragments of prion protein behave differently to its monomeric unit. *Dalton Trans.*, pp. 1284–93.

Valensin, D., Mancini, F. M., Luczkowski, M., Janicka, A., Wisniewska, K., Gaggelli, E., Valensin, G., Lankiewicz, L. and Kozlowski, H. (2004b). Identification of a novel high affinity copper binding site in the APP(145–155) fragment of amyloid precursor protein. *Dalton Trans.*, pp. 16–22.

Verity, M. A. (1999). Manganese neurotoxicity: a mechanistic hypothesis. *Neurotoxicology*, 20, pp. 489–97.

Viles, J. H., Cohen, F. E., Prusiner, S. B., Goodin, D. B., Wright, P. E. and Dyson, H. J. (1999). Copper binding to the prion protein: Structural implications of four identical cooperative binding sites. *Proc. Natl. Acad. Sci. USA*, 96, pp. 2042–7.

Zarranz, J. J., Alegre, J., Gomez-Esteban, J. C., Lezcano, E., Ros, R., Ampuero, I., Vidal, L., Hoenicka, J., Rodriguez, O., Atares, B., Llorens, V., Gomez Tortosa, E., del Ser, T., Munoz, D. G. and de Yebenes, J. G. (2004). The new mutation, E46K, of α-synuclein causes parkinson and Lewy body dementia. *Ann. Neurol.*, 55, pp. 164–173.

Chapter 11

An Attempt to Treat Amyotrophic Lateral Sclerosis by Intracellular Copper Modification Using Ammonium Tetrathiomolybdate and/or Metallothionein: Fundamentals and Perspective

Shin-Ichi Ono,[a,b] Ei-Ichi Tokuda,[a,c] Eriko Okawa,[a,d] and Shunsuke Watanabe[a,e]

[a]*Laboratory of Clinical Medicine, School of Pharmacy, Nihon University, Funabashi, Chiba 274-8555, Japan*

[b]*Division of Neurology, Akiru Municipal Medical Center, Tokyo 197-0834, Japan*

[c]*Department of Medical Biosciences, Clinical Chemistry, Umeå University, Umeå, Sweden*

[d]*Department of Pharmacy, St. Marianna University School of Medicine Hospital, Kawasaki, Kanagawa, 216-8511, Japan*

[e]*Department of Pharmacy, International University of Health and Welfare, Atami Hospital, Atami, Shizuoka, 413-0012, Japan*

Mutation in superoxide diamutase1 (SOD1) is a cause of hereditary form of amyotrophic lateral sclerosis (ALS). Novel acquired toxicity (gain-of-function) is believed to play a crucial role. We propose that

Brain Diseases and Metalloproteins
Edited by David R. Brown

ISBN 978-981-4316-01-9 (Hardcover), 978-981-4364-07-2 (eBook)
www.panstanford.com

the nature of mutant SOD1 toxicity is disruption of intracellular Cu homeostasis. We provide evidences that copper transporters and chaperons are geared to accumulate Cu ion in the cells, and its excretion is downregulated with mutant SOD1 ("intracellular copper dysregulation" theory). Intracellular Cu modification using a Cu chelator and/or metallothionein resulted in a favorable outcome in an experimental study with a rodent model for hereditary form of ALS.

11.1 Introduction

Amyotrophic lateral sclerosis (ALS) is an adult-onset lethal neurodegenerative disorder that affects both corticospinal tract pathways (upper motor neurons) and α-motoneurons (lower motoneurons) (Bruijn *et al.*, 2004; Cleveland *et al.*, 2001; Julien, 2001; Rowland *et al.*, 1995; Rowland *et al.*, 2001; Strong *et al.*, 1996). ALS is characterized by affecting the upper and lower motor neurons selectively (Bruijn *et al.*, 2004; Cleveland *et al.*, 2001; Julien, 2001; Rowland *et al.*, 1995; Rowland *et al.*, 2001; Strong *et al.*, 1996). The clinical manifestations of ALS are highlighted by a creeping muscle weakness and muscle atrophy that usually begins in the upper limbs or bulbar muscles. The disorder is known colloquially as Lou Gehrig's disease in the USA. The more the upper motoneurons are affected, the more exaggerated is the deep tendon reflex and the more prominent is the planter reflex (Babinski sign). Meanwhile, fasciculation is the hallmark of spinal motoneuronal injury. The respiratory muscles are also involved in advanced stages of ALS, resulting in respiratory failure. Patients with ALS are usually bedridden in 3-to-4 years after onset and later develop pneumonia complications that often result in death from respiratory failure and/or a systemic infectious process. ALS has a prevalence of about 5 per 100,000 individuals (Bruijn *et al.*, 2004; Cleveland *et al.*, 2001; Julien, 2001; Rowland *et al.*, 1995; Rowland *et al.*, 2001; Strong *et al.*, 1996).

ALS was first described by Charcot in 1874. Despite medical advances during the past century, knowledge of the pathogenesis of this disease and its treatment have, frankly, not visibly progressed, and for years ALS has been ironically referred to as "the representative of intractable neurologic disorders". Two phenotypes of ALS are known: sporadic (non-genetic or non-familial), and familial (genetic). Approximately 90% of all ALS cases are non-familial (Bruijn *et al.*, 2004; Cleveland *et al.*, 2001; Julien, 2001; Rowland *et al.*, 1995; Rowland *et al.*, 2001; Strong *et al.*, 1996). Some pathogeneses have been proposed for sporadic ALS. These etiologies can be grossly divided into environmental causes, glutamate toxicity, and alternative theories such as an autoimmune mechanism or a special form of infectious process, as well as miscellaneous causes. The remaining 10% are familial cases that are caused by a single gene mutation. Familial ALS is expected to provide a clue to resolve the pathogenesis of ALS, even the sporadic forms.

11.2 Causes of ALS

11.2.1 *Environmental Causes (Heavy Metals, Minerals, and Pesticide Exposures)*

The incidence and prevalence of ALS vary little worldwide, with the exception of being 50-fold higher in Guam and considerably higher in the Kii peninsula of Japan (Steele *et al.*, 1990; Strong *et al.*, 1996; Yase 1972). ALS patients in Guam had a characteristic clinical manifestation: accompanying dementia (Kuzuhara *et al.*, 2001; Palato *et al.*, 2002). Soldiers who survived the Gulf War reportedly have a 2-fold higher incidence and prevalence of ALS (Horner *et al.*, 2008; Kasarski *et al.*, 2009; Miranda *et al.*, 2009; Johnson *et al.*, 2009). This epidemiologic evidence suggests a possible relation between ALS and environment factors, and exposure to pesticide use (especially organophosphates), heavy and/or toxic metals (e.g., mercury, lead, and arsenic) and/or mineral imbalances in the local soil have been proposed as possible environment causes (Yase 1972; Spencer *et al.*, 1987; Ahlskog *et al.*, 1995; Cox *et al.*, 2002). Curiously, these higher incidences and prevalences have gradually decreased and become more uniform (Chen *et al.*, 2002; Blazer *et al.*, 2008). While these aspects suggest that environment factors may be a possible underlying cause of the pathogenesis of ALS, no convincing results to support this theory have been published. Neurologists do not ever consider that heavy metals and/or minerals are implicated in ALS pathogenesis. While such theories were not uncommon in the history of ALS research, they have been somewhat forgotten in the last few decades.

11.2.2 *Viral Infection and Prion Disease*

Persistent viral infection, including poliovirus, had been proposed as a possible cause of ALS, but this speculation has not been confirmed (Berger *et al.,* 2000; Swanson *et al.,* 1995). Furthermore, the concept of an amyotrophic form of Creutzfeldt-Jakob disease, a representative prion disease, has been raised (Salazar *et al.,* 1983; Worrall *et al.,* 2000). However, prions are generally accepted to be an unlikely cause of ALS.

11.2.3 *Autoimmunity*

Some disorders mimic the clinical manifestations of sporadic ALS, including chronic inflammatory demyelinating polyneuropathy (CIDP), multifocal motor neuropathy with persistent conduction block (Lewis-Sumner syndrome), and others (Appel *et al.,* 1995). Sporadic ALS could reasonably be caused by an autoimmune mechanism, similar to these other disorders. Unfortunately, the ineffectiveness of immunosuppressive therapy, such as corticosteroids, gamma-globulins, plasma exchange, or immunosuppressive agents, makes an autoimmune mechanism unlikely (Vincent *et al.,* 1996).

11.2.4 *Genetic Causes*

Approximately 10% of ALS cases are familial, while the remaining cases are believed to be sporadic. In familial ALS, almost 20% of all cases are attributable to a point mutation in Cu/Zn superoxide dismutase (SOD1), which was first identified in 1993 (Rosen *et al.,* 1993). More than 100 different SOD1 mutations have been identified (Gaudette *et al.,* 2000; Andersen 2001). Apart from SOD1 mutations, other causative genes (ALS2 [encoding alsin], ALS3, ALS4, ALS6, ALS7, and ALS8) have been identified (Kunst 2004). Among the SOD1 mutations, the substitution of valine for alanine at position 4 (A4V) is the most common in the U.S.A. (Cudkowicz *et al.,* 1997). Another mutation, G93A, results in rapid disease progression and is the origin of a commercially available rodent model of familial ALS (described in detail in the section titled, "Characterization of G93A mutant SOD1 mouse"). Mutations in the SOD1 gene result, to some extent, in a reduction in SOD1 enzymatic activity; however, the magnitude of this reduction varies, and no correlation has been found between its enzymatic activity and longevity (Cudkowicz *et al.*, 1997). Knockout of the SOD1 gene in mice did not lead to the development of ALS-like symptoms (Reaume *et al.*, 1996; Bruijn *et al.*, 1998). In view of these observations, the loss or reduction of SOD1 enzymatic activity is unlikely to cause ALS pathogenesis. Instead, SOD1 mutations are generally accepted to gain a novel cytotoxic function (gain-of-function theory). The nature of this novel cytotoxic function remains unclear; thus, further elucidation of this function might lead to a better understanding of the pathogenesis of ALS and suggest novel pharmacological treatments, at least for familial ALS.

11.2.5 *Glutamate Toxicity*

Glutamate acts as an excitatory neurotransmitter at spinal motoneurons. Glutamate receptor activation allows calcium ions entry into motoneurons, but over-activation results in alterations in the cytosolic free calcium ion level, leading to motoneuronal death. The glutamate transporter EAAT2 controls the synaptic glutamate level. A decrease in glutamate EAAT2 has been reported in the affected regions (motor cortex and spinal cord) of non-familial ALS patients (Rothstein *et al.*, 1995). Based on these observations, the anti-glutamate agent riluzole has been used therapeutically (Bensimon *et al.*, 1994; Gurney *et al.*, 1996).

11.2.6 *Miscellaneous (Recent Hypotheses)*

Recent research has revealed that spinal motoneurons are more vulnerable to calcium influx than other neurons because of the dysediting of glutamate AMPA receptor subunit (GluR2) RNA (Kawahara *et al.*, 2003). D-serine has also been implicated in the pathogenesis of ALS (Sasabe *et al.*, 2007). A recent discovery of mutations in the DNA/RNA-binding proteins known as transactive response DNA-binding protein 43 (TDP-43) and fused in sarcoma (FUS) in familial ALS and the cytosolic sequestration of TDP-43 in sporadic ALS have opened new avenues in ALS research and have suggested that alterations in RNA processing may be involved in the pathogenesis of ALS (Neumann *et al.*, 2006; Arai *et al.*, 2006; Kwiatkowski Jr *et al.*, 2009; Vance C *et al.*, 2009).

11.3 Characterization of G93A Mutant SOD1 Mouse

A human mutant *SOD1* (G93A) was genetically transfected into a mouse strain (Gurney *et al.,* 1994) and is now commercially available from the Jackson Laboratory (Bar Harbor, ME). These mice develop motor paralysis beginning in their hind limbs at around 12 weeks of age (Gurney *et al.,* 1994; Ono *et al.,* 2006; Tokuda *et al.,* 2007a). The motor activity of the mice, as assessed using an infrared beam test, gradually decreases thereafter with aging, and the mice eventually die at 17–18 weeks of age (Gurney *et al.,* 1994; Ono *et al.,* 2006; Tokuda *et al.,* 2007a). Although this strain of mice is a rodent model of G93A SOD1-linked familial ALS, it has become the most popular rodent model of ALS.

Some potential mechanisms of pathogenesis have been proposed, as described above, but none of these hypotheses have adequately explained the intractable nature of the disease mechanism or have led to satisfactory therapeutics. Although this strain of transgenic mice is a rodent model of familial ALS, understanding how mutant SOD1 leads to selective motoneuron damage might also be useful for elucidating the pathogenesis of sporadic ALS.

11.4 Metallothionein, Copper Ions and ALS

Metallothionein (MT) is a low molecular weight, cysteine-rich, metal-binding cytoplasmic protein (Kägi *et al.,* 1988; Aschner *et al.,* 1997; Hidalgo *et al.,* 2001). Four isoforms of MT have been identified. The major isoforms, MT-I/MT-II, are genetically encoded in a coordinate manner and are ubiquitously expressed in all mammalian tissues including the CNS(Kägi *et al.,* 1988; Aschner *et al.,* 1997; Hidalgo *et al.,* 2001). MT-I/MT-II are upregulated by numerous stimuli including exposure to metals, chemical substances (steroid hormones, non-steroidal anti-inflammatory drugs, lipopolysaccharides, and so on), radiation exposure, and physical stress such as starvation and physical exercise (Kägi *et al.,* 1988; Aschner *et al.,* 1997; Hidalgo *et al.,* 2001; Ono *et al.,* 1998a, Hashimoto *et al.,* 2009). MT-I/MT-II are essentially believed to have beneficial roles in the detoxification of toxic heavy metals (i.e., mercury and cadmium), the regulation of essential trace element bioavailability (i.e., zinc and copper), and protection against reactive-oxygen species (ROS) (Kägi *et al.,* 1988;

Aschner *et al.*, 1997; Hidalgo *et al.*, 2001; Ono *et al.*, 1997; Ono *et al.*, 1998b). The other isoforms are rather tissue- and/or organ-specific; MT-III is found in the CNS and urogenital organs, and MT-IV is found in stratified epithelia (Uchida *et al.*, 1991; Quaife *et al.*, 1994). Unlike MT-I/MT-II, MT-III is not easily induced (Ono *et al.*, 2000, Ono *et al.*, 2007). MT-III is deeply involved in zinc storage (Kägi *et al.*, 1988; Aschner *et al.*, 1997; Hidalgo *et al.*, 2001). MT-III is likely induced in areas of critical tissue damage and is thought to be involved in tissue repair (Hozumi *et al.*, 1996; Hozumi *et al.*, 1998; Ono *et al.*, 2007); however, the definitive role of the MT-III isoform remains unclear. The expression of the fourth isoform, MT-IV, in the epithelia remains of uncertain significance. Because of the unique functions of MTs, the roles of MTs in human diseases have attracted attention (Simpkins 2000).

Considering the classic function of MT-I/MT-II, an examination of the possible role of MT-I/MT-II in ALS pathogenesis might challenge researchers to refocus on a forgotten theme. Sillevis Smitt *et al.* (1992a; 1994) first reported an increase in MT proteins (MT-I/MT-II) in the spinal cords of patients with ALS. MT immunoreactivity was prominent in the nucleus and cytoplasm of astrocyte glias (Sillevis Smitt *et al.*, 1992a; Sillevis Smitt *et al.*, 1994). Blaauwgeers *et al.* (1996) confirmed their observations by examining the mRNA level. Curiously, Hozumi *et al.* (2008) found a decrease in MT-I/MT-II protein in postmortem spinal cords.

Koto *et al.* (1997) found that hyaline inclusions in the spinal cords of familial ALS patients were positive for MT-I immunostaining. Although the discrepancy in MT changes should be resolved, we cannot offer a reasonable answer at this time. In the frontal cortex, which is another region affected by ALS, MT-I mRNA expression was reportedly upregulated (Lederer *et al.*, 2007).

Puttaparthi *et al.* (2002) found that the crossing of an ALS model mouse (G93A SOD1 mouse) with an MT-I/MT-II or MT-III gene knockout mouse resulted in a reduced survival and an accelerated disease onset and disease progression. Nagano *et al.* (2001) also generated a double transgenic mouse (an ALS model mouse [G93A SOD1 mouse] with an MT-I/MT-II gene knockout mouse) and showed similar results to those reported by Puttaparthi.

Concerning the MT-III isoform, Ishigaki *et al.* (2002a) reported a decrease in spinal MT-III mRNA expression in sporadic ALS patients. Hozumi *et al.* (2008) observed a reduction of MT-III immunoreactivity in the spinal cords of sporadic ALS patients who required a respirator. In contrast, Gong *et al.* (2000) and Olsen *et al.,* (2001) demonstrated an increase in MT-III in mutant SOD1 mice. While the discrepancy in the MT-I/MT-II or MT-III expression level should be elucidated, MT-I/MT-II and MT-III are likely associated with ALS pathogenesis and might have a potentially therapeutic role (Hozumi *et al.,* 2004; Tokuda *et al.,* 2007a).

A few reports have examined the copper ion levels in ALS patients, with descriptions of an increase in the frontal lobe (Gellein *et al.,* 2003) and a decrease in the serum (Domzal *et al.,* 1983). However, these investigations were performed sporadically and on a case-report basis. No integrated studies examining the significance of MT-I/MT-II or MT-III and copper and/or zinc to ALS pathogenesis have been conducted (Carrì *et al.,* 2001). Considering the conventional functions of MT, the previously mentioned changes in MTs, and the changes in copper ion levels, the forgotten themes of ALS research should be reconsidered. Of note, the generation of double transgenic mice by crossing a strain of mutant SOD1 mice with a rodent model of Menkes disease (a congenital copper deficient disorder) unexpectedly resulted in a longer survival period than that of either of the original transgenic mice (Kiaei *et al.,* 2004). A hybrid of two different strains of mice that has a longer survival period than the original strains is a remarkable finding. Therefore, we systematically examined the possible relation between MT and copper ions in the pathogenesis of ALS using a mouse model of ALS, the most popular animal model of ALS.

11.4.1 *Changes in MT mRNA and Protein Levels in the Spinal Cord*

The spinal cord is an affected region of ALS, since the lower motor neurons originate in the spinal cords. MT-I mRNA (Ono *et al.,* 2006) and protein expression (Tokuda *et al.,* 2007a) were already significantly upregulated in mutant SOD1 mice at as early as 4 weeks, when motor paralysis was not yet evident. Thereafter MT-I protein level exhibited an age-dependent increase and remained significantly elevated at 16 weeks. Although the MT-III mRNA and protein levels were not altered at an early stage (8 weeks of age), they were also elevated at advanced stages (16 weeks), and the

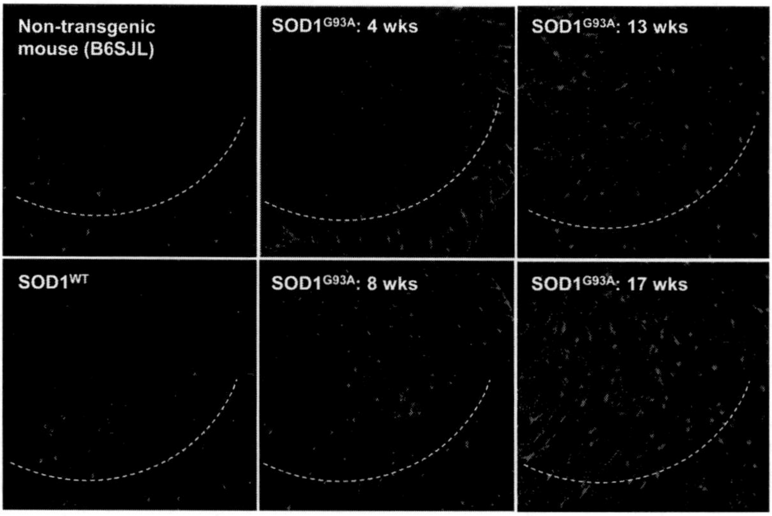

Figure 11.1 Immunodetection and confocal imaging showed spinal metallothionein-1 protein expressed was increased in an age dependent manner in the anterior horn. See also Colour Insert.

increase in the protein level was significant (Ono *et al.*, 2006; Tokuda *et al.*, 2007a).

An immunohistochemical analysis revealed that MT-I protein expression increased in an age-dependent fashion. Of note, MT-I protein expression and its changes were prominent in the anterior horn, where the motoneurons originate (Fig. 11.1). These changes in the MT mRNA and protein levels were not observed in a non-responsible region, the cerebellum (used as a control). Therefore, the exclusive changes in the MT mRNA and protein levels in the spinal cord strongly suggest a close association between ALS pathogenesis and MTs.

11.4.2 *MT Changes in Other Organs*

In humans, Sillevis Smitt *et al.* (1992b) first reported that MT-I was increased in the liver and kidney as well as in the spinal cord. In mutant SOD1 mice, we observed that MT-I mRNA expression was significantly increased in the liver and kidney not during an advanced stage (16 weeks), but during a pre-symptomatic stage (8 weeks old). No change in the MT-I mRNA level was found in the pancreas at either a pre-symptomatic or an advanced stage (Ono *et al.*, 2006).

The liver and kidney play central roles in metabolism, detoxification, and secretion. From the stand-point of the classic MT functions, these findings suggest that ALS might be attributable to unknown environmental causes. The CNS is vulnerable to both xenobiotic and/or endogenous toxic substances. Damage and injury to motoneurons might appear first because of the vulnerability of the CNS. This hypothesis would seem to make ALS a selective motor neuron disease. In other words, ALS might be a systemic disorder, with motor paralysis the most prominent and/or earliest clinical manifestation.

11.4.3 *Copper Ion Changes and SOD1 Enzymatic Activity in the Spinal Cord*

Copper ion concentrations were measured using ICP-MS (inductively-coupled plasma mass spectrometry). As shown in Fig. 11.2, the spinal Cu levels increased in an age-dependent manner (Tokuda *et al.*, 2009a) as did MT-I. Our findings were in agreement with the previous reports, showing an elevation of Cu in the spinal cords (Li *et al.*, 2006; Ahtoniemi *et al.*, 2007). In the present study, we used two other strains of mice as controls:

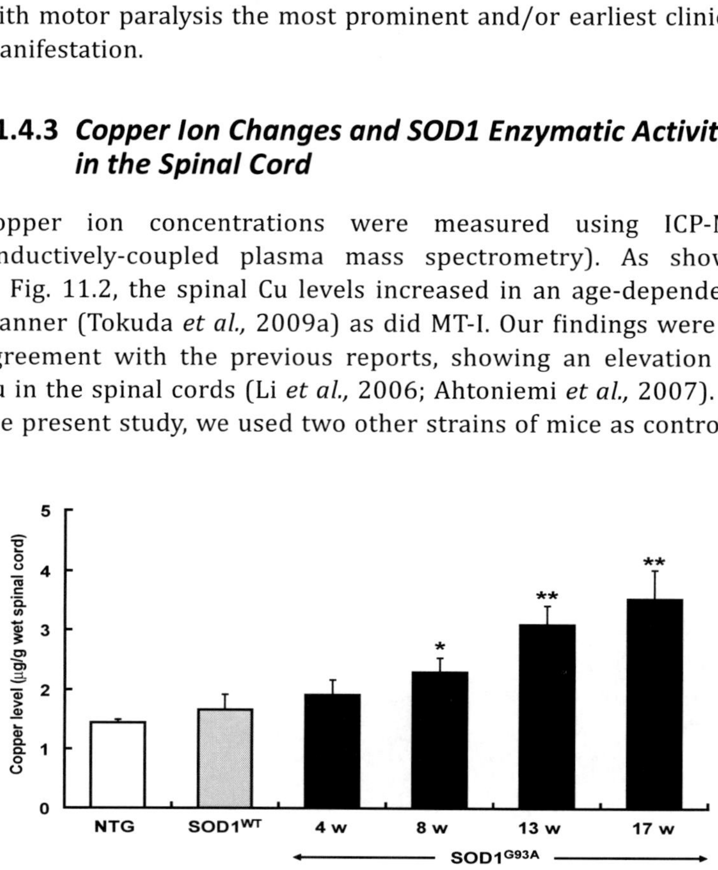

Figure 11.2 Spinal copper ion levels measured by inductively coupled plasma mass spectroscopy (ICP-MS) were increased in an age dependent manner in mutant SOD1 mice. NTG – non transgenic mice, $SOD1^{WT}$ – wild-type SOD1 mice, $SOD1^{G93A}$ – mutants SOD1 mice, W – weeks of age. * $p < 0.05$, ** $p < 0.01$.

wild-type SOD1 transgenic (SOD1^{WT}) mice and non-transgenic (SOD1^{WT}) mice (B6SJL strain). The SOD1^{WT} mice were used as a control for the mutant SOD1 mice. The NTG mice were used as a control for the two transgenic strains of mice irrespective of the wild-type or mutant status, since these transgenic mice carry an unphysiologically high copy number of human SOD1 genes. This might have affected the Cu levels in the SOD1^{WT} mice. The absence of an increase in Cu ions in the NTG mice eliminated the possibility that the Cu increase was not brought about by the unphysiologically high copy number of the SOD1 gene, but was a characteristic finding of the mutant SOD1 gene. We performed Rhodanine staining to identify cell populations in the spinal cord, but failed to observe any notable populations. The sensitivity of Rhodanine staining might not be sufficient to visualize Cu ion deposits for this purpose, since the lowest visualization limit is on the order of 100 μg/g of tissue (Watanabe *et al.*, 2001). The form of the Cu ions (free or bound) might be another reason for the lack of staining.

Higher Cu levels were probably associated with a higher SOD1 enzymatic activity level, since Cu ions are essential for the chemical composition and enzymatic activity of SOD1. Unexpectedly, despite the fact that the protein expression level of SOD1 in SOD1^{WT} mice was much higher than that in NTG mice, the Cu ion level in the SOD1^{WT} mice was not proportionally elevated and instead remained at the same level as that observed in the NTG mice (Tokuda *et al.*, 2009a). Because Cu is essential for SOD1 activity, a discrepancy between the SOD1 expression level and the Cu level in SOD1^{WT} seemed to exist. Therefore, we measured the SOD1 enzymatic activity in the spinal cord of mice. SOD1 activity and the human SOD1 protein expression level were well correlated in both the SOD1^{WT} and mutant SOD1 mice (Fig. 11.3) (Tokuda *et al.*, 2009a).

These results suggest that the unphysiological overexpression of human wt SOD1 did not affect spinal Cu homeostasis. In other words, mutant SOD1 probably has a higher affinity to Cu ions, resulting in an increase in the spinal Cu ion levels (Figs. 11.2 and 11.3). Along with an age-dependent increase in MT-I, the present observations of the changes in Cu ion concentration provide an important clue to the pathogenesis of ALS.

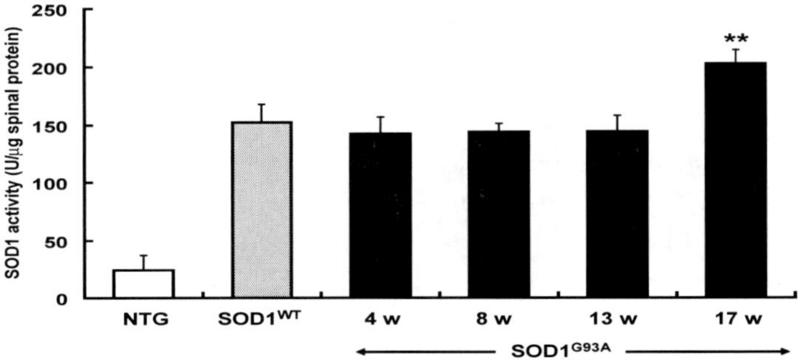

Figure 11.3 Spinal SOD1 activity was measured by a tetrazolium blue method. Irrespective of an age-dependent increase in spinal Cu levels in mutant SOD1 mice, SOD1 activity was almost at the same level as that in wild-type SOD1 mice, except at 17 weeks of age. NTG – non transgenic mice, $SOD1^{WT}$ – wild-typeSOD1 mice, $SOD1^{G93A}$ – mutants SOD1 mice, W – weeks of age. $^{**}p < 0.01$.

11.4.4 *Cu Ion Regulation in Cells*

Copper is an essential trace element that can form the backbone of chemical structures and is used as a co-factor to enable chemical reactions (Prohaska *et al.,* 2004). On the other hand, Cu overload is harmful to cells, since it may work as a pro-oxidant to accelerate oxidative stress and/or act as a ligand for the Fas receptor in the initiation of the apoptotic process (Levenson 2005).

While the intracellular Cu concentration is tightly maintained, the trafficking pathway of Cu ions remains poorly understood (Choi *et al.,* 2009; Linder *et al.,* 1996). Briefly, Cu ions are reduced by Steap2 (six transmembrane epithelial antigen of prostate 2) and enter the cells; only Cu^+ is available inside the cells (Ohgami *et al.,* 2006; Kuo *et al.,* 2001; Lee *et al.,* 2001). Ctr1 (copper transporter 1) acts as the gate to the cell. CCS (copper chaperone for SOD1), Cox17 (copper chaperone for cytochrome c oxidase), and Atox1 (antioxidant protein 1) work as transporters and chaperones to deliver Cu ions to destinations inside the cell (Larin *et al.,* 1999; Wong *et al.,* 2000; Takahashi *et al.,* 2002). In the CNS, Atp7a is a key enzyme in the excretion of Cu ions. The Atp7a N-terminus (Atp7a(N)) acts as a receiver of Cu ions (Lutsenko *et al.,* 1997), and the translocation of the C-terminus (Atp7a(C)) to the plasma

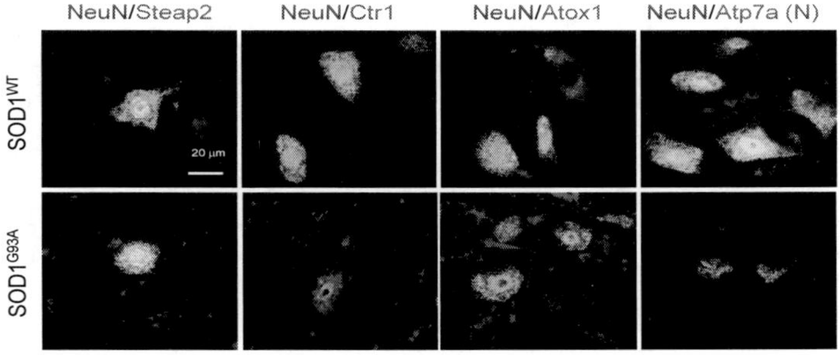

Figure 11.4 Confocal images revealed that up- and downregulation of copper transporters and chaperones were taking place in spinal motoneurons. SOD1^{WT} – wild-typeSOD1 mice, SOD1^{G93A} – mutants SOD1 mice, NeuN – marker for neurons by immunohistochemical staining, Steap2, six transmembrane epithelial antigen of the prostate, Ctr1 – copper transporter, Atox1 – antioxidant-1, a copper chaperone. See also Colour Insert.

membrane is necessary to excrete Cu ions (Petris *et al.,* 1996; Petris *et al.,* 1998) (See also Fig. 11.10). Therefore, we investigated the status of these Cu transporters and chaperons in SOD1 mice.

These transporters and chaperones (Steap2, Ctr1, CCS, and Cox17) were all upregulated with disease progression in an age-dependent fashion in mutant SOD1 mice, but not in SOD1^{WT} mice or NTG mice (the uniform upregulation of the CCS level was probably secondary to a high copy number of SOD1 in both SOD1^{WT} and mutant SOD1 mice). In contrast, the Atp7a(N) level was downregulated, suggestive of failure to receive Cu ion. Evidently, these changes promoted the accumulation of Cu ion within the cells. Moreover, these transporters and chaperons were mainly expressed in spinal motoneurons (Fig. 11.4). In another term, intracellular Cu ion accumulation takes place in spinal motoneurons.

Of note, these transporters and chaperones (Steap2, Ctr1, and Atox1), which play a crucial role in Cu influx and outflow, were co-aggregated with mutant SOD1, but not with SOD1^{WT} (Fig. 11.5 upper panel). The co-aggregation of these proteins might lead to the dysfunction of the proteins. Regardless of the accumulation of Cu ions inside the cells, the translocation of Atp7a (C) did not occur (Fig. 11.6). As expected, Atp7a(C) was also co-aggregated with mutant SOD1 (Fig. 11.7). Along with the decrease in the Atp7a(N)

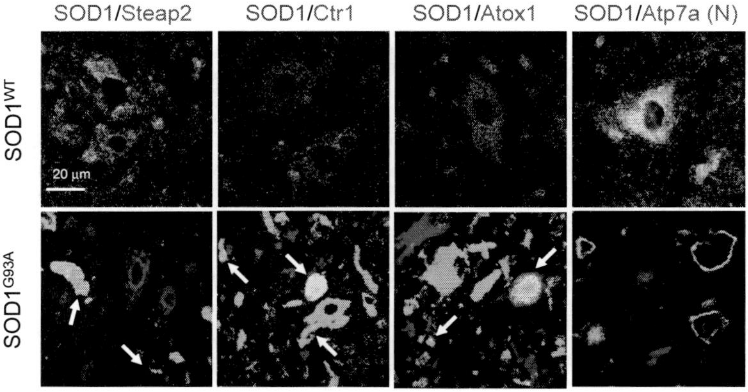

Figure 11.5 Co-aggregation with mutant SOD1 (Steap 2, Ctr1, Atox1) suggests dysfunction of these Cu transporters and chaperones in mutant SOD1 mice. See also Colour Insert.

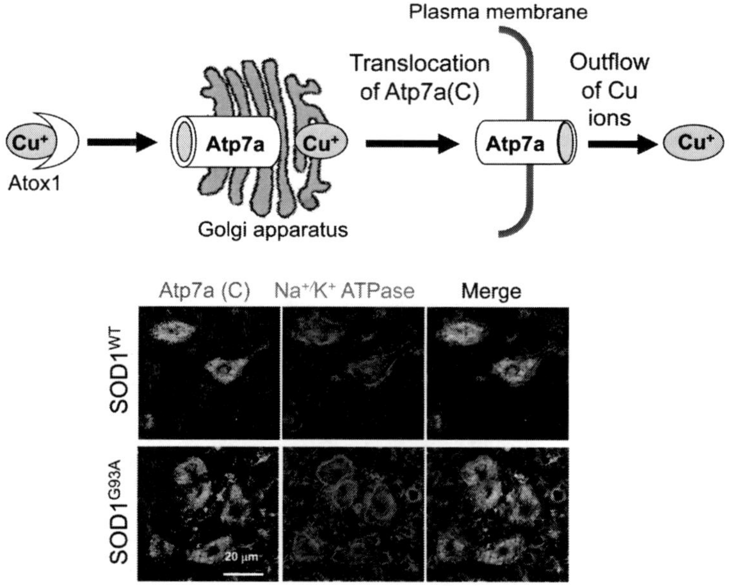

Figure 11.6 In response to Cu ions accumulation inside the cell, translocation of Atp7a(C) toward plasma membranes occurs. This enables Cu ions efflux from the cells and regulates intracellular Cu ions level (upper illustrations). However, no translocation of Atp7a(C) was observed in SOD1 G93A, suggesting that this protein did not excrete Cu ions, regardless of Cu overload (lower panel). Atp7a(C) – Atp7a C-terminus, Na^+/K^+ ATPase is a surrogate marker for plama membrane. See also Colour Insert.

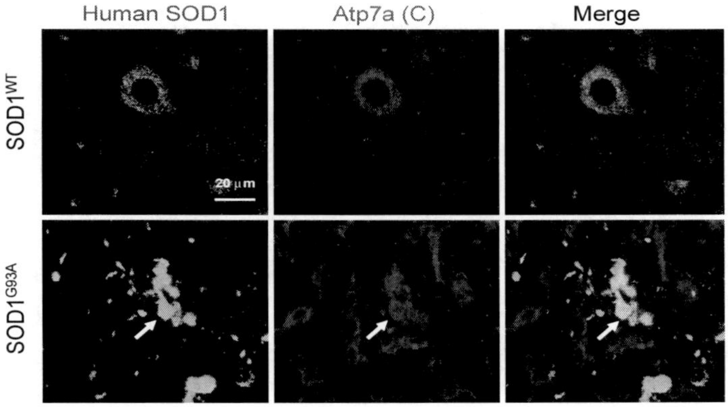

Figure 11.7 Confocal images revealed that Atp7a (C) was co-aggegated with mutant SOD1 but not with wild-type SOD1. This suggests that translation of Atp7a was disturbed by intracellular copper accumulation. $SOD1^{WT}$ – wild-typeSOD1 mice, $SOD1^{G93A}$ – mutants SOD1 mice. See also Colour Insert.

expression level, this status resulted in the further accumulation of Cu ion. These findings could explain the selective motoneuronal degeneration in ALS.

SOD1 has four histidine residues (His46, His48, His63, and His120) to which Cu ions are ligated (Valentine *et al.,* 2005). The deletion of all four of these histidine residues or the knockout of CCS, which inserts Cu ions into the histidyl site of SOD1, did not inhibit the development of ALS-like symptoms in mutant SOD1 mice (Wang *et al.,* 2003; Subramaniam *et al.,* 2002). Another Cu-binding site, a cysteine residue at position 111 (Cys111) that is located on the surface of SOD1, has been identified; this site does not contribute to the enzymatic activity of SOD1 and has been shown to play a crucial role in the interaction between Cu ions and mutant SOD1 (Li *et al.,* 2000; Watanabe *et al.,* 2007). Cys111 might be associated with the higher affinity of mutant SOD1 to Cu ions (described in the section entitled, "Copper ion changes and SOD1 enzymatic activity in the spinal cord") and its co-aggregation with Cu chaperons and transporters. Kishigami *et al.* (2010) recently reported that the modification of Cys111 increased the affinity of monomeric SOD1 to Cu ions. Along with the higher affinity of mutant SOD1 to Cu ions, the present observations are further supported by the presence of Cu ion deposits in spinal motoneurons.

It seems reasonable and acceptable that sufficiently large Cu ion deposits would provoke oxidative stress and/or the apoptotic process in spinal motoneurons (described in "Cu ion regulation in the cell"). In other words, the dysregulation of the Cu trafficking pathway provoked by mutant SOD1 might explain the nature of SOD1 cytotoxicity (i.e., its gain-of-toxic function). We would like to propose a hypothesis; "intracellular Cu dysregulation" as a major role in the pathogenesis of mutant SOD1-linked ALS.

11.5 Cellular Damage by Copper Overload

Although we were unable to identify cellular populations in the spinal cords, we provided evidence to support the hypothesis that a process leading to apoptotic cell death is accelerated in both a mitochondrial pathway via caspase-9 and a Fas pathway via caspase-8 as well as a final common pathway via caspase-3 (Tokuda *et al.,* 2007b). The apoptotic process is balanced between caspases and IAPs (inhibitor of apoptosis proteins). We found that the protein expression of an IAP, survivin, was already decreased before the onset of motor paralysis (8 weeks) in the spinal cords of mutant SOD1 mice.

Survivin is thought to have a crucial role in tissue regeneration and repair (Deguchi *et al.,* 2002; Blanc-Brude *et al.,* 2002). The protein expression of another IAP, XIAP (X-linked inhibitor of apoptosis protein), was decreased at an advanced stage (16 weeks), but not at a pre-symptomatic stage (8 weeks). These results are nearly consistent with previous reports (Inoue *et al.,* 2003; Ishigaki *et al.,* 2002b; Yoshihara *et al.,* 2002; Wootz *et al.,* 2006). These results were obtained exclusively in the spinal cord, and not in cerebella (as a control region). Thus, these animal experiments imply that the dysequilibrium of the apoptotic process and its inhibitors is involved in the pathogenesis of ALS. Interestingly, XIAP plays an important role in intracellular Cu ion homeostasis (Burstein *et al.,* 2004), and its level was greatly reduced in Wilson's disease, a representative congenital metabolic disorder of Cu metabolism (Mufti *et al.,* 2006). The excess level of intracellular Cu ions results in a conformational change in XIAP, leading to its degradation and a subsequent decrease in its ability to inhibit caspase activities (Mufti *et al.,* 2006). A higher level of Cu ion might trigger the Fas pathway of apoptosis, since Cu ions act as a ligand for Fas (Levenson, 2005). Taking these findings into account, the dysequilibrium of the apoptotic process could be another result of Cu ion accumulation in cells, presumably motoneurons. However, no evidence of such an apoptotic process has been reported in humans with ALS.

Cu overloading increases the risk of oxidative stress via a Fenton-like reaction (Carrì *et al.,* 2001). We also observed that the production of lipid peroxides (LPOs), measured as malondialdehyde substances (MDARS), was significantly increased in the spinal cord of mutant SOD1 mice at an advanced stage of disease (16 weeks) (Tokuda *et al.,* 2007a). 8-hydroxy-2'-deoxyguanosine (8-OHdG), a marker for DNA damage, was translocated from mitochondrial fraction (original position) to cytosol fraction at advanced stages, suggesting the presence of DNA damage and dysfunction of its repair (Tokuda *et al.,* 2007c; Warita *et al.,* 2001). While no cellular populations have been identified as of yet, Cu ion-mediated oxidative stress and DNA damage might be involved in neuronal degeneration in ALS.

11.6 Therapeutic Strategy in Mutant SOD1 Mice Based on "Intracellular Cu Dysregulation" Hypothesis: Intracellular Copper Removal Using Ammonium Tetrathiomolybdate

Based on our hypothesis that the dysregulation of the Cu trafficking pathway is involved in the pathogenesis of ALS, Cu removal from the spinal cord might be beneficial in terms of the survival period, disease progression, and disease duration. We conducted an experimental treatment using ammonium tetrathiomolybdate (ATTM) in mutant SOD1 mice. ATTM is a selective Cu chelator, can pass through the blood-brain barrier, and has the ability to remove both intracellular and extracellular Cu ions (Ogura *et al.,* 1996; Ogura *et al.,* 1998; Ogura *et al.,* 1999). The present animal experiment was designed as a placebo-controlled, randomized blind study corresponding to a randomized placebo-controlled trial (RCT), the most reliable type of clinical trial. A group of 62 mice (either NTG or mutant SOD1) was allocated for the survival study and another group of 32 SOD1 mice (either NTG or mutant SOD1) was assigned to a biochemical analysis. A daily intraperitoneal dose of ATTM (5 mg/kg of body weight) was administered to the mice beginning at an age of 4 weeks (pre-symptomatic stage). We defined the end-point as the inability of the mouse to right itself within 30 seconds after being pushed onto its side (a surrogate end-point to its death) (Gurney *et al.*, 1996). The identities of the SOD1 mice were revealed at the end-point. The ATTM administration significantly delayed the progression of weight loss and motor performance in the mutant SOD1 mice group (Tokuda *et al.*, 2008a). The onset of disease (19.8%), survival (24.6%), and progression of disease (42.3%) were all significantly improved, as indicated in Figs. 11.8a–c. Unlike ATTM, the efficacy of D-penicillamine and trientine, which are incapable of removing Cu ions inside cells or from histidyl sites, was limited and increased the survival period by only 5–8% (Andreassen *et al.*, 2001; Hottinger AF, *et al.*, 1997; Nagano *et al.*, 2003). Riluzole, the sole drug approved for therapeutic use in patients with ALS, provided a modest increase (9–10%) in the survival period of the mutant SOD1 mice (Bensimon *et al.*, 1994; Gurney *et al.*, 1996). Of note, the efficacy of ATTM was about 2-fold higher than that of riluzole.

ALS is diagnosed based on its clinical manifestations and supplementary investigations, such as electromyography, and by ruling out other diseases that share its clinical characteristics, since there is no biomarker to support a diagnosis of ALS as there are for other neurodegenerative disorders. Of note, ATTM administration started even after the onset of motor paralysis resulted in similar results to those obtained when administration was started before the onset of motor paralysis (Ono *et al.*, 2009).

Also of note, ATTM treatment inhibited SOD1 enzymatic activity by up to 50% in mutant SOD1 mice (Fig. 11.9). However, the mutant SOD1 protein expression level was not altered in mutant SOD1 mice after ATTM administration (Fig. 11.9).

ATTM is known to remove Cu ions in a coordinated manner by acting on cysteine first and then on histidine (Ogura *et al.*, 1998). ATTM administration resolved SOD1 oligomer formation in vitro in mutant SOD1 (Tokuda *et al.*, 2008b). ATTM likely ameliorated the ALS-like symptoms of mutant SOD1 mice by chelating Cu ions from the Cys111 site, and not by suppressing SOD1 enzymatic activity. The present results provide strong evidence that Cys111 is relevant to Cu neurotoxicity ("intracellular Cu dysregulation").

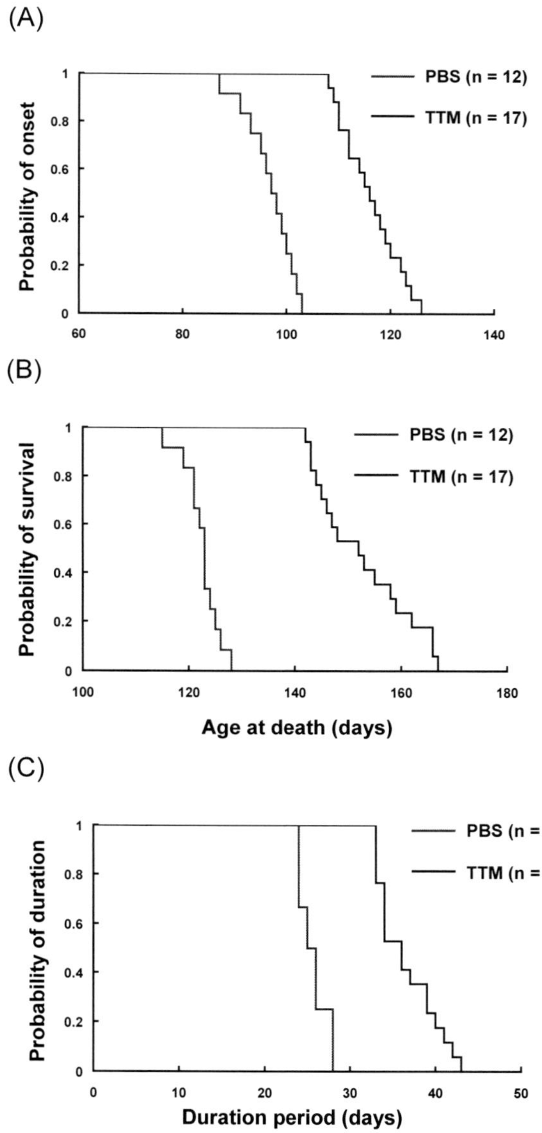

Figure 11.8 Ammonium tetrathiomolybdate administration significantly delayed onset (96.8 ± 4.9 versus 116.1± 5.6 days, $p = 1.06 \times 10^{-7}$) (A) prolonged survival (112.5 ± 3.3 versus 152.7 ± 8.8 days, $p = 2.48 \times 10^{-7}$) (B) and slowed progression of disease (25.7 ± 1.6 versus 35.6 ± 3.5 day, $p = 1.49 \times 10^{-6}$). TTM = ammonium tetrathiomolybdate, PBS = phosphate buffered saline. See also Colour Insert.

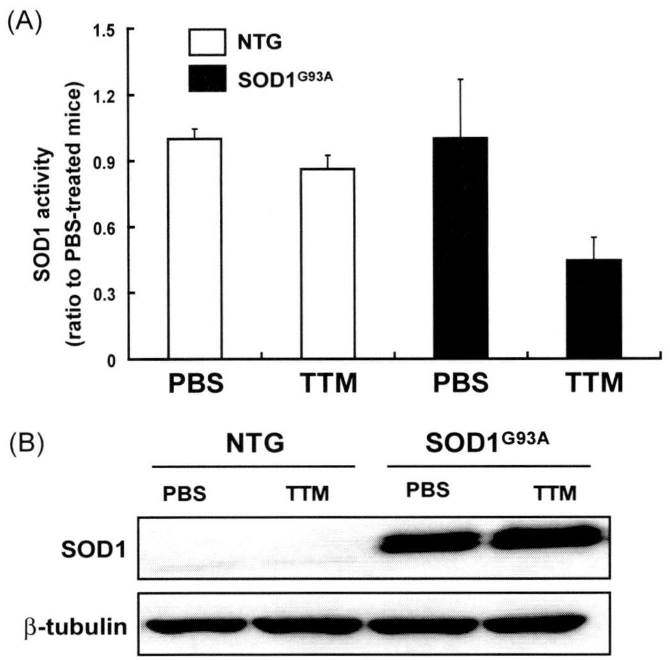

Figure 11.9 Ammonium tetrathiomolybdate administration inhibited SOD1 enzymatic activity up to 50% (A) without affecting its protein expression (B). PBS- phosphate-buffered saline, TTM – ammonium tetrathiomolybdate, non transgenic mice, SOD1^{G93A} – mutants SOD1 mice.

11.7 Therapeutic Strategy in Mutant SOD1 Mice Based on "Intracellular Cu Dysregulation" Hypothesis: Intracellular Copper Modification Using a Metallothionein-I Isoform

The excellent outcome of copper removal using ATTM confirmed the presence of "intracellular copper dysregulation". The "intracellular copper dysregulation" hypothesis seems to explain the pathogenesis of mutant SOD1-linked ALS. Taking the observation that MT-I gene knockout in mutant SOD1 mice resulted in a shorter survival period (Nagano *et al.*, 2001; Puttaparthi *et al.*, 2002) and the classic functions of MT-I into account, the MT-I isoform appears to have a protective function and might be beneficial for the treatment of ALS. We first attempted to generate double transgenic mice by crossing a transgenic strain of MT-I overexpressing mice (MT-I*, The Jackson Laboratory, Palmiter *et al.,* 1993) with mutant SOD1 mice to evaluate the effect of congenital MT-I overexpression. Of note, previous authors have pointed out that the mouse strain itself could influence the survival of mutant SOD1 mice. The decline in motor performance, as measured using the Rota-rod test, was significantly slowed in mutant SOD1/MT-I* mice, compared with in mutant SOD1 mice. The survival period was also significantly prolonged in the mutant SOD1/MT-I* mice (Tokuda *et al.,* 2009b).

Based on these favorable results, we next examined acquired MT-I/MT-II induction using a chemical agent. MT-I/MT-II can be easily induced by various stimuli including chemical, physical, emotional, irradiation, and physical exertion (Ono *et al.,* 1998a, Ono *et al.,* 2007; Hashimoto *et al.,* 2009). Of these possible stimuli, dexamethasone (Dex) is a popular chemical inducer of MT-I/MT-II (Palmiter *et al.,* 1992; Méndez-Armenta *et al.,* 2003). An intraperitoneal daily injection of Dex (2 mg/kg of body weight) was administered to the mice every 2 days beginning at an age of 4 weeks (pre-symptomatic stage) significantly delayed onset by 9.8%, prolonged survival period by 16.4%, and slowed disease progression by 36.7% (Figs. 11.10a-c) (Watanabe *et al.,* 2008). An immunohistochemical analysis revealed that MT-I/MT-II was prominently induced in glial cells, but not in motoneurons, in the anterior horn. Of note, MT-I/MT-II induction after disease onset prolonged the survival period by 11.7% (120 ± 2.7 days vs. 134 ± 2.8 days) and slowed the progression of the disease by 88% (25 ± 2.7 days vs. 47 ± 6.8 days) (Watanabe *et al.,* 2008; Ono *et al.,* 2009).

We also generated another double transgenic mouse crossing a transgenic strain of MT-I/MT-II knockdown (MT-null) mice with mutant SOD1 mice to confirm the above effect of MT-I/MT-II induction. In those mice, an intraperitoneal daily injection of Dex (2 mg/kg of body weight), beginning at an age of 4 weeks, altered none of the outcomes (disease onset, survival period, and rate of progression) (Watanabe *et al.,* 2008; Ono *et al.,* 2009). These results verified that the excellent outcomes by Dex administration were brought by MT-I/MT-II themselves, but not by Dex.

In view of the classic function of MT-I/MT-II, it is reasonable and acceptable that excess amounts of copper ion inside cells were captured by induced MT-I/MT-II as ATTM administration removed.

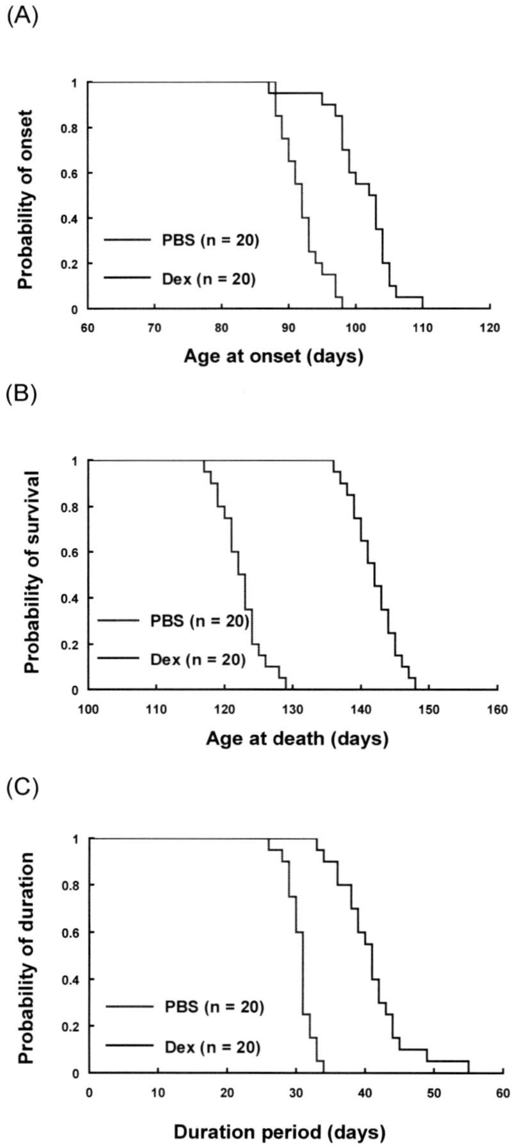

Figure 11.10 Metallothionein-1 induction by dexamethasone administration significantly delayed onset (92 ± 3.1 versus 101 ± 4.9 days, $p = 1.42 \times 10^{-6}$) (A), prolonged survival (122 ± 2.9 versus 142 ± 3.2 days, $p = 1.29 \times 10^{-8}$) (B) and slowed progression of diseases (30 ± 1.8 versus 41 ± 5.1 days, $p = 1.43 \times 10^{-7}$) (C) Dex = dexamethasone, PBS = phosphate buffered saline. See also Colour Insert.

Several lines of evidence have shown that MT has antiapoptotic properties (Wang *et al.*, 2001; Kang *et al.*, 2003; Gurel *et al.*, 2007). We also observed antiapoptotic effects of MT-I/MT-II in spinal cords of mutant SOD1 mice (Tokuda *et al.*, 2006).

The expression of mutant SOD1 in motoneurons is reportedly required for the development of this disease, while its expression in glial cells contributes to disease progression (Boilles *et al.*, 2006; Clement *et al.*, 2003; Yamanaka *etal.*, 2008). The effect of Dex on disease progression was greater than on survival in the present study. One possible explanation for this difference might be that MT-I/MT-II is primarily expressed and induced in glial cells (Kägi *et al.*, 1988; Aschner *et al.*, 1997; Hidalgo *et al.*, 2001). Intriguingly, MT-I induction enhanced mutant SOD1 solubility *in vitro* (Tokuda *et al.*, 2009b). A reduction in the aggregation of mutant SOD1 is expected *in vivo*.

No effect of corticosteroids (Werdelin *et al.*, 1990) or immunosuppressive agents (Vincent *et al.*, 1996) on ALS has been reported. Humans are exposed to various stimuli during their lifetimes. In humans, MT proteins are likely induced to a considerable level in response to various stimuli during normal aging. In view of this aspect, the administration of an exogenous MT-I isoform, such as recombinant MT-I administration might be worthwhile regardless of the negative clinical results.

11.8 Conclusion

A recent promising clinical trial examining minomycin resulted in disappointment (Gordon *et al.*, 2007). Although lithium carbonate previously provided an excellent outcome (Fornai *et al.*, 2008), recent examinations have failed to reproduce the same efficacy (Gill *et al.*, 2008). Treatment with edaravone, an ROS scavenger (Ito *et al.*, 2008), hepatocyte growth factor (Ishigaki *etal.*, 2007), ceftriaxone (Rothstein *et al.*, 2005), sodium valproate (Piepers *et al.*, 2009), and a combined therapy with lithium carbonate and sodium valproate (Feng *et al.*, 2008) all exhibited a promising efficacy in mutant SOD1 mice.

Regardless of these discoveries in animal models, which might be relevant only to the SOD1-linked familial form, the pace of therapy for ALS remains frustratingly slow. According to the recommendation of the American Academy of Neurology, candidate

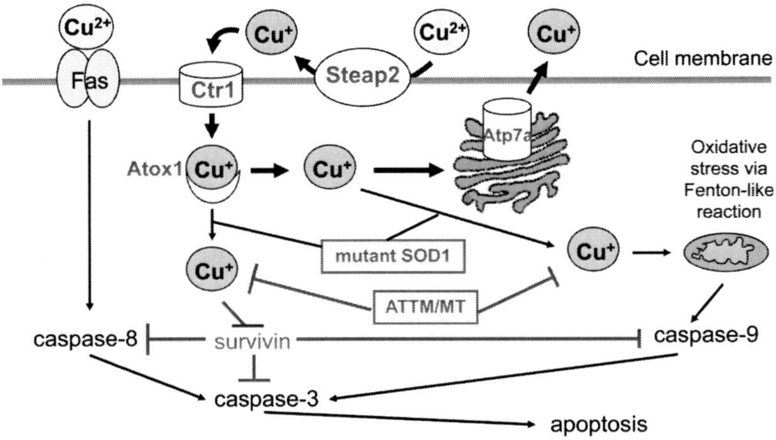

Figure 11.11 In normal conditions intracellular Cu ion concentration is maintain in a narrow range by Cu transporters (Ctr1) and chaperones such as Atox1 and Atp7a (bold arrows). Mutant SOD1 disturbs this regulation by co-aggregating with these transporters, chaperones and Atp7a. This increases intracellular Cu accumulation. Mutant SOD1 has higher affinity for Cu ions than wild-type SOD1. Excess Cu ions mediate fention like reactions and facilitate apoptosis via the mitochondrial pathway (caspases 9) and suppress antioxidant proteins (such as surviving). Extracellular Cu^{2+} activates caspase-8 via the Fas receptor. Ammonium tetrathiolmolybdate (ATTM) and metallothionein (MT) chelate or capture excess amounts of intracellular Cu to inhibit intracellular Cu dysregulation. See also Colour Insert.

drugs for the treatment of ALS should fulfill the following criteria: a favorable preclinical outcome, low toxicity, and good CNS penetration (Traynor *et al.,* 2006). The excellent results obtained using ATTT for copper removal and the administration of MT-I for Cu modification seem to validate the "intracellular Cu dysregulation" theory, and vice versa. Intracellular copper control seems to be a rational therapeutic strategy (Fig. 11.11). Our observations may be sufficient to expect that MT-I could modify the course of ALS.

Acknowledgements

Figures 2, 3, 4, 5, 6 and 7 were reproduced with or without modification from *Journal of Neurochemistry* 111, 181–191, 200 (Tokuda *et al.*) with permission (license No. 2391781085664). Figures 7, 8 and 9 were reproduced with or without modification from *Experimental Neurology* 213, 122–128, 2008 (Tokuda *et al.*) with permission (license No. 2366220502242).

Present work was partially supported by Grants-in-aid for scientific research #21659222 in Japan.

References

Ahlskog, J. E., Waring, S. C., Kurland, L. T., Petersen, R. C., Moyer, T. P., Harmsen, W. S., Maraganore, D. M., O'Brien, P. C., Esteban-Santillan, C., and Bush, V. (1995). Guamanian neurodegenerative disease: Investigation of the calcium metabolism/heavy metal hypothesis. *Neurology* 45, pp. 1340–1344.

Ahtoniemi, T., Goldsteins, G., Keksa-Goldsteine, V., Malm, T., Kanninen, K., Salminen A., and Koistinaho, J. (2007). Pyrrolidine dithiocarbamate inhibits induction of immunoproteasome and decreases survival in a rat model of amyotrophic lateral sclerosis. *Mol Pharmacol* 71, pp. 30–37.

Andersen, P. M. (2001). Genetics of sporadic ALS. *Amyotroph Lateral Scler Other Motor Neuron Disord* 2(Suppl 1), pp. S37–S41.

Andreassen, O. A., Dedeoglu, A., Friedlich, A., Ferrante, K. L., Hughes, D., Szabo, C., and Beal, M. F. (2001). Effects of an inhibitor of poly(ADP-Ribose) polymerase, desmethylselegiline, trientine, and lipoic acid in transgenic ALS mice. *Exp Neurol* 168, pp. 419–424.

Appel, S. H., Smith, R. G., Alexianu, M. E., Engelhardt, J. L., and Stefani, E. (1995). Autoimmunity as an etiological factor in sporadic amyotrophic lateral sclerosis. *Adv Neurol* 68, pp. 47–57.

Arai, T., Hasegawa, M., Akiyama, H., Ikeda, K., Nonaka, T., Mori, H., Mann, D., Tsuchiya, K., Yoshida, M., Hashizume, Y., and Oda, T. (2006). TDP-43 is a component of ubiquitin-positive tau-negative inclusions in frontotemporal lobar degeneration and amyotrophic lateral sclerosis. *Biochem Biophys Res Commun* 351, pp. 602–611.

Aschner, M., Cherian, M. G., Klaassen, C. D., Palmiter, R. D., and Erickson, J. C., and Bush A. I. (1997). Metallothionein in brain. The role of physiology and pathology. *Toxicol Appl Pharmacol* 142, pp. 229–242.

Bensimon, G., Lacomblez, L., Meininger, V., and the ALS/Riluzole Study Group (1994). A control trial of Riluzole in amyotrophic lateral sclerosis. *N Engl J Med* 330, pp. 585–591.

Berger, M. M., Kopp, N., Vital, C., Redl, B., Aymard, M., and Lina, B. (2000). Detection and cellular localization of enterovirus RNA sequences in spinal cord of patients with amyotrophic lateral sclerosis. *Neurology* 54, pp. 20–25.

Blaauwgeers, H. G., Anwar, Chand, M., van den berg, F. M., Vianney de Jong, J. M., and Troost, D. (1996). Expression of different metallothionein messenger ribonucleic acids in motor cortex, spinal cord and liver from patients with amyotrophic lateral sclerosis. *J Neurol Sci* 142, pp. 39–44.

Blanc-Brude, O. P., Yu, J., Simosa, H., Conte, M. S., Sessa, W. C., and Altieri, D. C. (2002). Inhibitor of apoptosis protein survivin regulates vascular injury. *Nat Med* 8, pp. 987–994.

Blazer, D., Gray, G. C., Hotopf, M., Macfarlane, G., Sim, M., Smith, T. C., and Wessely, S. (2008). Acetylcholinesterase inhibition and Gulf War illnesses. Conclusions are not supported by independent reviews of the same evidence. *Proc Natl Acad Sci USA* 105, pp. 20.

Boillée, S., Yamanaka, K., Lobsiger, C. S., Copeland, N. G., Jenkins, N. A., Kassiotis, G., Kollias, G., and Cleveland, D. W. (2006). Onset and progression in inherited ALS determined by motor neurons and microglia. *Science* 312, pp. 1389–1392.

Bruijn, L. I., Houseweart, M. K., Kato, S., Anderson, K. L., Anderson, S. D., Ohoma, E., Reaume, A. G., Scott, R. W., and Cleveland, D. W. (1998). Aggregation and motor neuron toxicity of an ALS-linked SOD1 mutant independent from wild-type SOD1. *Science* 281, pp. 1851–1854.

Bruijn, L. I., Miller, T. M., and Cleveland, D. W. (2004). Unraveling the mechanisms involved in motor neuron degeneration in ALS. *Ann Rev Neurosci* 27, pp. 723–749.

Burstein, E., Ganesh, L., Dick, R. D., van De Sluis, B., Wilkinson, J. C., Klomp, L. W., Wijmenga, C., Brewer, G. J., Nabel, G. J., and Duckett, C. S. (2004). A novel role for XIAP in copper homeostasis through regulation of MURR1. *EMBO J* 23, pp. 244–254.

Carrì, M. T., Ferri, A., Casciati, A., Celsi, F., Ciriolo, M. R., and Rotilio, G. (2001). Copper-dependent oxidative stress, alteration of signal transduction and neurodegeneration in amyotrophic lateral sclerosis. *Funct Neurol* 16(Suppl 4), pp. 181–188.

Chen, K. M., Craig, U. K., Lee, C. T., and Haddock, R. (2002). Cycad neurotoxin, consumption of flying foxes, and ALS/PDC disease in Guam. *Neurology* 59, pp. 1664.

Choi, B.-S. and Zheng, W. (2009). Copper transport to the brain by the blood-brain barrier and blood-CSF barrier. *Brain Res* 1248, pp. 14–21.

Clement, A. M., Nguyen, M. D., Roberts, E. A., Garcia, M. L., Boillée, S., Rule, M., McMahon, A. P., Doucette, W., Siwek, D., Ferrante, R. J., Brown, R. H. Jr, Julien, J. P., Goldstein, L. S., and Cleveland, D. W. (2003). Wild-type nonneuronal cells extend survival of SOD1 mutant motor neurons in ALS mice. *Science* 302, pp. 113–117.

Cleveland, D. W. and Rothstein, J. D. (2001). From Charcot to Lou Gehrig. Deciphering selective motor neuron death in ALS. *Nat Rev Neurosci* 2, pp. 806–819.

Cox, P. A. and Sacks, O. W. (2002). Cycad neurotoxins, consumption of flying foxes, and ALS-PDC disease in Guam. *Neurology* 58, pp. 956–959.

Cudkowicz, M. E., McKenna-Yasek, D., Sapp, P. E., Chin, W., Geller, B. S., Hayden, D. L., Schoenfeld, D. A., Hosler, B. A., Horvitz, H. R., and Brown, R. H. (1997). Epidemiology of mutations in superoxide dismutase in amyotrophic lateral sclerosis. *Ann Neurol* 41, pp. 210–221.

Deguchi, M., Shiraki, K., Inoue, H., Okano, H., Ito, T., Yamanaka, T., Sugimoto, K., Sakai, T., Ohmori, S., Murata, K., Furusaka, A., Hisatomi, H., and Nakano, T. (2002). Expression of survivin during liver regeneration. *Biochem Biophys Res Commun* 297, pp. 59–64.

Domzał, T., Radzikowska, B., Domzał, T., and Radzikowska, B. (1983). Ceruloplasmin and copper in the serum of patients with amyotrophic lateral sclerosis. *Neurol Neurochir Pol* 17, pp. 343–346.

Feng, H. L., Leng, Y., Ma, C. H., Zhang, J., Ren, M., and Chuang, D. M. (2008). Combined lithium and valproate treatment delays disease onset, reduces neurological deficits and prolongs survival in an amyotrophic lateral sclerosis mouse model. *Neuroscience* 155, pp. 567–572.

Fornai, F., Longone, P., Cafaro, L., Kastsiuchenka, O., Ferrucci, M., Manca, M. L., Lazzeri, G., Spalloni, A., Bellio, N., Lenzi, P., Modugno, N., Siciliano, G., Isidoro, C., Murri, L., Ruggieri, S., and Paparelli, A. (2008). Lithium delays progression of amyotrophic lateral sclerosis. *Proc Natl Acad Sci USA* 105, pp. 2052–2057.

Gaudette, M., Hirano, M., and Siddique, T. (2000). Current status of SOD1 mutations in familial amyotrophic lateral sclerosis. *Amyotroph Lateral Scler Other Motot Neuron Disord* 1, pp. 83–89.

Gellein, K., Garruto, R. M., Syversen, T., Sjøbakk, T. E., and Flaten, T. P. (2003). Concentrations of Cd, Co, Cu, Fe, Mn, Rb, V, and Zn in formalin-fixed brain tissue in amyotrophic lateral sclerosis and Parkinsonism-dementia complex of Guam determined by High-resolution ICP-MS. *Biol Trace Elem Res* 96, pp. 39–60.

Gill, A., Kidd, J., Vieira, F., Thompson, K., and Perrin, S. (2008). No benefit from chronic lithium dosing in a sibling-matched, gender balanced, investigator-blinded trial using a standard mouse model of familial ALS. *PLoS One* 4, pp. e6489.

Gong, Y. H. and Elliot, J. H. (2000). Metallothionein expression is altered in a transgenic murine model of familial amyotrophic lateral sclerosis. *Exp Neurol* 162, pp. 27–36.

Gordon, P. H., Moore, D. H., Miller, R. G., Florence, J. M., Verheijde, J. L., Doorish, C., Hilton, J. F., Spitalny, G. M., MacArthur, R. B., Mitsumoto, H., Neville, H. E., Boylan, K., Mozaffar, T., Belsh, J. M., Ravits, J., Bedlack, R. S., Graves, M. C., McCluskey, L. F., Barohn, R. J., Tandan, R., and Western ALS Study Group (2007). Efficacy of minocycline in patients with amyotrophic lateral sclerosis: a phase III randomised trial. *Lancet Neurol* 6, pp. 1045–1053.

Gurel, Z., Ozcelik, D., and Dursun, S. (2007). Apoptotic rate and metallothionein levels in the tissues of cadmium- and copper-exposed rats. *Biol Trace Elem Res* 116, pp. 203–217.

Gurney, M. E., Cutting, F. B., Zhai, P., Doble, A., Taylor, C. P., Andrus, P. K., and Hall, E. D. (1996). Benefit of vitamin E, Riluzole, and gabapentin in a transgenic model of familial amyotrophic lateral sclerosis. *Ann Neurol* 39, pp. 147–157.

Gurney, M. E., Pu, H., Chiu, A. Y., Dal Canto, M. C., Polchow, C. Y., Alexander, D. D., Caliendo, J., Hentati, A., Kwon, Y. W., Deng, H. X., Chen, W., Zhai, P., Sufit, R. L., and Siddique, T. (1994). Motoer neuro degeneration in mice that express a human Cu/Zn superoxide dismutase mutation. *Science* 264, pp. 1772–1775.

Hashimoto, K., Hayashi, Y., Inuzuka, T., and Hozumi, I. (2009). Exercise induces metallothioneins in mouse spinal cord. *Neuroscience*, 163, pp. 244–251.

Hidalgo, J., Aschner, M., Zatta, P., and Vašák, M. (2001). Roles of the metallothionein family of proteins in the central nervous system. *Brain Res Bull* 55, pp. 133–145.

Horner, R. D., Grambow, S. C., Coffman, C. J., Lindquist, J. H., Oddone, E. Z., Allen, K. D., and Kasarskis, E. J. (2008). Amyotrophic lateral Sclerosis among 1991 Gulf War veterans. Evidence for a time-limited outbreak. *Neuroepidemiology* 31, pp. 28–32.

Hottinger, A. F., Fine, E. G., Gurney, M. E., Zurn, A. D., and Aebischer, P. (1997). The copper chelator d-penicillamine delays onset of disease and extends survivial in a transgenic mouse model of familial amyotrophic lateral sclerosis. *Eur J Neurosci*, pp. 1458–1451.

Hozumi, I., Inuzuka, T., Ishiguro, H., Hiraiwa, M., Uchida, Y., and Tsuji, S. (1996). Immunoreactivity of growth inhibitory factor in normal rat brain and after stab wounds. An immunocytochemical study using confocal laser scan microscope. *Brain Res* 741, pp. 197–204.

Hozumi, I., Inuzuka, T., and Tsuji, S. (1998). Brain injury and growth inhibitory factor (GIF). A minireview. *Neurochem Res* 23, pp. 319–328.

Hozumi, I., Asanuma, M., Yamada, M., and Uchida, Y. (2004). Metallothionein and neurodegenerative diseases. *J Health Sci* 50, pp. 323–331.

Hozumi, I., Yamada, M., Uchida, Y., Ozawa, K., Takahashi, H., and Inuzuka, T. (2008). The expression of metallothioneins is diminished in the spinal cords of patients with sporadic ALS. *Amyotroph Lateral Scler* 9, pp. 294–298.

Inoue, H., Tsukita, K., Iwasato, T., Suzuki, Y., Tomioka, M., Tateno, M., Nagao, M., Kawata, A., Saido, T. C., Miura, M., Misawa, H., Itohara, S., and Takahashi, R. (2003). The crucial role of caspase-9 in the disease progression of a transgenic ALS mouse model. *EMBO J* 22, pp. 6665–6674.

Ishigaki, S., Niwa, J., Ando, Y., Yoshihara, T., Sawada, K., Doyu, M., Yamamoto, M., Kato, K., Yotsumoto, Y., and Sobue, G. (2002a). Differentially expressed genes in sporadic amyotrophic lateral sclerosis spinal cords--screening by molecular indexing and subsequent cDNA microarray analysis. *FEBS Lett* 531, pp. 354–358.

Ishigaki, S., Liang, Y., Yamamoto, M., Niwa, J., Ando, Y., Yoshihara, T., Takeuchi, H., Doyu, M., and Sobue, G. (2002b). X-Linked inhibitor of apoptosis protein is involved in mutant SOD1-mediated neuronal degeneration. *J Neurochem* 82, pp. 576–584.

Ishigaki, A., Aoki, M., Nagai, M., Warita, H., Kato, S., Kato, M., Nakamura, T., Funakoshi, H., and Itoyama, Y. (2007). Intrathecal delivery of hepatocyte growth factor from amyotrophic lateral sclerosis onset suppresses disease progression in rat amyotrophic lateral sclerosis model. *J Neuropathol Exp Neurol* 66, pp. 1037–1044.

Ito, H., Wate, R., Zhang, J., Ohnishi, S., Kaneko, S., Ito, H., Nakano, S., and Kasukabe, H. (2008). Treatment with edaravone, initiated at symptom onset, slows motor decline and decreases SOD1 deposition in ALS mice. *Exp Neurol* 213, pp. 448–455.

Johnson, F. L. and Atchison, W. D. (2009). The role of environmental mercury, lead and pesticide exposure in development of amyotrophic lateral sclerosis. *Neurotoxicology* 30, pp. 761–768.

Julien, J.-P. (2001). Amyotrophic lateral sclerosis. Unfolding the toxicity of the misfolded. *Cell* 104, pp. 581–591.

Kägi, J. H. R. and Schäffer, A. (1988). Biochemistry of metallothionein. *Biochemistry* 27, pp. 8509–8515.

Kang, Y. J., Li, Y., Sun, X., and Sun, X. (2003). Antiapoptotic effect and inhibition of ischemia/reperfusion-induced myocardial injury in metallothionein-overexpressing transgenic mice. *Am J Pathol* 163, pp. 1579–1586.

Kasarskis, E. J., Lindquist, J. H., Coffman, C. J., Grambow, S. C., Feussner, J. R., Allen, K. D., Oddone, E. Z., Kamins, K. A., Horner, R. D., and ALS Gulf War Clinical Review Team (2009). Clinical aspects ALS in Gulf War veterans. *Amyotroph Lateral Scler* 10, pp. 35–41.

Kato, S., Hayashi, H., Nakashima, K., Nanba, E., Kato, M., Hirano, A., Nakano, I., Asayama, K., and Ohama, E. (1997). Pathological characterization of astrocytic hyaline inclusions in familial amyotrophic lateral sclerosis. *Am J Pathol* 151, pp. 611–620.

Kawahara, Y., Kwak, S., Sun, H., Ito, K., Hashida, H., Aizawa, H., Jeong, S. Y., and Kanazawa, I. (2003). Human spinal motoneurons express low relative abundance of GluR2 mRNA . An implication for excitotoxicity in ALS. *J Neurochem* 85, pp. 680–689.

Kiaei, M., Bush, A. I., Morrison, B. M., Morrison, J. H., Cherny, R. A., Volitakis, I., Beal, M. F., and Gordon, J. W. (2004). Genetically decreased spinal cord copper concentration prolongs life in a transgenic mouse model of amyotrophic lateral sclerosis. *J Neurosci* 24, pp. 7945–7950.

Kishigami, H., Nagano, S., Bush, A. I., and Sakota, S. (2010). Monomerized Cu, Zn-superoxide dismutase induces oxidative stress though aberrant Cu binding. *Free Radic Biol Med* 48, pp. 945–952.

Kunst, C. B. (2004). Complex genetics of amyotrophic lateral sclerosis. *Am J Hum Genet* 75, pp. 933–947.

Kuo, Y. M., Zhou, B., Cosco, D., and Gitschier, J. (2001). The copper transporter CTR1 provides an essential function in mammalian embryonic development. *Proc Natl Acad Sci* 98, pp. 6836–6841.

Kuzuhara, S., Kokubo, Y., Sasaki, R., Narita, Y., Yabana, T., Hasegawa, M., and Iwatsubo, T. (2001). Familial amyotrophic lateral sclerosis and parkinsonism-dementia complex of the Kii Peninsula of Japan: clinical and neuropathological study and tau analysis. *Ann Neurol* 49, pp. 501–511.

Kwiatkowski, Jr. T. J., Bosco, D. A., LeClerc, A. L., Tamrazian, E., Vanderburg, C. R., Russ, C., Davis, A., Gilchrist, J., Kasarskis, E. J., Munsat, T., Valdmanis, P., Rouleau, G. A., Hosler, B. A., Cortelli, P., de Jong, P. J., Yoshinaga, Y., Haines, J. L., Pericak-Vance, M. A., Yan, J., Ticozzi, N., Siddique, T., McKenna-Yasek, D., Sapp, P. C., Horvitz, H. R., Landers, J. E., Brown, Jr. R. H. (2009). Mutations in the FUS/TLS gene on chromosome 16 casue familial amyotrophic lateral sclerosis. *Science* 323, pp. 1205–1208.

Larin, D., Mekios, C., Das, K., Ross, B., Yang, A. S., and Gilliam, T. C. (1999). Characterization of the interaction between the Wilson and Menkes disease proteins and cytoplasmic copper chaperone, HAH1p. *J Biol Chem* 274, pp. 28497–28504.

Lederer, C. W., Torrisi, A., Pantelidou, M., Santama, N., and Cavallaro, S. (2007). Pathways and genes differentially expressed in the motor cortex of patients with sporadic amyotrophic lateral sclerosis. *BMC Genomics* 8, pp. 26.

Lee, J., Prohaska, J. R., and Thiele, D. J. (2001). Essential role for mammalian copper transporter Ctr1 in copper homeostasis and embryonic development. *Proc Natl Acad Sci* 98, pp. 6842–6847.

Levenson, C. W. (2005). Trace metal regulation of normal apoptosis: From genes to behavior. *Physiol Behav* 86, pp. 399–406.

Li, H., Zhu, H., Eggers, D. K., Nersissian, A. M., Faull, K. F., Goto, J. J., Ai, J., Sanders-Loehr, J., Gralla, E. B., and Valentine, J. S. (2000). Copper (2+) binding to the surface residue cysteine111 of His46Arg human copper-zinc superoxide dismutase. A familial amyotrophic lateral sclerosis mutant. *Biochemistry* 39, pp. 8125–8132.

Li, Q. X., Mok, S. S., Laughton, K. M., McLean, C. A., Volitakis, I., Cherny, R. A., Cheung, N. S., White, A. R., and Masters, C. J. (2006). Overexpression of Aβ is associated with acceleration of onset of motor impairment and superoxide dismutase 1 aggregation in an amyotrophic lateral sclerosis mouse model. *Aging Cell* 5, pp. 153–165.

Linder, M. C. and Hazegh-Azam, M. (1996). Copper biochemistry and molecular biology. *Am J Clin Nutr* 63, pp. 797S–811S .

Lutsenko, S., Petrukhin, K., Cooper, M. J., Gilliam, C. T., and Kaplan, J. H. (1997). N-terminal domains of human copper-transporting adenosine triphosphatases (the Wilson's and Menkes disease proteins) bind copper selectively *in vivo* and *in vitro* with stoichiometry of one copper per metal-binding repeat. *J Biol Chem* 272, pp. 18939–18944.

Méndez-Armenta, M., Villeda-Hernández, J., Barroso-Moguel, R., Nava-Ruíz, C., Jiménez-Capdeville, M. E., and Ríos, C. (2003). Brain regional lipid peroxidation and metallothionein levels of developing rats exposed to cadmium and dexamethasone. *Toxicol Lett* 144, pp. 151–157.

Miranda, M. L., Overstreet, A., Galeano, M., Tassone, E., Allen, K. D., and Horner, R. D. (2009). Spatial analysiss of the epidemiology of amyotrophic lateral sclerosis among 1991 Gulf War veterans. *Neurotoxicology* 29, pp. 964–970.

Mufti, A. R., Burstein, E., Csomos, R. A., Graf, P. C., Wilkinson, J. C., Dick, R. D., Challa, M., Son, J. K., Bratton, S. B., Su, G. L., Brewer, G. J., Jakob, U., and Duckett, C. S. (2006). XIAP is a copper binding protein deregulated in Wilson's disease and other copper toxicosis disorders. *Mol Cell* 21, pp. 775–785.

Nagano, S., Satoh, M., Sumi, H., Fujimura, H., Tohyama, C., Yanagihara, T., and Sakoda, S. (2001). Reduction of metallothioneins promotes the disease expression of familial amyotrophic lateral sclerosis mice in a dose-dependent manner. *Eur J Neurosci* 13, pp. 1363–1370.

Nagano, S., Fujii, Y., Yamamoto, T., Taniyama, M., Fukada, K., Yanagihara, T., and Sakoda, S. (2003). The efficacy of trientine or ascorbate alone compared to that of the combined treatment with these two agents in familial amyotrophic lateral sclerosis model mice. *Exp Neurol* 179, pp. 176–180.

Neumann, M., Sampathu, D. M., Kwong, L. K., Truax, A. C., Micsenyi, M. C., Chou, T. T., Bruce, J., Schuck, T., Murray Grossman, M., Clark, C. M., McCluskey, L. F., Miller, B. L., Masliah, E., Mackenzie, I. R., Feldman, H., Feiden, W., Kretzschmar, H. A., Trojanowski, J. Q., and Lee, V. M. Y. (2006). Ubiquitinated TDP-43 in frontotemporal lobar degeneration and amyotrophic lateral sclerosis. *Science* 314, pp. 130–133.

Ogura, Y., Ohmichi, M., and Suzuki, K. T. (1996). Mechanism of selective copper removal by tetrathiomolybdate from metallothionein in LEC rats. *Toxicology* 106, pp. 75–78

Ogura, Y. and Suzuki, K. T. (1998). Targeting of tetrathiomolybdate on the copper accumulating in the liver of LEC rats. *J Inorg Biohem* 70, pp. 49–55.

Ogura, Y., Komada, Y., and Suzuki, K. T. (1999). Comparative mechanism and toxicity of tetra- and dithiomolybdate in the removal of copper. *J Inorg Biochem* 75, pp. 199–204.

Ohgami, R. S., Campagna, D. R., McDonald, A., and Fleming, M. D. (2006). The steap proteins are metalloreductases. *Blood* 108, pp. 1388–1394.

Olsen, M. K., Roberds, S. L., Ellerbrock, B. R., Fleck, T. J., McKinley, D. K., and Gurney, M. E. (2001). Disease mechanisms revealed by transcription profiling in SOD1-G93A transgenic mouse spinal cord. *Ann Neurol* 50, pp. 730–740.

Ono, S. I., Koropatnick, D. J., and Cherian, M. G. (1997). Regional brain distribution of metallothionein, zinc and copper in toxic milk and transgenic mice. *Toxicology* 124, pp. 1–10.

Ono, S. I., Cai, Lu, and Cherian, M. G. (1998a). Effects of gamma radiation on levels of brain metallothionein and lipid peroxidation in transgenic mice. *Rad Res* 159, pp. 52–57.

Ono, S. I., and Cherian, M. G. (1998b). Changes in brain metallothionein and zinc during development in transgenic mice. *Biol Trace Elem Res* 61, pp. 41–49.

Ono, S. I., Cai, L., Koropatnick, J., and Cherian, M. G. (2000). Radiation exposure does not alter metallothionein III isoform expression in mouse brain. *Biol Trace Elem Res* 74, pp. 23–29.

Ono, S. I., Endo, Y., Tokuda, E., Ishige, K., Tabata, K., Asami, S., Ito, Y., and Suzuki, T. (2006). Upregulation of metallothionein-I mRNA expression in a rodent model for amyotrophic lateral sclerosis. *Biol Trace Elem Res* 113, pp. 93–103.

Ono, S. I., Ishizaki, Y., Tokuda, E., Tabata, K., Asami, S., and Suzuki, T. (2007). Different patterns in the induction of metallothionein mRNA synthesis amoung isoforms after acute ethanol administration. *Biol Trace Elem Res* 115, pp. 147–156.

Ono, S. I., Tokuda, E., Watanabe, S., and Okawa, E. (2009). An attempt to treat amyotrophic lateral sclerosis by metallothionein. Its fundamentals and perspective. The 7th Japanese meeting on metallothionein and metal-biosciences. Oct. 17, 2009, Tokyo (Abstract in Japanese)

Palato, C. C., Galasko, D., Garruto, R. M., Plato, M., Gamst, A., Craig, U. K., Torres, J. M., and Wiederholt, W. (2002). ALS and PDC of Guam: forty-year follow-up. *Neurology* 58, pp. 765–773.

Palmiter, R. D., Findley, S. D., Whitmore, T. E., and Durnam, D. M. (1992). MT-III, a brain-specific member of the metallothionein gene family. *Proc Natl Acad Sci USA* 89, pp. 6333–6337.

Palmiter, R. D., Sandgren, E. P., Koeller, D. M., and Brinster, R. L. (1993). Distal Regulatory Elements from the mouse metallothionein locus stimulate gene expression in transgenic mice. *Mol Cell Biol* 13, pp. 5266–5275.

Petris, M. J., Mercer, J. F., Culvenor, J. G., Lockhart, P., Gleeson, P. A., and Camakaris, J. (1996). Ligand-regulated transport of the Menkes copper P-type ATPase efflux pump from the Golgi apparatus to the plasma membrane: a novel mechanism of regulated trafficking. *EMBO J* 15, pp. 6084–6095.

Petris, M. J., Camakaris, J., Greenough, M., LaFontaine, S., and Mercer, J. F. (1998). A C-terminal di-leucine is required for localization of the Menkes protein in the trans-Golgi network. *Hum Mol Genet* 7, pp. 2063–2071.

Piepers, S., Veldink, J. H., de Jong, S. W., van der Tweel, I., van der Pol, W. L., Uijtendaal, E. V., Schelhaas, H. J., Scheffer, H., de Visser, M., de Jong, J. M., Wokke, J. H., Groeneveld, G. J., and van den Berg, L. H. (2009). Randomized sequential trial of valproic acid in amyotrophic lateral sclerosis. *Ann Neurol* 66, pp. 227–234.

Prohaska, J. R. and Gybina, A. A. (2004). Intracellular copper transport in Mammals. *J Nutr* 134, pp. 1003–1006.

Puttaparthi, K., Gitomer, W. L., Krishnan, U., Son, M., Rajendran, B., and Elliott, J. L. (2002). Disease progression in a transgenic model of familial amyotrophic lateral sclerosis is dependent on both neuronal and non-neuronal zinc binding proteins. *J Neurosci* 22, pp. 8790–8796.

Quaife, C. J., Findley, S. D., Erickson, J. C., Froelick, G. J., Kelly, E. J., Zambrowicz, B. P., and Palmiter, R. D. (1994). Induction of a new metallothionein isoform (MT-IV) occurs during differentiation of stratified squamous epithelia. *Biochemistry* 33, pp. 7250–7259.

Reaume, A. G., Elliot, J. L., Hoffmann, E. K., Kowall, N. W., Ferrante, R. J., Siwek, D. F., Wilcox, H. M., Flood, D. G., Beal, M. F., Brown, Jr, R. H., Scott, R. W., and Snider, W. D. (1996). Motor neurons in Cu/Zn superoxide dismutase-deficient mice develop normally but exhibit enhanced cell death after axonal injury. *Nat Genet* 13, pp. 43–47.

Rosen, D. R., Siddique, T., Patterson, D., Figlewicz, D. A., Sapp, P., Hentati, A., Donaldson, D., Goto, J., O'Regan, J. P., Deng, H. X., Rahmani, Z., Krizus, A., McKenna-Yasek, D., Cayabyab, A., Gaston, SM., Berger, R., Tanzi, R. E., Halperin, J. J., Herzfeldt, B., Van den Berg, R., Hung, W.-Y., Bird, T., Deng, G., Mulder, D. W., Smyth, C., Laing, N. G., Soriano, E., Pericak-Vance, M. A., Haines, J., Rouleau, G. A., Gusella, J. S., Horvitz, H. R., and Brown, Jr. RH. (1993). Mutations in Cu/Zn superoxide dismutase gene are associated with familial amyotrophic lateral sclerosis. *Nature* 362, pp. 59–62.

Rothstein, J. D., Van Kammen, M., Levey, A. I., Martin, L. J., and Kuncl, R. W. (1995). Selective loss of glial glutamate transporter GLT-1 in amyotrophic lateral sclerosis. *Ann Neurol* 38, pp. 128–131.

Rothstein, J. D., Patel, S., Regan, M. R., Haenggeli, C., Huang, Y. H., Bergles, D. E., Jin, L., Dykes, Hoberg, M., Vidensky, S., Chung, D. S., Toan, S. V., Bruijn, L. I., Su, Z. Z., Gupta, P., and Fisher, P. B. (2005). Beta-lactam antibiotics offer neuroprotection by increasing glutamate transporter expression. *Nature* 433, pp. 73–77.

Rowland, L. P. (1995). Hereditary and acquired motor neuron diseases. In *Merritt's textbook of neurology*. 9th ed. Ed by Rowland, L. P., Willians and Wilkins, Baltimore, pp. 742–749.

Rowland, L. P. and Shneider, N. A. (2001). Amyotrophic lateral Sclerosis. *N Engl J Med* 344, pp. 1688–1700.

Salazar, A. M., Masters, C. L., Gajdusek, D. C., and Gibbs, C. J. (1983). Syndrome of amyotrophic lateral sclerosis and dementia. Relation to transmissible Creutzfelt-Jacob disease. *Ann Neurol* 14, pp. 17–26.

Sasabe. J., Chiba, T., Yamada, M., Okamoto, K., Nishimoto, I., Matsuoka, M., and Aiso, S. (2007). D-serine is a key determinant of glutamate toxicity in amyotrophic lateral sclerosis. *EMBO J* 26, pp. 4149–4159.

Sillevis Smitt, P. A., Blaauwgeers, H. G., Troost, D., and de Jong, J. M. (1992a). Metallothionein immunoreactivity is increased in the spinal cord of patients with amyotrophic lateral sclerosis. *Neurosci Lett* 144, pp. 107–110.

Sillevis Smitt, P. A., van Beek, H., Baars, A. J., Troost, D., Louwerse, E. S., Krops-Hermus, A. C., de Wolff, F. A., and de Jong, J. M. (1992b). Increased metallothionein in the liver and kidney of patients with amyotrophic lateral sclerosis. *Arch neurol* 49, pp. 721–724.

Sillevis Smitt, P. A., Mulder, T. P., Verspaget, H. W., Blaauwgeers, H. G., Troost, D., and de Jong, J. M. (1994). Metallothionein in amyotrophic lateral sclerosis. *Biol Signals* 3, pp. 193–197.

Simpkins, C. O. (2000). Metallothionein in human diseases. *Cell Mol Biol* 46, pp. 465–488.

Spencer, P. S., Ohta, M., and Palmer, V. S. (1987). Cycad use and motor neurone disease in Kii peninsula of Japan. *Lancet* 2, pp. 1462–1463.

Steele, J. C., Guzman, T. Q., Driver, M. G., Zolan, W., Heitz, L. F., Kilmer, F. H., Parker, C. M., Standal, B. R., Pobutsky, A. M., and McLachan, D. R. (1990). Nutritional factors in amyotrophic lateral sclerosis on Guam. Observations from Umatac. In *Amyotrophic lateral sclerosis. Concepts in pathogenesis and etiology*. Ed by Hudson, A. J., University of Toronto press, Toronto, pp. 193–223.

Strong, M. J. and Garruto, R. M. (1996). Motor Neuron disease. In *Mineral and metal toxicology*, Ed by Yasui, M., Strong, M. J, Ota, K., and Verity, M. A. CRC Press, Florida, pp. 107–112.

Subramaniam, J. R., Lyons, W. E., Liu, J., Bartnikas, T. B., Rothstein, J., Price, D. L., Cleveland, D. W., Gitlin, J. D., and Wong, P. C. (2002). Mutant SOD1 causes motor neuron disease independent of copper chaperone-mediated copper loading. *Nat Neurosci* 5, pp. 301–307.

Swanson, N. R., Fox, S. A., and Mastaglia, F. L. (1995). Search for patient infection with poliovirus or other enteroviruses in amyotrophic lateral sclerosis-motor neuron disease. *Neuromuscl Disord* 5, pp. 457–465.

Takahashi, Y., Kako, K., Kashiwabara, S., Takehara, A., Inada, Y., Arai, H., Nakada, K., Kodama, H., Hayashi, J., Baba, T., and Munekata, E. (2002). Mammalian copper chaperone Cox17 has an essential role in activation of cytochrome *c* oxidase and embryonic development. *Mol Cell Biol* 22, pp. 7614–7621.

Tokuda, E., Ono, S. I., Ishige, K., Ito, Y., Naganuma, A., and Suzuki, T. (2006). Metallothionein restores dysequilibrium between caspases and their inhibitors in a mouse model for amyotrophic lateral sclerosis. 125th Annual meeting of the Pharmaceutical Society of Japan, Sendai, Japan, pp. 103. (Abstract in Japanese)

Tokuda, E., Ono, S. I., Ishige, K., Naganuma, A., Ito, Y., and Suzuki, T. (2007a). Metallothionein proteins expression, copper and zinc concentrations, and lipid peroxidation level in a rodent model for amyotrophic lateral sclerosis. *Toxicology* 229, pp. 33–41.

Tokuda, E., Ono, S. I., Ishige, K., Watanabe, S., Okawa, E., Ito, Y., and Suzuki, T. (2007b). Dysequilibrium between caspases and their inhibitors in a mouse model for amyotrophic lateral sclerosis. *Brain Res* 1148, pp. 234–242.

Tokuda, E., Ono, S. I., Ishige, K., Ito, Y., and Suzuki, T. (2007c). Oxidative damage to mitochondrial DNA in spinal cord is associated with disease onset in a rodent model for amyotrophic lateral sclerosis. 126th Annual meeting of the Pharmaceutical Society of Japan, Toyama, Japan, pp. 33. (Abstract in Japanese)

Tokuda, E., Ono, S. I., Ishige, K., Watanabe, S., Okawa, E., Ito, Y., and Suzuki, T. (2008a). Ammonium tetrathiomolybdate delays onset, prolongs survival, and slows progression of disease in a mouse model for amyotrophic lateral sclerosis. *Exp Neurol* 213, pp. 122–128.

Tokuda, E., Okawa, E., Suzuki, T., and Ono, S. I. (2008b). Ammonium tetrathiomolybdate, a copper-chelating drug, suppresses SOD1 aggregation and has therapeutic effects in a model mouse of amyotrophic lateral sclerosis. *Amyotroph Lateral Scler* 9(Suppl 1), p. 65. (Abstract)

Tokuda, E., Okawa, E., and Ono, S. I. (2009a). Dysregulation of intracellular copper trafficking pathway in a mouse model of mutant copper/zinc superoxide dismutase-linked familial amyotrophic lateral sclerosis. *J Neurochem* 111, pp. 181–191.

Tokuda, E. and Ono, S. I. (2009b). Overexpression of glial metallothionein slows disease progression in a mouse model of mutant SOD1-linked familial ALS. *Amyotroph Lateral Scler* 10(Suppl 1), p. 85. (Abstract)

Traynor, B. J., Bruijin, L., Conwit, R., Beal F, O'Neill, G., Fagan, S. C, and Cudkowicz, M. E. (2006). Neuroprotective agents for clinical trials in ALS. A systemic assessment. *Neurology* 67, pp. 20–27.

Uchida, Y., Takio, K., Titani, K., Ihara, Y., and Tomonaga, M. (1991). The growth inhibitory factor that is deficient in the Alzheimer's disease brain is a 68 amino acid metallothionein-like protein. *Neuron* 7, pp. 337–347.

Valentine, J. S., Doucette, P. A., and Potter, S. Z. (2005). Copper-zinc superoxide dismutase and amyotrophic lateral sclerosis. *Ann Rev Biochem* 74, pp. 563–593.

Vance, C., Rogelj, B., Hortobágyi, T., De Vos, K. J., Nishimura, A. L., Sreedharan, J., Hu, X., Smith, B., Rubby, D., Wright, P., Ganesalingam, J., Williams, K. L., Tripathi, V., Al-Saraj, S., Al-Chalabi, A., Leigh, P. N., Blair, I. P., Nicholson, G., de Belleroche, J., Gallo, J. M., Miller, C. C., Shaw, C. E. (2009). Mutations in FUS, an RNA processing protein, cause familial amyotrophic lateral sclerosis type 6. *Science* 323, pp. 1208–1211.

Vincent, A. and Drachman, D. B. (1996). Amyotrophic lateral sclerosis and antibodies to voltage-gated calcium channels. New doubts. *Ann Neurol* 40, pp. 691–693.

Wang, G. W., Klein, J. B., and Kang, Y. J. (2001). Metallothionein inhibits doxorubicin-induced mitochondrial cytochrome c release and caspase-3 activation in cardiomyocytes. *J Pharmacol Epx Ther* 298, pp. 461–468.

Wang, J., Slunt, H., Gonzales, V., Fromholt, D., Coonfield, M., Copeland, N. G., Jenkins, N. A., and Borchelt, D. R. (2003). Copper-binding-site-null SOD1 causes ALS in transgenic mice. Aggregates of non-native SOD1 delineate a common feature. *Hum Mol Genet* 12, pp. 2753–2764.

Warita, H., Hayashi, T., Murakami, T., Manabe, Y., and Abe, K. (2001). Oxidative damage to mitochondrial DNA in spinal motoneurons of transgenic ALS mice. *Mol Brain Res* 89, pp. 147–152.

Watanabe, K., Miyakawa, O., and Kobayashi, M. (2001). New method for quantitative mapping of metallic elements in tissue section by electron probe microanalyser with wavelength dispersive spectrometers. *J Electron Microsc* 50, pp. 77–82.

Watanabe, S., Nagano, S., Duce, J., Kiaei, M., Li, Q. X., Tucker, S. M., Tiwari, A., Brown, R. H. Jr, Beal M. F., Hayward, L. J., Culotta, V. C., Yoshihara, S., Sakoda, S., and Bush, A. I. (2007). Increased affinity for copper mediated by cysteine 111 in forms of mutant superoxide dismutase

1 linked to amyotrophic lateral sclerosis. *Free Radic Biol Med* 42, 1534–1542.

Watanabe, S., Tokuda, E., and Ono, S. I. (2008). Metallothionein-I/II induction improves clinical manifestation in a mouse model for amyotrophic lateral sclerosis. *Rinsho Shinkeigaku* (*Clinical Neurology*) 48, p. 1242. (Abstract in Japanese)

Werdelin, L., Boysen, G., Jensen, T. S., and Mogensen, P. (1990). Immunosuppressive treatment of patients with amyotrophic lateral sclerosis. *Acta Neurol Scand* 82, pp. 132–134.

Wong, P. C., Waggoner, D., Subramanian, J. R., Tessarollo, L., Bartnikas, T. B., Culotta, V. C., Price, D. L., Rothstein, J., and Giltin, J. D. (2000). Copper chaperone for superoxide dismutase is essential to activate mammalian Cu/Zn superoxide dismutase. *Proc Natl Acad Sci* 97, pp. 2886–2891.

Wootz, H., Hansson, I., Korhonen, L., and Lindholm, D. (2006). XIAP decreases caspase-12 cleavage and calpain activity in spinal cord of ALS transgenic mice. *Exp Cell Res* 312, pp. 1890–1898.

Worrall, B. B., Rowland, L. P., Chin, S. S., and Mastrianni, J. A. (2000). Amyotrophy in prion diseases. *Arch Neurol* 57, pp. 33–38.

Yamanaka, K., Chun, S. J., Boillée, S., Fujimori-Tonou, N., Yamashita, H., Gutmann, D. H., Takahashi, R., Misawa, H., and Cleveland, D. W. (2008). Astrocyte as determinants of disease progression in inherited amyotrophic lateral sclerosis. *Nat Neurosci* 11, pp. 251–253.

Yase, Y. (1972). The pathogenesis of amyotrophic lateral sclerosis. *Lancet* 2, pp. 292–296.

Yoshihara, T., Ishigaki, S., Yamamoto, M., Liang, Y., Niwa, J., Takeuchi, H., Doyu, M., and Sobue, G. (2002). Differential expression of inflammation- and apoptosis-related genes in spinal cords of a mutant SOD1 transgenic mouse model of familial amyotrophic lateral sclerosis. *J Neurochem* 80, pp. 158–167.

Index

Colour Insert

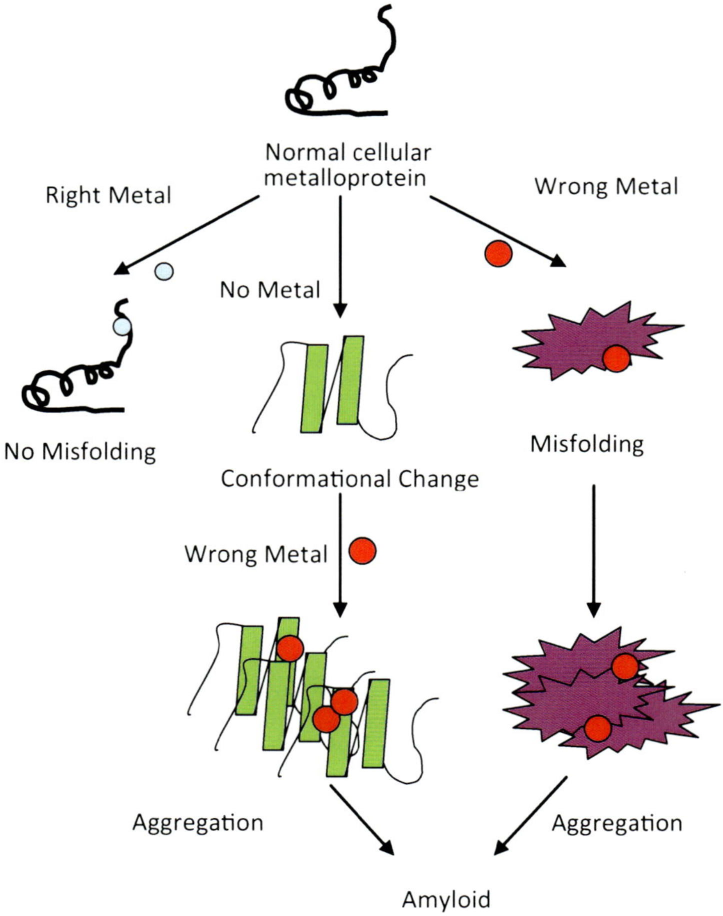

Figure 1.1

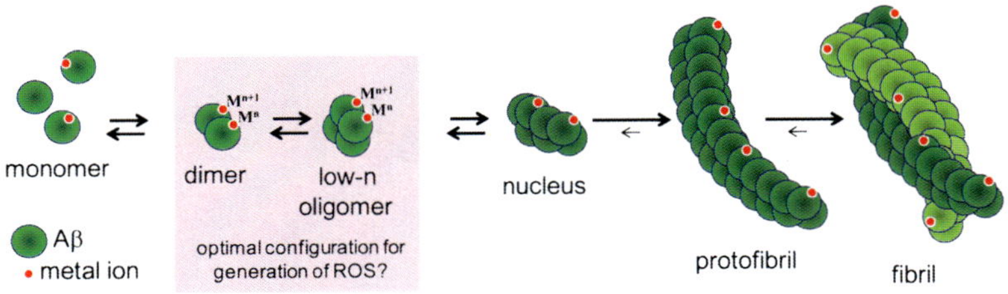

Figure 2.2

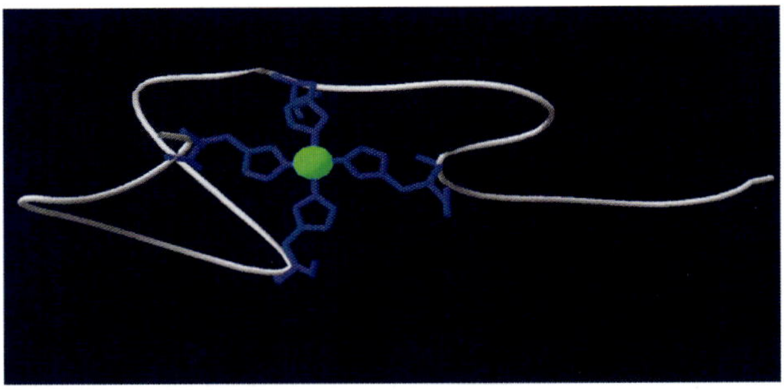

Figure 3.2

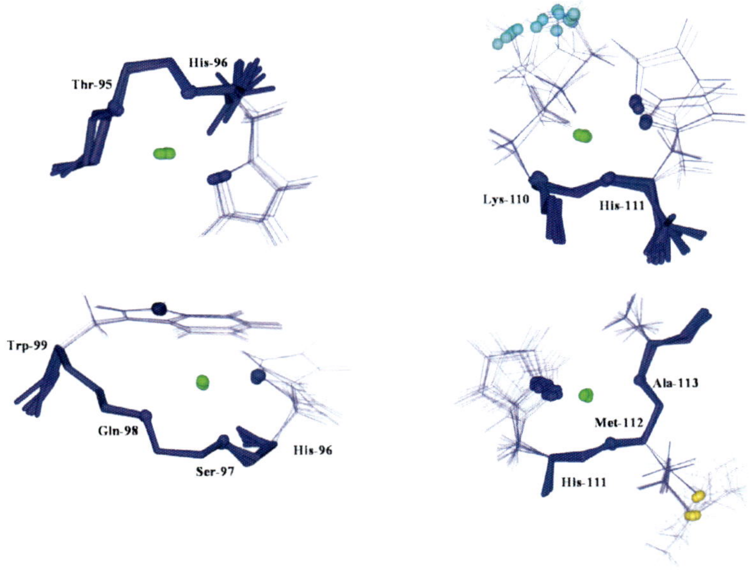

Figure 3.5

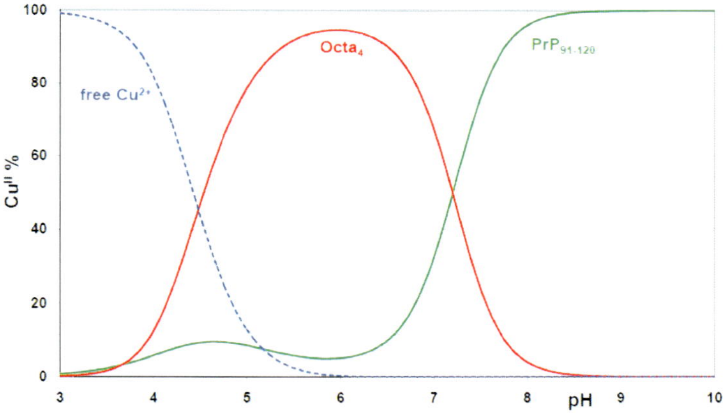

Figure 3.9

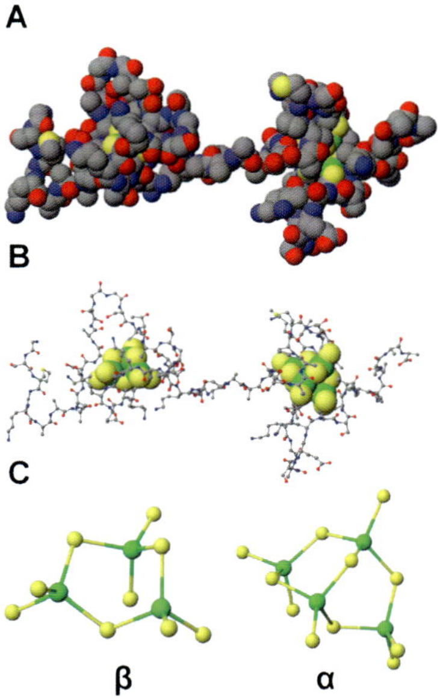

Figure 4.2

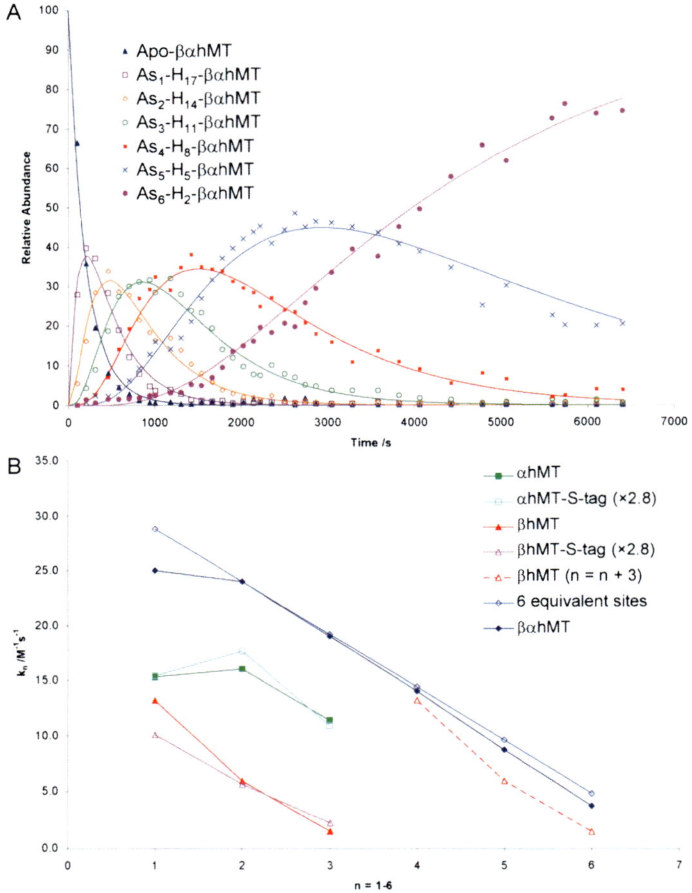
A
Apo-βαhMT
As1-H17-βαhMT
As2-H14-βαhMT
As3-H11-βαhMT
As4-H8-βαhMT
As5-H5-βαhMT
As6-H2-βαhMT
Relative Abundance
Time /s
B
αhMT
αhMT-S-tag (×2.8)
βhMT
βhMT-S-tag (×2.8)
βhMT (n = n + 3)
6 equivalent sites
βαhMT
kn /M-1 s-1
n = 1-6

Figure 4.11

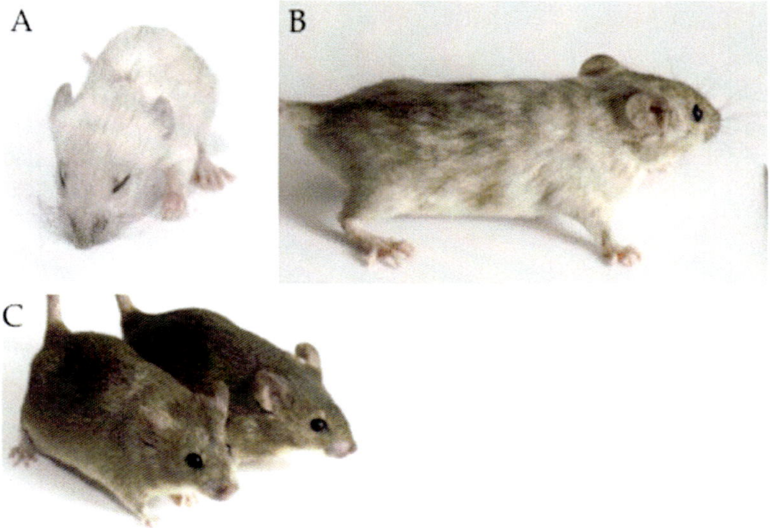

Figure 5.2

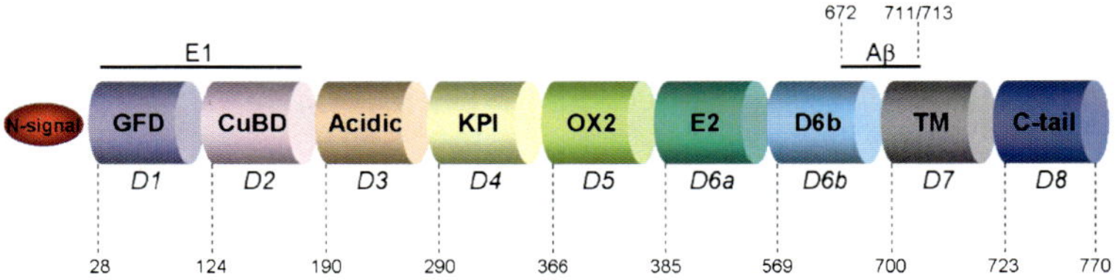

Figure 6.1

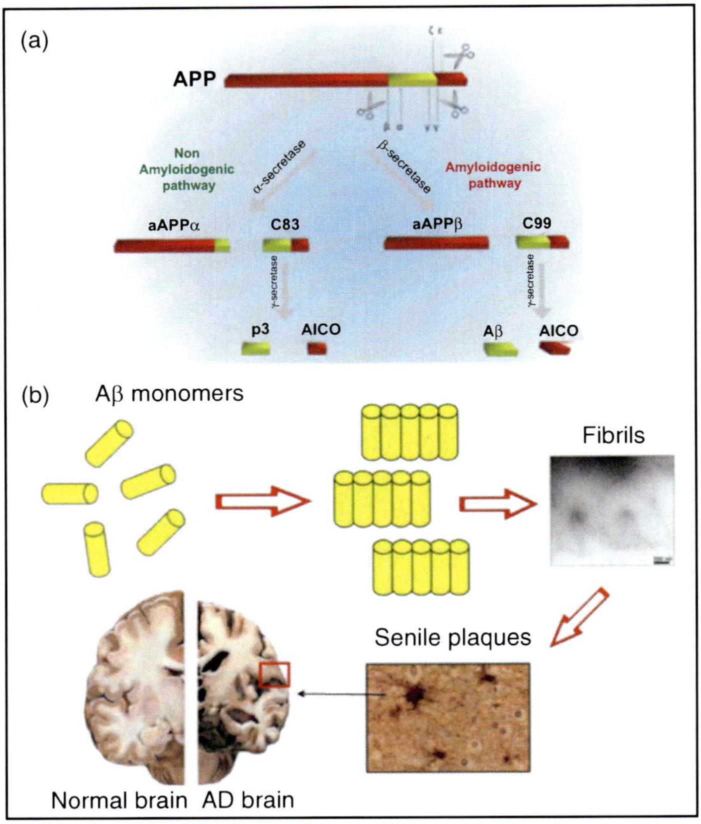

Figure 7.1

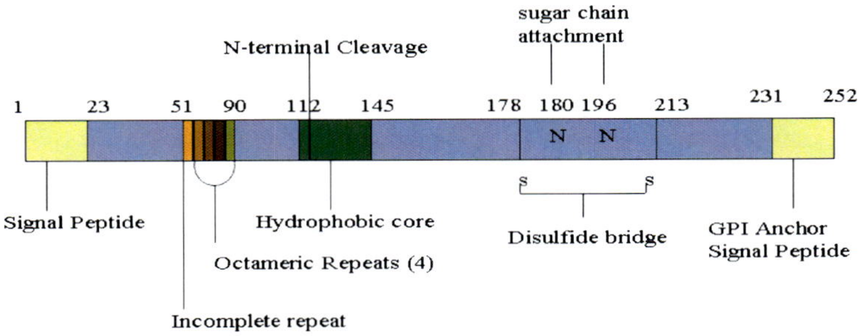

Figure 8.1

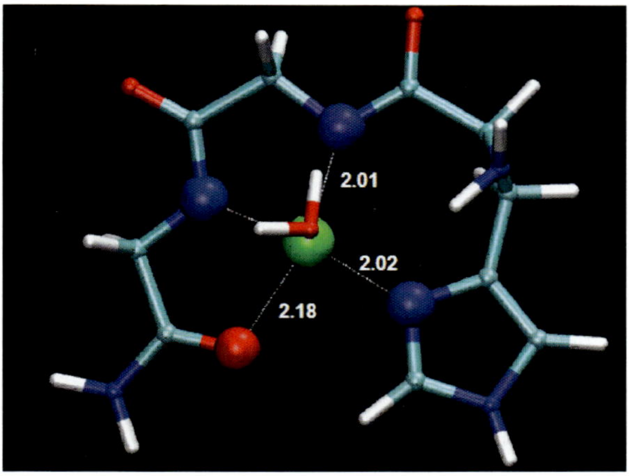

Figure 8.2

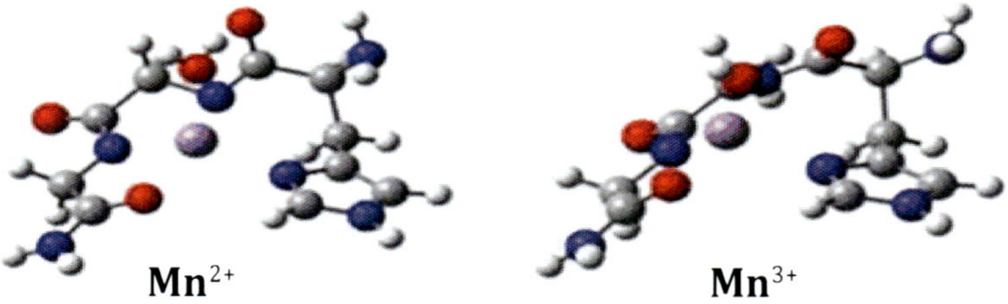

Figure 8.3

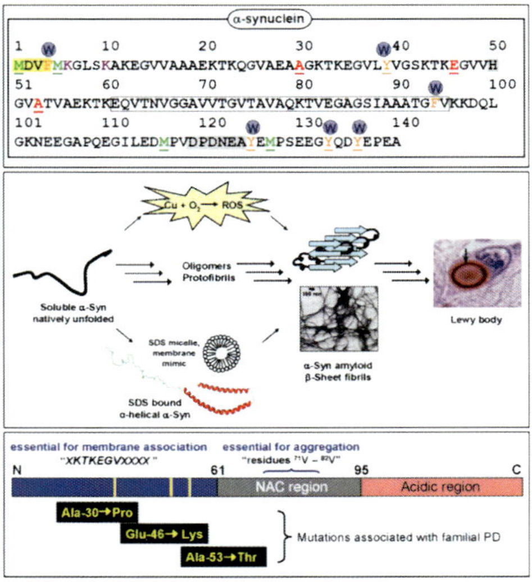

Figure 9.1

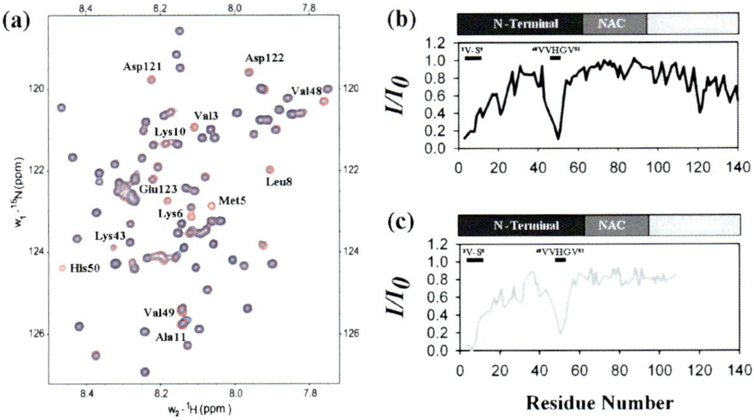

Figure 10.4

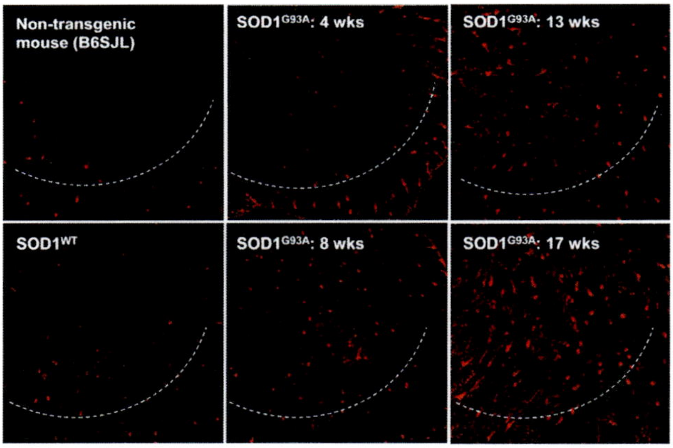

Figure 11.1

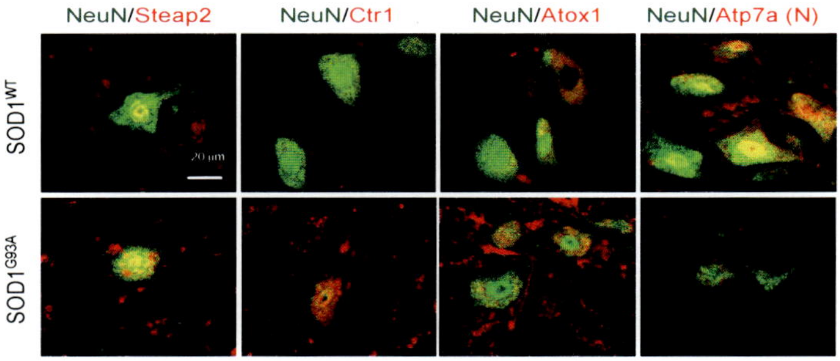

Figure 11.4

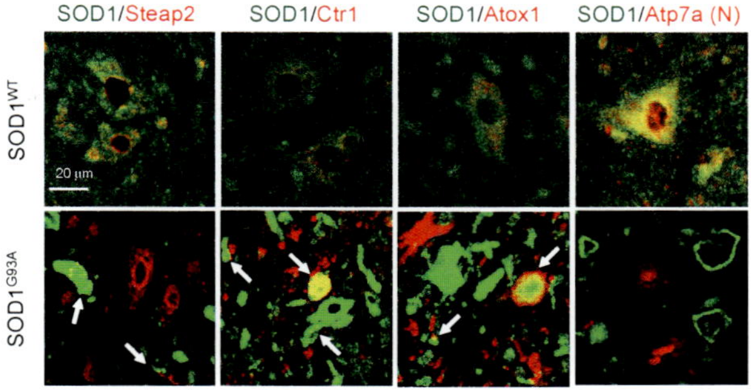

Figure 11.5

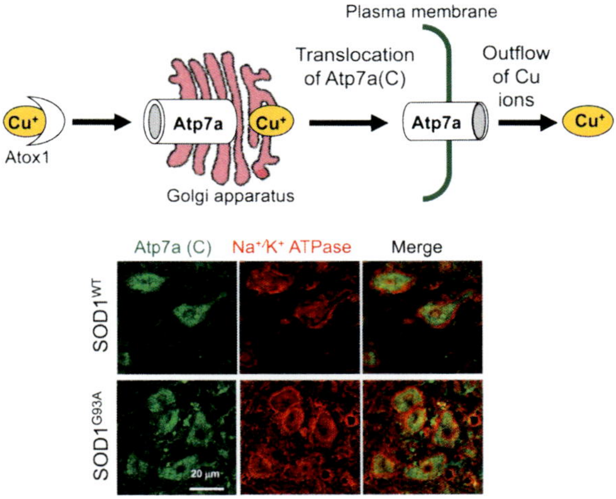

Figure 11.6

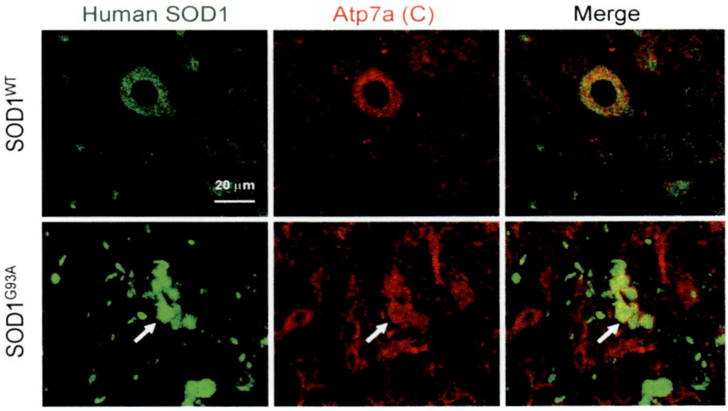

Figure 11.7

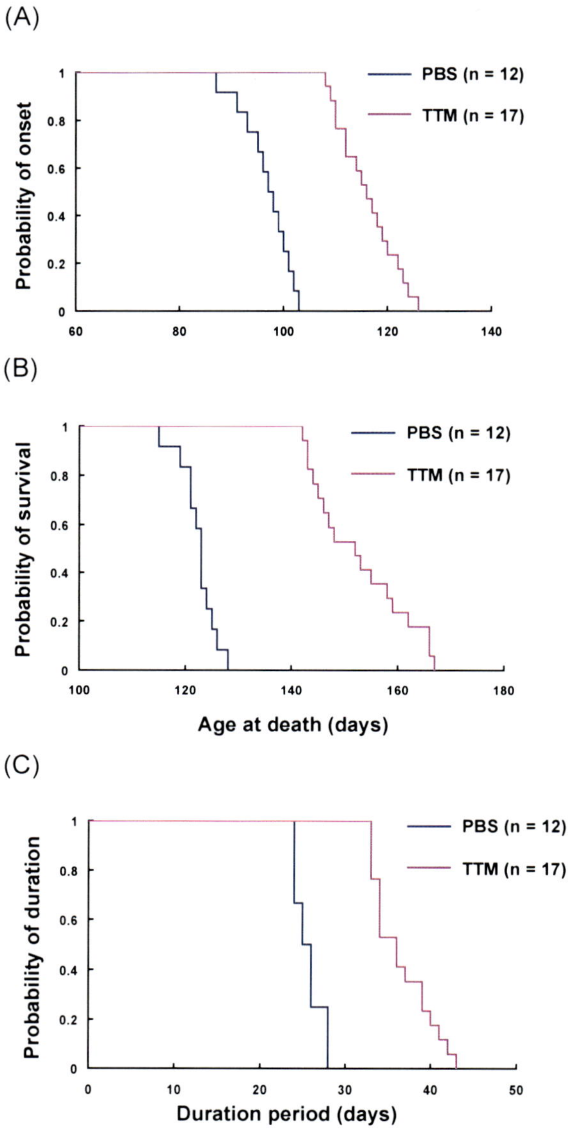
(A)
Probability of onset
1
0.8
0.6
0.4
0.2
0
60
80
100
120
140
PBS (n = 12)
TTM (n = 17)
(B)
Probability of survival
1
0.8
0.6
0.4
0.2
0
100
120
140
160
180
Age at death (days)
PBS (n = 12)
TTM (n = 17)
(C)
Probability of duration
1
0.8
0.6
0.4
0.2
0
0
10
20
30
40
50
Duration period (days)
PBS (n = 12)
TTM (n = 17)

Figure 11.8

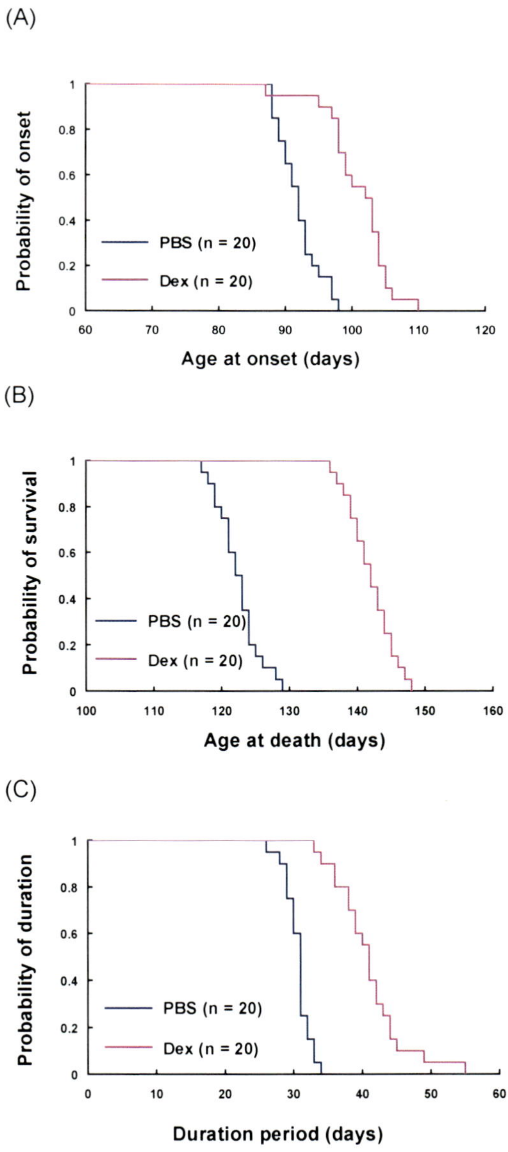
(A)
Probability of onset
PBS (n = 20)
Dex (n = 20)
Age at onset (days)
(B)
Probability of survival
PBS (n = 20)
Dex (n = 20)
Age at death (days)
(C)
Probability of duration
PBS (n = 20)
Dex (n = 20)
Duration period (days)

Figure 11.10

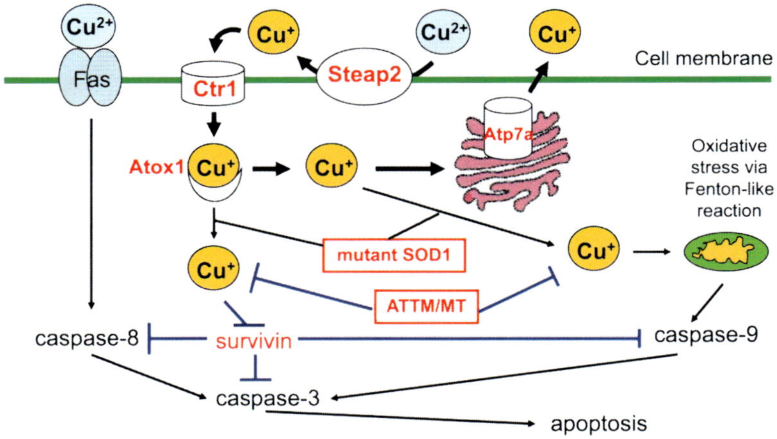
Cu2+
Fas
Cu+
Ctr1
Steap2
Cu2+
Cu+
Cell membrane
Atp7a
Atox1
Cu+
Cu+
Oxidative stress via Fenton-like reaction
mutant SOD1
Cu+
Cu+
ATTM/MT
caspase-8
survivin
caspase-9
caspase-3
apoptosis

Figure 11.11